AF388919

Sustainable Construction

Sustainable Construction

Editors

Víctor Yepes
Tatiana García-Segura

MDPI • Basel • Beijing • Wuhan • Barcelona • Belgrade • Manchester • Tokyo • Cluj • Tianjin

Editors
Víctor Yepes
Institute of Concrete Science
and Technology (ICITECH),
Universitat Politècnica de
València
Spain

Tatiana García-Segura
Department of Construction
Engineering and Civil
Engineering Projects, Universitat
Politècnica de València
Spain

Editorial Office
MDPI
St. Alban-Anlage 66
4052 Basel, Switzerland

This is a reprint of articles from the Special Issue published online in the open access journal *Sustainability* (ISSN 2071-1050) (available at: https://www.mdpi.com/journal/sustainability/special_issues/sustainable_construction).

For citation purposes, cite each article independently as indicated on the article page online and as indicated below:

LastName, A.A.; LastName, B.B.; LastName, C.C. Article Title. *Journal Name* **Year**, *Volume Number*, Page Range.

ISBN 978-3-0365-0482-7 (Hbk)
ISBN 978-3-0365-0483-4 (PDF)

Contents

About the Editors

Víctor Yepes is a full professor of Construction Engineering; he holds a Ph.D. in civil engineering. He serves at the Department of Construction Engineering, Universitat Politecnica de Valencia, Valencia, Spain. He has been the Academic Director of the M.S. studies in concrete materials and structures since 2007 and a Member of the Concrete Science and Technology Institute (ICITECH). He is currently involved in several projects related to the optimization and life-cycle assessment of concrete structures, as well as optimization models for infrastructure asset management. He currently teaches courses in construction methods, innovation, and quality management. He has authored more than 250 journals and conference papers, including more than 100 published in journals quoted in JCR. He acted as an expert for project proposal evaluation for the Spanish Ministry of Technology and Science, and he is a main researcher in many projects. He currently serves as an Editor-in-Chief for the *International Journal of Construction Engineering and Management* and a member of the editorial board of 12 other international journals (*Structure and Infrastructure Engineering, Structural Engineering and Mechanics, Mathematics, Sustainability, Revista de la Construcción, Advances in Civil Engineering, Advances in Concrete Construction*, among others).

Tatiana García-Segura is an Assistant Professor at the Department of Construction Engineering, Universitat Politècnica de València, Spain. She obtained her International Doctorate with outstanding "cum laude" in 2016. She received the IALCCE (International Association for Life-Cycle Civil Engineering) international award for his scientific contributions and two awards for her Master's Final Paper on sustainable construction (AEIPRO award and first prize of the Cemex-Sustainability Chair). She has published 24 articles in JCR journals, 30 articles in scientific congresses, has participated in several research projects (one as a PI), and one innovation project.

Preface to "Sustainable Construction"

Construction is one of the main sectors that generates greenhouse gases. This industry consumes large amounts of raw materials, such as stone, timber, water, etc. Additionally, infrastructure should provide service over many years without safety problems. Therefore, their correct design, construction, maintenance, and dismantling are essential to reducing economic, environmental, and societal consequences. That is why promoting sustainable construction has recently become extremely important.

To help address and resolve these types of questions, this book is comprised of twelve chapters that explore new ways of reducing the environmental impacts caused by the construction sector, as well to promote social progress and economic growth. The chapters collect papers included in the "Sustainable Construction" Special Issue of the *Sustainability* journal.

We would like to thank both the MDPI publishing and editorial staff for their excellent work, as well as the 43 authors who collaborated in its preparation. The papers cover a wide spectrum of issues related to the use of sustainable materials in construction, the optimization of designs based on sustainable indicators, the life-cycle assessment, the decision-making processes that integrate economic, social, and environmental aspects, and the promotion of durable materials that reduce future maintenance.

Víctor Yepes, Tatiana García-Segura
Editors

Article

Life Cycle Cost Assessment of Preventive Strategies Applied to Prestressed Concrete Bridges Exposed to Chlorides

Ignacio J. Navarro [1], Víctor Yepes [2] and José V. Martí [2,*]

[1] Department of Construction Engineering, Universitat Politècnica de València, 46022 Valencia, Spain; ignamar1@cam.upv.es

[2] Institute of Concrete Science and Technology (ICITECH), Universitat Politècnica de València, 46022 Valencia, Spain; vyepesp@upv.es

* Correspondence: jvmartia@upv.es; Tel.: +34-963-879-563

Received: 23 February 2018; Accepted: 13 March 2018; Published: 16 March 2018

Abstract: This paper applies Life Cycle Assessment methodology to aid in the decision making to select the preventive measure against chloride corrosion in concrete structures that works best for the socio-economic context of the structure. The assumed model combines the concepts of Life Cycle Cost Analysis and Social Life Cycle Analysis to assess the impacts on users derived from the maintenance activities associated with each alternative analyzed in terms of economic costs. The model has been applied to a prestressed concrete bridge to obtain a preventive measure that can reduce the total costs incurred over the period of analysis by up to 58.5% compared to the cost of the current solution.

Keywords: Life Cycle Assessment; Social Life Cycle Analysis; reinforced concrete; chloride corrosion; preventive measures

1. Introduction

Corrosion of reinforcing steel in concrete structures is one of the most important durability problems associated with this material. Currently, construction companies of different countries state that the refurbishment and maintenance of buildings represents up to 30% of the activity of the construction sector [1]. The poor durability of many concrete structures, which results in short structural service lives, is not sustainable [2] neither in social nor in economic terms. In recent times, it has been common practice to deal with concrete deterioration mechanisms once the problem is detected and not before it arises. Such kinds of strategies result in greater socio-economic impacts, since they require more material in the long term than a design based on prevention. Although there are several mechanisms that may degrade concrete in severe environments, experience demonstrates that the most critical threat to concrete structures exposed to marine environments is chloride-induced corrosion of the reinforcing steel bars [3–5]. Research has been carried out on this specific mechanism for many years [6–9], leading to the development of different preventive measures to increase resistance to corrosion from the beginning of the structure life cycle, thus resulting in less maintenance-demanding solutions.

Some of the measures developed to prevent chloride corrosion focus on the reinforcement itself and others seek to prevent corrosion by reducing the porosity of the concrete cover. Corrosion can also be prevented by isolating the structure from the environment by means of surface protection treatments or by altering the kinetics of the reactions or electrochemical potential of the affected metals. Although the degree of knowledge associated with some of these measures is still precarious, the use of preventive measures such as those mentioned above is common when a concrete structure is exposed to chlorides. It is the task of the designer to find the solution that entails the lowest cost and consumption of resources [10–13]. Regarding durability, decision support techniques, such as Life

Cycle Cost Assessment (LCCA) can be used to find a durable solution with the minimum associated costs [14].

Cost comparison is the usual procedure for selecting the best design alternative. However, when only considering the costs derived from implementing a particular solution, it may happen that the costs associated to the maintenance operations of the structure can exceed the initial investment, thus tilting the balance in favor of other alternatives with higher initial investment costs [15] but lesser maintenance. This leads to the consideration of LCCA techniques in order to evaluate the costs generated throughout every stage of the life cycle of a structure. In addition, the economic costs deferred over time have associated social costs that can also be evaluated by means of Social Life Cycle Assessment (SLCA) techniques. When applied to the choice of the most appropriate prevention measure, it is common practice to overlook the social impacts generated. In urban environments, this impact may lead to adopting preventive measures that are more expensive in economic terms, but that require fewer interventions and, consequently, generate less social costs. Thus, the integration of social criteria in decision making is presented as an effective step towards a sustainable structural design [16,17].

It is possible to integrate both methodologies when choosing between different prevention measures. The present paper proposes an LCCA based methodology for decision-making regarding the most appropriate preventive measure for concrete structures exposed to chlorides, taking into account economic and social criteria. The economic and social costs considered in the proposed methodology are described below.

2. Materials and Methods

2.1. Bridge Description

In the present paper, some of the most usual preventive strategies against chloride corrosion are applied to a particular bridge deck. The subject in this study is the bridge of Illa de Arosa, in Galicia, Spain. A cross-section of the bridge deck is shown in Figure 1. The input data regarding the durability and geometry characterization of this structure has been obtained from the literature [18]. The concrete mix of the bridge deck has a cement content of 485 kg/m^3, and a water/cement ratio w/c = 0.45. The concrete cover of the deck is 30 mm. A steel amount of 100 kg/m^3 of concrete has been assumed, as is usual for these types of prestressed structures. This quantity does not include the steel of the prestressing tendons. The deck has a width of 13 m and a section depth of 2.3 m. The deck, with a span of 50 m, is located 9.6 m over the high tide sea water level. It is worth noting that according to the Spanish regulations for marine environments, the deck is designed for no cracking of concrete, i.e., concrete remains uncracked.

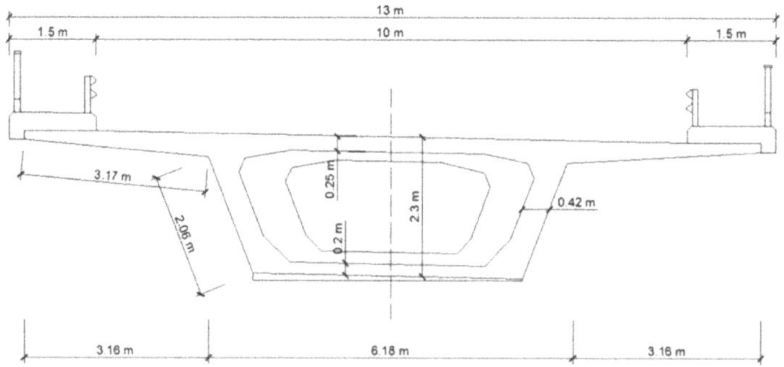

Figure 1. Cross-section of the Arosa's concrete bridge deck.

The particular preventive measures evaluated in the present study are as follows. Firstly, an increase in the reinforcement concrete cover to 35 mm, 45 mm, and to 50 mm (measures R35, R40, and R50, respectively, henceforth) are considered, taking into account that as more concrete cover is considered, the steel amount needed to guarantee the proper structural behavior of the bridge deck is greater. The steel amounts considered are 112 kg/m for R35, 136 kg/m for R45, and 147 kg/m for R50. A second group of measures consists of the addition to the existing concrete mixture of fly ash, silica fume, or polymers. The resulting concrete mixes have been assumed to be applied to the whole deck, although only the properties of the cover will affect the durability performance of the design alternative. Additions of 10% and 20% of fly ash (measures CV10 and CV20), additions of 5% and 10% of silica fume (measures HS5 and HS10), and additions of 10% and 20% of styrene-butadiene rubber (SBR) latex (measures HMP10 and HMP20) have been assumed in the analysis. The mentioned percentages are expressed in relation to the cement content of the reference concrete mix design. In the cases where fly ash or silica fume are added, the amount of cement is partially substituted by those components, as they contribute to the resistance development of the resulting concrete. The cement amount considered in the mix proportions of those alternatives is reduced according to the efficiency factor associated to the specific addition. In the present study, efficiency factors of $K = 0.3$ and $K = 2$ have been assumed for fly ash and silica fume additions, respectively. It is worth noting that the addition of silica fume may reduce the critical chloride threshold [19]. This effect, which is a consequence of the decrease in the chloride binding capacity of the resulting concrete when such additions are considered, has been taken into account in the present study. Thirdly, a decrement in the water/cement ratio to $w/c = 0.40$ and to $w/c = 0.35$ (measures AC40 and AC35) is considered. When the water/cement ratio is reduced, it is common practice to add special additives in order to increase concrete workability. As these products may increase the economic impacts of the measure, the addition of superplasticisers has been considered in the definition of measures AC40 and AC35. The concrete mixes corresponding to the design alternatives presented above are shown in Table 1. It shall be noted that, in order to make alternatives comparable, the alternative designs shall not only guarantee the same service life under an appropriate maintenance, but the resulting design should also have the same mechanical strength as the reference design. According to the mix proportions reported by León et al. [18], the reference design has a mean compressive strength f_{cm} equal to 40 MPa, with a modulus of elasticity E_c equal to 29 GPa. The alternative concrete mixes have been designed in order to reach the reference compressive strength and elastic modulus.

Table 1. Concrete mix design for the different alternatives.

	Cement	Water	w/b	Fly Ash	Silica Fume	Polymer (SBR)	Plasticiser	Gravel	Sand
	(kg/m^3)	(L/m^3)	(%)	(kg/m^3)	(kg/m^3)	(kg/m^3)	(kg/m^3)	(kg/m^3)	(kg/m^3)
REF [1]	485.6	218.5	0.45	-	-	-	-	926.7	827.9
AC40	500	200	0.40	-	-	-	7.5	948.0	844.1
AC35	500	175	0.35	-	-	-	10	976.7	882.8
CV10	471	218.5	0.45	48.6	-	-	-	926.7	798.3
CV20	456.4	218.5	0.45	97.1	-	-	-	926.7	768.7
HS5	437	218.5	0.45	-	24.3	-	-	926.7	849.1
HS10	388.4	218.5	0.45	-	48.6	-	-	926.7	870.2
HMP10	485.6	218.5	0.45	-	-	48.6	-	926.7	827.9
HMP20	485.6	218.5	0.45	-	-	97.1	-	926.7	827.9

[1] This mix is also considered in alternatives R35, R45, R50, INOX, GALV, HIDRO and SEAL.

Next, the replacement of the existing ordinary steel with galvanized steel (measure GALV) and with stainless steel (measure INOX) are considered. Finally, treatment the exposed deck surface with a generic silane-based hydrophobic impregnation (measure HIDRO) and with a silicate-based sealant product (measure SEAL) is considered. Fifteen preventive measures are considered. It shall be said that there are other ways to deal with corrosion in severe environments, such as cathodic protection. These other relevant measures have been excluded from the present analysis, as their performance is hard to assess in the same terms as the presented strategies.

2.2. Durability Performance of the Preventive Measures

The assessment of the durability of a structure requires a criterion indicating the time at which it becomes necessary to perform a maintenance operation. In the case of prestressed concrete bridges, it is usual to consider the time to corrosion initiation (Figure 2) as proposed by Tuutti [20] as the time where maintenance activities shall be held. The time to corrosion initiation is the time where chlorides reach a concentration high enough to start the corrosive process of the reinforcement. Consequently, this maintenance criterion guarantees that, when maintenance operations are performed, the reinforcing steel is still not damaged by corrosion, and it is not necessary to replace it.

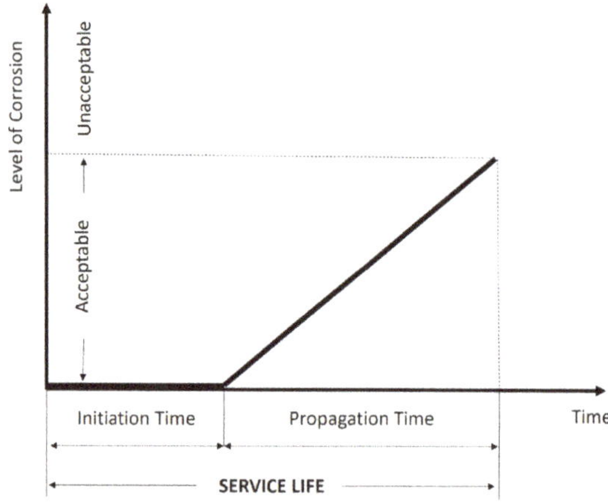

Figure 2. Service life definition based on Tuutti's model of corrosion.

The calculation of the initiation time requires a physical model that describes how chloride ions move through the concrete cover. Existing models for the prediction of the required time to initiate corrosion are based on Fick's second law of diffusion, and they assume that the porous concrete cover is a homogeneous material in which ions migrate through a diffusion process. A deterministic solution of the Fick's equation for the diffusion of chlorides in the concrete cover is used in this analysis, namely, the one proposed in Fib Bulletin 34 [21] that assumes a constant, time independent surface chloride concentration. So, the chloride concentration to be expected in the concrete cover at a specific depth x and in a particular time *t* is expressed as:

$$C(x,t) = C_0 + (C_{s,\Delta x} - C_0) \times \left(1 - erf \frac{x - \Delta x}{2\sqrt{D_{app,C} \times t}} \right)$$

(1)

where $C(x,t)$ is the chloride concentration (wt.%/binder) at concrete depth x (mm) and time t (years); $C_{s,\Delta x}$ is the chloride concentration at depth Δx (wt.%/binder); Δx is the depth of the convection zone (mm), which is the surface layer depth for which the process of chloride penetration differs from Fick's second law of diffusion; *erf(.)* is the Gauss error function; and $D_{app,C}$ is the apparent coefficient of chloride diffusion through concrete (mm^2/years). Note that if Δx is considered to be zero, the term $C_{s,\Delta x} = C_s$ is the chloride concentration at the concrete surface. The model proposed by Fib Bulletin 34 [21] assumes that the chloride front advances in only one direction. However, this hypothesis is not true when specific bars are exposed to two simultaneously advancing fronts, as is the case of reinforcing bars located at the edges of the studied section. In such cases, the use of one-dimensional models results in inaccurate, overestimated service lives. In the present study, the Fib model has been slightly modified in order to consider the two-dimensional advance of chlorides, the so called corner effect:

$$C(x,y,t) = C_0 + (C_s - C_0) \times \left(1 - erf \frac{x - \Delta x}{2\sqrt{D_{app,C,x} \times t}} \times erf \frac{y - \Delta y}{2\sqrt{D_{app,C,y} \times t}} \right) \qquad (2)$$

The concrete cover in the y-direction (r_y) is assumed to be constant and equal to 50 mm for every alternative analyzed, while the cover in the x-direction (r_x) is assumed to vary between 30 mm and 50 mm depending on the prevention alternative studied. The apparent diffusion coefficient is obtained from the experimental non-steady state migration coefficient using the equation proposed by Fib [21]:

$$D_{app,C} = exp\left(b_e \times \left(\frac{1}{T_{ref}} - \frac{1}{T_{real}} \right) \right) \times D_{RCM,0} \times k_t \times \left(\frac{t_0}{t} \right)^\alpha \qquad (3)$$

where b_e is a regression variable (constant); T_{ref} is the standard test temperature (°C); T_{real} is the temperature of the structural element (°C); $D_{RCM,0}$ is the non-steady state chloride migration coefficient (mm^2/years); k_t is a transfer parameter (constant); t_0 is a reference point of time (years); and α is an age factor, which is assumed to be equal to 0.5 according to the Spanish concrete code EHE-08 [22]. In the present study, T_{ref} and T_{real} are assumed to be the same, and the transfer variable is considered to be $k_t = 1$ as suggested by Fib [21]. The age factor α determines the way the diffusion coefficient varies with the time. As reference time, $t_0 = 0.0767$ years (namely 28 days) has been considered. Table 2 shows the value of the parameters that allow for the characterization of the analyzed measures in terms of durability.

Table 2. Durability characterization of the analyzed preventive strategies.

Design Alternative	Description	$D_{RCM,0}$ ($\times 10^{-12}$ m^2/s)	C_{crit} (%)	r_x (mm)	Service Life (Years)	Reference
REF	Current bridge design	10	0.6	30	7.1	[22]
R35		10	0.6	35	11.2	
R45		10	0.6	45	21.1	
R50		10	0.6	50	26.5	
AC40	$w/c = 0.40$	6.15	0.6	30	18.8	[23,24]
AC35	$w/c = 0.35$	4.32	0.6	30	43.1	
INOX	Stainless steel reinforcement bars	10	5	30	>100	[25]
GALV	Galvanized steel reinforcement bars	10	1.2	30	23.4	[26]
HMP10	Polymer modified concrete (10%)	7.32	0.6	30	13.2	[27,28]
HMP20	Polymer modified concrete (20%)	3.04	0.6	30	75.6	
HS5	5% Silica fume	3.31	0.38	30	36.2	[19,29]
HS10	10% Silica fume	1.38	0.22	30	>100	
CV10	10% Fly ash	6.16	0.6	30	18.6	[30]
CV20	20% Fly ash	5.23	0.6	30	26.1	
HIDRO	Hydrophobic surface treatment	7.73	0.6	30	5 [1]	[31,32]
SEAL	Sealant surface treatment	4.87	0.6	30	5 [1]	[33]

[1] In the present study, the service life of surface treatments (HIDRO and SEAL) is limited to five years according to manufacturer specifications.

It is assumed that the surface concentration of chlorides and the age coefficient is the same for all the alternatives evaluated. Back to the Tuutti model presented above, the time to corrosion initiation can be obtained by equalizing the chloride concentration at the rebar depth $C(r_x, r_y, t)$ to the critical chloride concentration for each specific measure. In the calculations, and on the basis of the distance between the bridge deck bottom surface and the mean sea water level, a surface chloride content of $C_s = 3.34\%$ is assumed for the evaluation of the bridge deck. Table 2 shows the resulting expected service lives for the analyzed prevention alternatives considering the durability parameters assumed in the present study. It shall be noted that the effectiveness of the surface treatments depends greatly on the porosity of the substrate [34]. Consequently, the diffusion coefficients presented in Table 2 for HIDRO and SEAL have been obtained considering the *w/c* ratio of the REF alternative (current bridge design), namely *w/c* = 0.45. However, the durability performance of these treatments is very sensitive to ageing and weathering as derived from the existing literature [33,35–37]. In particular, conventional treatments are very affected by microcracking of the concrete cover, for when those cracks are deeper than the treatment thickness, barrier properties are lost and the impregnation becomes ineffective [38,39]. Periodic reapplication of surface treatments is, therefore, desirable in order to reestablish the protective effect of these measures. For these reasons, the present study limits the service life of surface treatments to five years according to manufacturer specifications.

2.3. Life Cycle Cost Analysis (LCCA)

In general, the economic costs associated with a structure can be divided into initial investment and maintenance costs. The investment costs correspond to the costs incurred at the moment of the initial economic analysis. Four main categories are usually distinguished chronologically: planning costs, land acquisition costs, construction costs, and interruption costs. The present study only considers the costs derived from the construction activities. It shall be noted that many of the investment costs are very similar between the different alternatives to be evaluated, for example, the costs associated to excavation or piles construction. Only those costs that are different between alternatives are considered in the present paper, as they are the ones that will have an influence on the final decision.

The maintenance costs, which are generated throughout the service life of the structure, are necessary for the infrastructures to be in operating condition throughout the required service life of the project, and they directly depend on the durability performance of the structure against the dominating degradation mechanism. Maintenance operations include the activities of hydrodemolition of the concrete cover, cleaning of the outermost reinforcement, and shotcreting with the corresponding concrete mixture to restore the original cover. In the case of surface treatment, maintenance operations consist only of the reapplication of the treatment.

The unitary costs assumed in the present study for the basic materials considered in the evaluation of both initial investment and maintenance costs are shown in Table 3. These costs are usual for the Spanish construction sector.

Table 3. Parameters considered in Life Cycle Cost Analysis (LCCA).

Parameter	Cost	Unit
Ordinary Portland Cement	87.77	€/t
Sand	13.98	€/t
Gravel	16.36	€/t
Fly Ash	38	€/t
Silica Fume	1.14	€/kg
SBR Latex	4.7	€/l
Superplastiziser	1.38	€/kg
Reinforcing Steel—B 500 S	0.8	€/kg
Reinforcing Steel—Stainless	4.5	€/kg
Hydrophobic Surface Treatment	19.01	€/m²
Sealant Surface Treatment	29.04	€/m²

2.4. Social Life Cycle Assessment (SLCA)

The social costs of a construction over its lifetime are difficult to quantify in monetary terms. These costs affect both users of the infrastructure and third parties who do not make use of it. For the road bridge studied, the main social costs affecting users [40,41] of the infrastructure are quantified. The user related costs considered in the present study are the Vehicle Operating Costs (VOC) and the Vehicle Delay Costs (VDC). The Vehicle Operating Costs are those costs arising from the normal use of the vehicles that must be borne by its users, such as fuel consumption, tire wear, and maintenance.

These costs can be quantified as the cost increase derived from the circulation in zones affected by the construction or maintenance of the bridge with regard to the costs associated to the circulation in normal, unaffected travelling conditions. According to Seshadri and Harrison [42], the traffic behavior along a zone affected by maintenance or construction works allows us to identify five different sections, as shown in Figure 3.

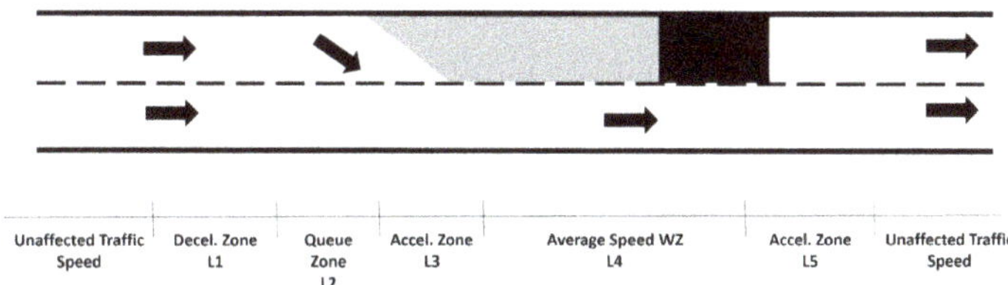

Figure 3. Traffic behavior along a Working Zone (WZ) according to Seshadri and Harrison [42].

The value of the vehicle's operating costs can be defined by the following equation [43]:

$$VOC = \sum_{j=1}^{24} \left(\sum_{k=1}^{5} \left(L_k - \frac{S_{ak}}{S_n} \cdot L_k \right) \cdot HT \cdot \sum_{i=1}^{4} (VOC_i \cdot p_i) \right) \tag{4}$$

where L_k is the length of the affected zone k depending on the behavior of the traffic involved, as shown in Figure 3, S_{ak} is the traffic speed in the zone k affected by maintenance works, S_{an} stands for the traffic speed under normal, unaffected conditions, HT is the average hourly traffic, p_i is the percentage of class *i* vehicles with respect to the total vehicle flow, and VOC_i represents the operating costs associated to class *i* vehicles, defined as the sum of the costs derived from fuel consumption, tire consumption, and maintenance costs. In the present paper, the assumed parameter values regarding the characterization of the traffic behavior along the Working Zone are shown in Table 4. In the analysis, a traffic speed under normal conditions of S_{an} = 80 km/h has been considered. The vehicle classes assumed in the present methodology are described in Table 5.

Table 4. Characterization of the Working Zone traffic behavior.

Section	Section Length (m)	Traffic Speed (km/h)
L1—Deceleration Zone	100	60
L2—Queue Zone	75	20
L3—Acceleration Zone	50	30
L4—Average Speed	250	40
L5—Acceleration Zone	50	50

Table 5. Cost per hour of a person travelling, with regard to their objective, according to De Rus [44].

Vehicle Class	Class Description	Cost of Driver's Working Time (€/h)	Cost of Passenger's Leisure Time (€/h)	Occupation Rate
Class 1	Motorcycles and vehicles with vertical height from the first axle of 1.10 m	22.34	7.15	2
Class 2	Vehicles with two axles and a vertical height greater than 1.10 m	22.34	-	1
Class 3	Vehicles with three axles and a vertical height greater than 1.10 m	17.93	5.13	24
Class 4	Vehicles with more than three axles and a vertical height greater than 1.10 m	17.93	-	1

Note: For every vehicle class, it is assumed that 1 passenger is working. For vehicle classes 1 and 3, an additional 1 and 23 passengers are assumed to be "not working" respectively, according to Gervásio [45].

On the other hand, the economic quantification of the delays generated by maintenance operations on road users (VDC) is calculated by assessing the difference between the costs derived from the time spent by the driver on crossing the section affected by the works and those resulting from the time spent under normal operating conditions. In addition, the reason for the driver's travel is also assessed and different estimates are made if the journey is for work or for other reasons. The value of the vehicle's delay costs (VDC) can be defined by the following equation:

$$VDC = \sum_{j=1}^{24} \left(\sum_{k=1}^{5} \left(\frac{L_k}{S_{ak}} - \frac{L_k}{S_n} \right) \times HT \times \sum_{i=1}^{4} DTC_i \times p_i \right) \tag{5}$$

where the meaning of the different variables is the same as that presented in the definition of VOC. In Equation (5), DTC_i is the hourly cost associated to a class i vehicle user, and it depends on the user's reason for travelling. This concept can be evaluated as:

$$DTC_i = \sum_{m}(TC_m \times OR_{i,m}) \tag{6}$$

where $OR_{i,m}$ is the occupation rate of passengers inside a class i vehicle that are travelling with a particular objective m, and TC_m is the hourly cost of a person travelling with a particular objective m as shown in Table 5. Two travel objectives are assumed, namely $m = 1$ for working reasons and $m = 2$ for non-working reasons. The relation between both costs, when no data is available, can be estimated as:

$$TC_{m=2} = 0.25 \times TC_{m=1} \tag{7}$$

Equations presented to evaluate VOC and VDC are based on the model developed by Gervásio [43]. For both the quantification of VOC and VDC, it is assumed that the maintenance operations only affect a 300 m long road section, and that the repair activities last six weeks. The average daily traffic is assumed to be 6000 vehicles per day, with an average traffic speed of 80 km/h. The costs derived from the use of vehicles are shown in Table 6.

Table 6. Costs associated to the assumed vehicle classes.

Vehicle Class	Class 1		Class 2	Class 3	Class 4
Traffic composition	11%		78%	3%	8%
Fuel type	Gasoline	Diesel	Diesel	Diesel	Diesel
Fuel consumption (l/100 km)	5.9	4.8	4.5	44	44
Fuel cost (€/l)	1.33	1.13	1.13	1.13	1.13
Average service life for vehicles (years) [1]	10		8	12	12
Service life of tires (km) [1]	40,000		40,000	75,000	200,000
Tire cost (€) [1]	73		62	333	463.6
Mean travel distance km/year [1]	20,000		30,000	70,000	85,000
Yearly vehicle maintenance cost (€) [1]	1575		1860	16,316	28,027
Vehicle depreciation (€) [1]	17,177		11,563	21,653	84,451

[1] This data has been obtained from Gervásio [45].

Both the economic and the social costs, all transformed into monetary terms, will occur in different time instants, depending on the initiation time resulting for each of the treatment alternatives considered. In the resolution of the proposed model, the temporality of the maintenance actions is taken into account through the following concepts: selection of an appropriate period of analysis, consideration of a residual benefit, and cost discounting to present values.

The analysis period is the time frame in which the different alternatives are compared. The choice of the analysis period greatly influences the results of the socio-economic evaluation; choosing a period equal to the shorter service life of the alternatives may not capture the differences in the behavior of the alternatives in the long term, penalizing those that have longer service lives. This period should be long enough to adequately reflect the existing performance differences between the alternatives being compared. The current LCCA includes the impacts derived for an analysis period of the first 100 years of bridge life. This is the required service life for bridge structures according to European Committee for Standardization [46]. The consideration of the durability performance shown in Table 2 results in the number of maintenance operations to be held during the analysis period.

The residual benefit represents the monetary value of the alternative at the end of the analysis period. This value should be taken into account in the evaluation of projects where the solution has a longer service life than the analysis period considered in the economic evaluation [47]. In a simplified way, the economic value of the structure in its optimum state corresponds to the value of the initial investment. From this point onwards, the structure loses value as the end of its useful life approaches, when its state becomes inadmissible from the point of view of durability and its residual value becomes zero. In cost accounting, this residual value of a solution when the end of the analysis period is reached is considered as a benefit, thus reducing the life cycle costs.

The criterion assumed to determine the residual value of an alternative is to take into account the advance of the critical chloride content. By calculating the depth reached by the critical chloride content at the end of the analysis period, it is possible to estimate the residual value of the alternative in question; if the critical chloride content has reached the reinforcement depth, the residual value of the structure is zero. If not, it is assumed that the residual value of the structure is a fraction of the installation costs of the alternative. In particular, this fraction is proportional to the penetration depth of the chlorides in relation to the concrete cover of the design.

Finally, the temporary aspect of costs and benefits remains to be addressed. Over time, the value of money varies depending on financial concepts such as the interest rates of the investment or the inflation rates. In order to properly compare two alternatives, it is necessary to convert the costs generated over time into comparable monetary values. As a criterion, the comparison is made in terms of present monetary value, using the discount rate. The equation for calculating costs in terms of present costs is as follows:

$$LCC = \sum_{t=t_0}^{t_{SL}} C_i \times \frac{1}{(1+d)^{t-t_0}} \tag{8}$$

where LCC is the Life Cycle Cost of the structure, C_i represents the economic costs associated to time t, t_0 is the time associated to the beginning of the analysis period, t_{SL} is the number of years considered in the analysis, and d is the discount rate. In Europe, the European Commission proposes a discount rate of between 3.5% and 5.5% for project evaluation. In the present paper, a discount rate of 5% is considered.

3. Results and Discussion

According to the methodology proposed, the resulting installation and maintenance costs per meter of bridge deck are presented in Figure 4 for the preventive strategies analyzed in this study. As explained above, since the assumed maintenance criterion is to repair the bridge when the initiation time is reached, the reinforcement will not have been affected by corrosion, and consequently, the maintenance operation will consist on removing the chloride contaminated concrete cover and executing it again. Regarding the surface treatments (HIDRO and SEAL), the maintenance activities consist in the reapplication of the treatment, without demolishing the existing concrete cover.

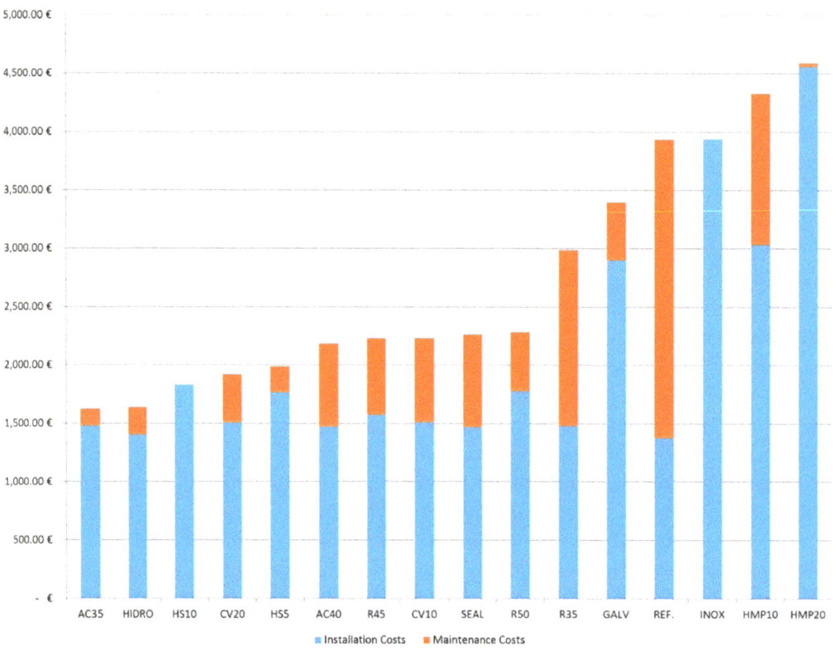

Figure 4. Installation and maintenance costs, in €/m of deck.

It is observed that there are several alternatives with initial costs very similar to the installation costs of the reference alternative. Those are the alternatives where the cost of the materials is very close to the reference case, namely the ones based on surface treatments (hydrophobic and sealant treatments), the addition of fly ash, the reduction of the water/cement ratio, and the increase of the concrete cover up to 35 mm.

The addition of silica fume in the concrete mix results in solutions with initial costs significantly greater than the those of the reference alternative, namely 28% to 33% greater for HS5 and HS10, respectively. It is observed that an excessive concrete cover (R45 and R50) leads to great initial costs as well, due to the associated increase in the reinforcing steel demand. On the other hand, those solutions where the reinforcement material is modified, namely by using galvanized or stainless steel reinforcements, result in almost the highest initial costs, due to the high costs of materials [48]. At last, it is observed that the most expensive solutions in terms of initial costs are those where the reference concrete is modified with polymers. This conclusion is in good accordance with Fowler [49], and is derived from the high costs of the polymer materials. Consequently, the economic limitation of polymer modified concretes or stainless steel used as reinforcement is only assumable in cases where good durability is required. As can be observed, the service life of both HMP10 and HMP20 solutions is considerable, and this results in lower or almost no maintenance costs. In the case of INOX, no maintenance is required.

Focusing on the costs generated during the service stage of the bridge, the results show that the alternatives with the least number of interventions are, in general, the ones with the lowest maintenance costs, as expected. An exception to the previous statement is made by alternatives involving surface treatments; as explained above, such alternatives require frequent maintenance. The costs associated with each of these maintenance operations are, however, small if compared to the cost of repair for any of the other alternatives. This means that, despite demanding a high number of repairs, they are preferable in terms of economic costs if compared to the other alternatives. It is also interesting to highlight the significant maintenance costs associated with the reference solution, which almost doubles the maintenance costs of the next more expensive alternative. However, it is worth noting that an increase of only 5 mm in the concrete cover may reduce the maintenance costs of the reference alternative by up to 40%.

By analyzing the social impact of the different alternatives in terms of user costs (Figure 5), it can be observed that those measures that require a greater number of interventions are the ones that generate the greatest social impact. Thus, alternatives HIDRO and SEAL, despite requiring faster maintenance operations, generate the greatest social costs, exceeding those resulting from the reference alternative. The reference alternative, which also requires intensive maintenance throughout its service life, results in high social costs as well. The rest of the analyzed measures show a significantly lower social impact.

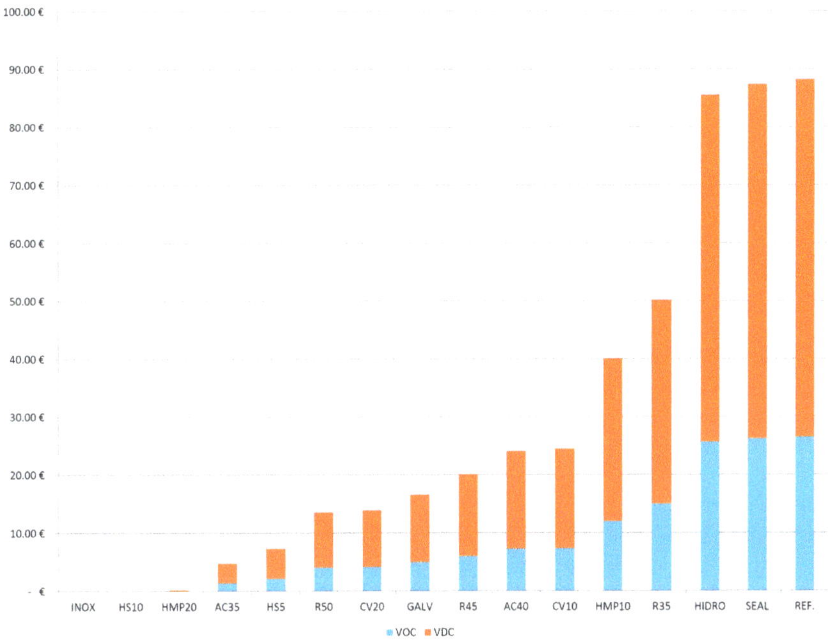

Figure 5. User costs of the analyzed preventive measures, in €/m of deck.

Table 7 shows the discounted total costs incurred by each of the different designs. It is observed that the user costs are significantly lower than the costs derived from maintenance. These costs depend on the socio-economic context of the infrastructure and on the traffic intensity registered. When considering both installation, maintenance ,and user costs, it is concluded that using polymer modified concrete is the most expensive prevention alternative, with a resulting cost that is even greater than the costs derived from the reference design. The difference between the cost of both strategies is, however, less than 15%. The use of stainless steel shows similar economic impact to the reference alternative. In this case, the same as in the case of HMP20, the total costs are mainly those associated with the installation phase of the structure, i.e., neither of them incurs maintenance costs (INOX) or maintenance costs are generated far into the future (HMP20).

Table 7. Total socio-economic costs of the analyzed alternatives, in €/m of deck. VOC = vehicle operating costs; VDC = vehicle delay costs.

Alternative	Investment Costs	Maintenance Costs	VOC	VDC	Residual Value	Total
AC35	1481.67 €	143.84 €	1.44 €	3.37 €	−7.66 €	1622.65 €
HIDRO	1406.91 €	230.92 €	25.61 €	59.93 €	−10.70 €	1712.68 €
HS10	1832.96 €	-	-	-	−2.24 €	1830.72 €
CV20	1511.39 €	409.19 €	4.19 €	9.80 €	−1.94 €	1932.63 €
HS5	1768.90 €	219.50 €	2.18 €	5.11 €	−3.20 €	1992.50 €
AC40	1475.82 €	707.39 €	7.22 €	16.88 €	−7.64 €	2199.66 €
R45	1579.74 €	654.04 €	6.02 €	14.09 €	−3.13 €	2250.76 €
CV10	1514.56 €	720.65 €	7.34 €	17.16 €	−7.18 €	2252.53 €
R50	1778.83 €	507.79 €	4.08 €	9.54 €	−3.06 €	2297.17 €
SEAL	1471.19 €	795.83 €	26.17 €	61.22 €	−11.19 €	2343.22 €
R35	1480.19 €	1506.24 €	15.02 €	35.14 €	−0.80 €	3035.79 €
GALV	2903.18 €	494.98 €	4.98 €	11.64 €	−16.04 €	3398.74 €
INOX	3941.28 €	-	-	-	−29.97 €	3911.31 €
REF.	1380.64 €	2555.78 €	26.43 €	61.83 €	−9.61 €	4015.07 €
HMP10	3033.84 €	1296.54 €	12.02 €	28.12 €	−9.79 €	4360.72 €
HMP20	4560.74 €	32.59 €	0.08 €	0.19 €	−23.49 €	4570.11 €

On the other hand, it shall be noted the high cost of maintenance derived from the reference design. For this alternative, the costs that are generated over the service life almost double the initial investment. Something similar happens for those alternatives in which the modification of the solution does not significantly affect the durability of the solution, as is the case with alternative R35. While this slight increase in the concrete cover significantly reduces maintenance costs, they still carry a significant weight on the final cost of the solution. Obviously, the more a solution contributes to improving the durability performance of the structure against corrosion, the lower the maintenance costs will be.

The prevention strategies that generate less costs throughout the life cycle of the structure, considering both economic and social costs on users, are those based on reducing the water to cement ratio of the original concrete mix (AC35), applying hydrophobic surface treatments on the deck surface (HIDRO), and adding of silica fume and fly ash on the concrete mix design (HS10, HS5, and CV20). As can be observed, the durability performance of both AC35 and HIDRO is far below the performance of other, more expensive solutions such as HMP20 or INOX. However, the resulting life cycle costs are the lowest, namely between 41.5% and 42.6% of the costs associated to the reference design. This is due to the fact that material costs in both cases are very competitive. Consequently, the combined effect of the inexpensive maintenance activities together with the more than acceptable durability performance in the case of AC35, and the low costs of maintenance in the case of HIDRO design, make them the most desirable strategies. It shall be highlighted that the life cycle costs of these alternatives only differ between 11% and 15% with the costs derived from alternatives HS10 and CV20, making them also very cost-efficient solutions in chloride laden environments. It has to be noted that the design HS10 does not incur any costs related to maintenance, as the service life associated to this alternative is greater than the required service life.

The results presented above are based on the assumption of a discount rate d = 5%. However, the assumed discount rate is highly uncertain when considering a period of analysis of 100 years. Taking into account that uncertainties about discount rate tends to be a key contributor on LCCA results when considered [50,51], a sensitivity analysis on this parameter is performed, taking into account three discount rates within the usual range for infrastructures in Europe, namely d = 3%, d = 4%, and d = 5%. For these three scenarios, the different preventive strategies have been ranked, based on the resulting life cycle total costs, including both economic and user impacts. Results are presented in Figure 6, where a rank value of 1 means that the alternative results in the lowest life cycle discounted costs.

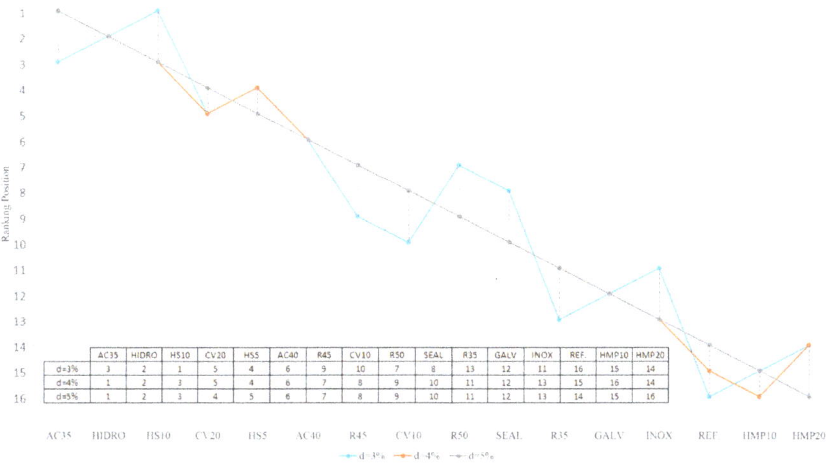

Figure 6. Preventive strategies ranking considering different discount rates.

It can be observed that the main conclusions derived above are robust, as they do not significantly depend on the chosen discount rate. There are, however, slight differences in the rankings. In Figure 6, a dashed line marks the ranking changes resulting from the analysis scenarios. It is observed that the worst alternatives, regardless of the discount rate assumed in the evaluation, are the reference alternative and the strategies based on polymer modified concrete. On the other hand, the best alternatives vary between measures HS10, AC35, and HIDRO. It is observed that, in general, the lower the chosen discount rate, the more preferable are those solutions with either less maintenance demand, such as HS10, or those with lower maintenance costs, such as hydrophobic surface treatments. However, it shall be highlighted that the differences between the best and worst of the alternatives HS10, AC35, and HIDRO in terms of total life cycle costs are found to be less than 10%. In consequence, it can be concluded that results presented in this paper are robust.

4. Conclusions and Future Lines of Research

LCCA methodology has been used to assess the different preventive measures against chloride corrosion of concrete reinforcement. These measures have been evaluated taking into account both their durability performance against corrosion and the economic and social costs derived during the life cycle of the structure. The exposed methodology has been applied to a prestressed concrete bridge in the Arosa Isle.

It has been shown that those alternatives with better performance in terms of durability against chloride corrosion, namely the use of stainless steel or polymer-modified concrete, are the ones that incur greater life cycle costs. Those alternatives that perform the worst, namely the reference alternative or the increase of the concrete cover up to 35 mm, show similar resulting life cycle costs. In these cases, in contrast to the previously described alternatives, the major part of the resulting costs is derived from maintenance, which can be up to three times greater than the installation costs, as the case of the reference design. From the results presented above, it seems reasonable to state that the optimum in economic terms is derived from a compromise solution between durability performance and material costs. For the bridge studied, the alternatives that perform best economically consist in reducing the water/cement ratio of the reference alternative up to $w/c = 0.35$ and treating the concrete surface by means of hydrophobic products. The addition of high amounts of silica fume and fly ash also provide very cost efficient solutions throughout the structure's life cycle. In particular, the addition of 10% silica fume results in a very cost-efficient, maintenance free solution.

It shall be highlighted that, although the LCCA methodology presented here is applicable in general terms to concrete bridge decks exposed to severe environments, the conclusions drawn are based on a particular bridge. The transferability of the obtained results is, therefore, dependent to the specific context of the structure to be evaluated. Further research is suggested to evaluate the combined effect of some of the presented preventive measures, and to bring to light both their compatibilities and their cost competitiveness.

Acknowledgments: The authors acknowledge the financial support of the Spanish Ministry of Economy and Competitiveness, along with FEDER funding (Project: BIA2017-85098-R).

Author Contributions: This paper represents a result of teamwork. Ignacio J. Navarro, Víctor Yepes, and José V. Martí jointly designed the research; Ignacio J. Navarro drafted the manuscript and José V. Martí revised the manuscript; José V. Martí and Víctor Yepes edited and improved the manuscript until all authors were satisfied with the final version.

Conflicts of Interest: The authors declare no conflict of interest.

References

1. Gil, E.; Vercher, J.; Mas, A.; Fenollosa, E. Seguridad remanente a flexión en forjados con corrosión en las viguetas. *Inf. Constr.* **2015**, *67*, e054. [CrossRef]
2. Gao, X.J.; Wang, X.Y. Impacts of Global Warming and Sea Level Rise on Service Life of Chloride-Exposed Concrete Structures. *Sustainability* **2017**, *9*, 460. [CrossRef]
3. Šavija, B.; Schlangen, E. Chloride ingress in cracked concrete—A literature review. In Proceedings of the International PhD Conference on Concrete Durability, Madrid, Spain, 19 November 2010; Andrade, C., Gulikers, J., Eds.; RILEM Publications: Madrid, Spain, 2012; pp. 133–142.
4. Maes, M.; De Belie, N. Resistance of concrete and mortar against combined attack of chloride and sodium sulphate. *Cem. Concr. Compos.* **2014**, *53*, 59–72. [CrossRef]
5. Miyazato, S.; Otsuki, N. Steel corrosion induced by chloride or carbonation in mortar with bending cracks or joints. *J. Adv. Concr. Technol.* **2010**, *8*, 135–144. [CrossRef]
6. Yang, K.H.; Singh, J.; Lee, B.Y.; Kwon, S.J. Simple technique for tracking chloride penetration in concrete based on the crack shape and width under steady-state conditions. *Sustainability* **2017**, *9*, 282. [CrossRef]
7. Zhu, X.; Zi, G.; Cao, Z.; Cheng, X. Combined effect of carbonation and chloride ingress in concrete. *Constr. Build. Mater.* **2016**, *110*, 369–380. [CrossRef]
8. Shaheen, F.; Pradhan, B. Effect of chloride and conjoint chloride–sulfate ions on corrosion of reinforcing steel in electrolytic concrete powder solution (ECPS). *Constr. Build. Mater.* **2015**, *101*, 99–112. [CrossRef]
9. Cheng, Y.; Zhang, Y.; Tan, G.; Jiao, Y. Effect of Crack on Durability of RC Material under the Chloride Aggressive Environment. *Sustainability* **2018**, *10*, 430. [CrossRef]
10. García-Segura, T.; Yepes, V.; Martí, J.V.; Alcalá, J. Optimization of concrete I-beams using a new hybrid glowworm swarm algorithm. *Lat. Am. J. Solids Struct.* **2014**, *11*, 1190–1205. [CrossRef]
11. Yepes, V.; García-Segura, T.; Moreno-Jiménez, J.M. A cognitive approach for the multi-objective optimization of RC structural problems. *Arch. Civ. Mech. Eng.* **2015**, *15*, 1024–1036. [CrossRef]
12. Martí, J.V.; García-Segura, T.; Yepes, V. Structural design of precast-prestressed concrete U-beam road bridges based on embodied energy. *J. Clean. Prod.* **2016**, *120*, 231–240. [CrossRef]
13. García-Segura, T.; Yepes, V. Multiobjective optimization of post-tensioned concrete box-girder road bridges considering cost, CO_2 emissions, and safety. *Eng. Struct.* **2016**, *125*, 325–336. [CrossRef]
14. Penadés-Plà, V.; Martí, J.V.; García-Segura, T.; Yepes, V. Life-cycle assessment: A comparison between two optimal post-tensioned concrete box-girder road bridges. *Sustainability* **2017**, *9*, 1864. [CrossRef]
15. Yepes, V.; Torres-Machí, C.; Chamorro, A.; Pellicer, E. Optimal pavement maintenance programs based on a hybrid greedy randomized adaptive search procedure algorithm. *J. Civ. Eng. Manag.* **2016**, *22*, 540–550. [CrossRef]
16. Sierra, L.A.; Pellicer, E.; Yepes, V. Social sustainability in the life cycle of Chilean public infrastructure. *J. Constr. Eng. Manag.* **2016**, *142*, 05015020. [CrossRef]
17. Penadés-Plà, V.; García-Segura, T.; Martí, J.V.; Yepes, V. An optimization-LCA of a prestressed concrete precast bridge. *Sustainability* **2018**, *10*, 685. [CrossRef]

18. León, J.; Prieto, F.; Rodríguez, F. Proyecto de rehabilitación del puente de la Isla de Arosa. *Hormig. Acero* **2013**, *270*, 75–89.
19. Manera, M.; Vennesland, Ø.; Bertolini, L. Chloride threshold for rebar corrosion in concrete with addition of silica fume. *Corros. Sci.* **2008**, *50*, 554–560. [CrossRef]
20. Tuutti, K. *Corrosion of Steel in Concrete*; Report No 4-82; Swedish Cement and Concrete Research Institute: Stockholm, Sweden, 1982.
21. Fib Bulletin 34. *Model Code for Service Life Design*; International Federation for Structural Concrete: Lausanne, Switzerland, 2006.
22. Spanish Ministry of Public Works. *EHE-08 Instrucción del Hormigón Estructural*; Secretaría General Técnica del Ministerio de Fomento: Madrid, Spain, 2008.
23. Cheewaket, T.; Jaturapitakkul, C.; Chalee, W. Concrete durability presented by acceptable chloride level and chloride diffusion coefficient in concrete: 10-year results in marine site. *Mater. Struct.* **2014**, *47*, 1501–1511. [CrossRef]
24. Vedalakshmi, R.; Saraswathy, V.; Song, H.W.; Palaniswamy, N. Determination of diffusion coefficient of chloride in concrete using Warburg diffusion coefficient. *Corros. Sci.* **2009**, *51*, 1299–1307. [CrossRef]
25. Bertolini, L.; Bolzoni, F.; Pastore, T.; Pedeferri, P. Behaviour of stainless steel in simulated concrete pore solution. *Br. Corros. J.* **1996**, *31*, 218–222. [CrossRef]
26. Darwin, D.; Browning, J.; O'Reilly, M.; Xing, L.; Ji, J. Critical Chloride Corrosion Threshold of Galvanized Reinforcing Bars. *ACI Mater. J.* **2009**, *106*, 176–183. [CrossRef]
27. Ohama, Y. *Handbook of Polymer-Modified Concrete and Mortars*; William Andrew: Amsterdam, The Netherlands, 1995.
28. Yang, Z.; Shi, X.; Creighton, A.T.; Peterson, M.M. Effect of styrene–butadiene rubber latex on the chloride permeability and microstructure of Portland cement mortar. *Constr. Build. Mater.* **2009**, *23*, 2283–2290. [CrossRef]
29. Frederiksen, J.M. Chloride threshold values for service life design. In Proceedings of the Second International RILEM Workshop on Testing and Modelling the Chloride Ingress into Concrete, Paris, France, 11–12 September 2000.
30. Otsuki, N.; Nishida, T.; Yi, C.; Nagata, T.; Ohara, H. Effect of Blast Furnace Slag Powder and Fly Ash on Durability of Concrete Mixed with Seawater. In Proceedings of the Fourth International Conference on the Durability of Concrete Structures, Purdue University, West Lafayette, IN, USA, 24–26 July 2014; pp. 229–241. [CrossRef]
31. Zhang, J.Z.; Buenfeld, N.R. Chloride profiles in surface-treated mortar specimens. *Constr. Build. Mater.* **2000**, *14*, 359–364. [CrossRef]
32. Liu, G.; Stavem, P.; Gjørv, O.E. Effect of Surface Hydrophobation for Protection of Early Age Concrete against Chloride Penetration. In Proceedings of the Fourth International Conference on Water Repellent Treatment of Building Materials; Silfwerbrand, J., Ed.; Aedificatio Publishers: Freiburg, Germany, 2005; pp. 93–104.
33. Medeiros, M.; Castro-Borges, P.; Aleixo, D.; Quarcioni, V.A.; Marcondes, C.; Helene, P. Reducing water and chloride penetration through silicate treatments for concrete as a mean to control corrosion kinetics. *Int. J. Electrochem. Sci.* **2012**, *7*, 9682–9696. [CrossRef]
34. Baltazar, L.; Santana, J.; Lopes, B.; Rodrigues, M.P.; Correia, J.R. Surface skin protection of concrete with silicate-based impregnations: Influence of the substrate roughness and moisture. *Constr. Build. Mater.* **2014**, *70*, 191–200. [CrossRef]
35. Christodoulou, C.; Goodier, C.I.; Austin, S.A.; Webb, J.; Glass, G.K. Long-term performance of surface impregnation of reinforced concrete structures with silane. *Constr. Build. Mater.* **2013**, *48*, 708–716. [CrossRef]
36. Schueremans, L.; Van Gemert, D.; Giessler, S. Chloride penetration in RC-structures in marine environment—Long term assessment of a preventive hydrophobic treatment. *Constr. Build. Mater.* **2007**, *21*, 1238–1249. [CrossRef]
37. Sadati, S.; Arezoumandi, M.; Shekarchi, M. Long-term performance of concrete surface coatings in soil exposure of marine environments. *Constr. Build. Mater.* **2015**, *94*, 656–663. [CrossRef]
38. Pan, X.; Shi, Z.; Shi, C.; Ling, T.C.; Li, N. A review on concrete surface treatment Part I: Types and mechanisms. *Constr. Build. Mater.* **2017**, *132*, 578–590. [CrossRef]
39. Dai, J.G.; Akira, Y.; Wittman, F.H.; Yokota, H.; Zhang, P. Water repellent surface impregnation for extension of service life of reinforced concrete structures in marine environments: The role of cracks. *Cem. Concr. Compos.* **2010**, *32*, 101–109. [CrossRef]

40. Torres-Machi, C.; Pellicer, E.; Yepes, V.; Chamorro, A. Towards a sustainable optimization of pavement maintenance programs under budgetary restrictions. *J. Clean. Prod.* **2017**, *148*, 90–102. [CrossRef]

41. García-Segura, T.; Yepes, V.; Frangopol, D.M.; Yang, D.Y. Lifetime reliability-based optimization of post-tensioned box-girder bridges. *Eng. Struct.* **2017**, *145*, 381–391. [CrossRef]

42. Seshadri, P.; Harrison, R. *Workzone Mobile Source Emission Prediction*; Center of Transportation Research, University of Texas at Austin: Austin, TX, USA, 1993.

43. Gervásio, H.; da Silva, L.S. Life-cycle social analysis of motorway bridges. *Struct. Infrastruct. Eng.* **2013**, *9*, 1019–1039. [CrossRef]

44. De Rus, G. *Manual de Evaluación Económica de Proyectos de Transporte*; Cedex Research Project PT2007-001-IAPP; Banco Interamericano de Desarrollo: Washington, DC, USA, 2010.

45. Gervásio, H. Sustainable Design and Integral Life-Cycle Analysis of Bridges. Ph.D. Thesis, University of Coimbra, Coimbra, Portugal, 2010.

46. European Committee for Standarization. *EN 1990:2002: Eurocode—Basis of Structural Design*; European Committee for Standarization: Brussels, Belgium, 2002; ISBN 0-580-40186-3.

47. Walls, J.; Smith, M. *Life-Cycle Cost Analysis in Pavement Design: In Search of Better Investment Decisions*; FHWA-SA-98-079; FHWA, U. S. Department of Transportation: Washington, DC, USA, 1998.

48. Mistry, M.; Koffler, C.; Wong, S. LCA and LCCA of the world's longest pier: A case study on nickel-containing stainless steel rebar. *Int. J. Life Cycle Assess.* **2016**, *21*, 1637–1644. [CrossRef]

49. Fowler, D.W. Polymers in concrete: A vision for the 21st century. *Cem. Concr. Compos.* **1999**, *21*, 449–452. [CrossRef]

50. Lee, E.B.; Kim, C.; Harvey, J.T. Value analysis using performance attributes matrix for highway rehabilitation projects: California Interstate 80 Sacramento case. *Transp. Res. Rec. J. Transp. Res. Board* **2011**, *2228*, 23–32. [CrossRef]

51. Harvey, J.T.; Rezaei, A.; Lee, C. Probabilistic approach to Life Cycle Cost Analysis of preventive maintenance strategies on flexible pavements. *Transp. Res. Rec. J. Transp. Res. Board* **2012**, *2292*, 61–72. [CrossRef]

Article

An Optimization-LCA of a Prestressed Concrete Precast Bridge

Vicent Penadés-Plà [1], Tatiana García-Segura [2], José V. Martí [1] and Víctor Yepes [1,*]

[1] Institute of Concrete Science and Technology (ICITECH), Universitat Politècnica de València, 46022 Valencia, Spain; vipepl2@cam.upv.es (V.P.-P.); jvmartia@cst.upv.es (J.V.M.)

[2] Construction Engineering Department, Universitat Politècnica de València, 46022 Valencia, Spain; tagarse@upv.es

* Correspondence: vyepesp@cst.upv.es; Tel.: +34-96-387-9563; Fax: +34-96-387-7569

Received: 30 January 2018; Accepted: 1 March 2018; Published: 2 March 2018

Abstract: The construction sector is one of the most active sectors, with a high economic, environmental and social impact. For this reason, the sustainable design of structures and buildings is a trend that must be followed. Bridges are one of the most important structures in the construction sector, as their construction and maintenance are crucial to achieve and retain the best transport between different places. Nowadays, the choice of bridge design depends on the initial economic criterion but other criteria should be considered to assess the environmental and social aspects. Furthermore, for a correct choice, the influence of these criteria during the bridge life-cycle must be taken into account. This study aims to analyse the life-cycle environmental impact of efficient structures from the economic point of view. Life-cycle assessment process is used to obtain all the environmental information about bridges. In this paper, a prestressed concrete precast bridge is cost-optimized and afterwards, the life-cycle assessment is carried out to achieve the environmental information about the bridge.

Keywords: sustainability; bridges; life-cycle assessment; optimization; ReCiPe

---

1. Introduction

The basis for the definition of sustainable development lies in the Brundtland Commission's report [1], which describes it as *"development that meets the needs of the present generation without compromising the needs of the future generation"*. This idea implies the consideration of different aspects of three main components: economic, environmental and social. Therefore, achieving sustainable development implies a consensus among these three main pillars, which usually have different goals. Wass et al. [2] stated that sustainable development implies that a decision-making strategy must be considered. Decision making is a process that can help to find a solution that provides a compromise between different aspects and therefore achieves a sustainable solution [3,4].

The construction sector is one of the most active sectors and one of the ones with a greater influence on the economic, environment and social aspects of the world. This indicates a need for a trend toward sustainability of buildings and structures. One of the most important structures in this sector is bridges. The construction and maintenance of bridges are crucial to generate and keep the best transport possible between different places. For this reason, the assessment of sustainable development during the whole life-cycle is necessary. Of the three main components of sustainable development, the social aspect is the least studied and there are more doubts about its assessment. On the contrary, the economic and environmental aspects have been studied more intensively and it is convenient to assume that their consideration is sufficient. Considering the evaluation of these two components to achieve sustainability of bridges, the objective is to design the bridge with the lowest cost and lowest environmental impact. Although these two pillars of sustainability have different goals, some works have stated that there is a relationship between the cost and CO_2 emissions of structures [5,6]. Therefore, reducing the cost implies a reduction of CO_2 emissions.

Obtaining the lowest cost or CO_2 emissions have been studied by several works. Optimization algorithms are most often used to reduce the cost or CO_2 emissions of structures. In some cases, this involves a mono-objective optimization of cost and CO_2 emissions [5–7], whereas other works carry out multi-objective optimization to achieve both objectives at the same time [8,9]. Despite the relationship between cost and CO_2 emissions, the environmental impact cannot be assessed by taking into account CO_2 emissions alone [10]. For this reason, the environmental impact assessment must achieve a complete environmental profile. This complete environmental profile can be obtained using the life-cycle assessment (LCA) process. LCA is one of the most important and accepted methods of assessing the environmental impacts [11–16], making it an excellent tool for assessing the environmental impact of a bridge.

In this paper, a prestressed concrete precast 40 m bridge is selected as the subject of an optimization-LCA. The optimization of the cost will reduce the cost of the bridge directly and the associated CO_2 emissions indirectly. This process makes it possible to obtain a cost-optimized bridge with a low environmental impact. After finishing the optimization, all the features of the cost-optimized bridge will be known, including its cost but the environmental impact will not yet have been obtained. The LCA makes it possible to obtain a complete environmental profile of this cost-optimized bridge. With this methodology, a bridge whose costs have been optimized directly and whose environmental impact has been improved is obtained and finally the LCA for the whole life-time can be performed. For this purpose, a hybrid memetic algorithm is used to carry out the cost-optimization of the bridge. Then, the Ecoinvent database [17] and the ReCiPe method [18] are used to conduct the LCA process of the bridge.

2. Optimization

The optimization process is used to achieve the best solution to a problem. This process is a clear alternative to designs based on experience. Optimization methods can be categorized into exact methods and heuristic methods. On one hand, the exact methods are based on mathematical algorithms that make it possible to obtain the global optimal solution [19]. On the other hand, the heuristic methods, which include a large number of algorithms [20], obtain an optimal solution starting from an initial solution. The exact methods are very useful in problems where there are a small number of variables, because the computing time becomes unworkable for a large number of variables. Structural optimization problems are defined for a large number of design variables and thus the heuristic method is the most useful for structural optimization. There are a large number of works that use heuristic algorithms for the optimization of different kinds of structures [8,9,21].

3. Life-Cycle Assessment

Life-cycle assessment (LCA) is one of the most important and accepted methods of evaluating the environmental impact of a product, process, or service during its whole life-cycle, taking into account all the activities involved, which are defined as inputs and outputs. The limits defined for these inputs and outputs are the boundaries of the system and represent the scheme to be considered. The LCA must be complete and thus it should consider all the activities needed for the achievement of the product, process, or service. Therefore, focusing on the construction sector, a full LCA of structures must consider all the activities from the acquisition of the raw material to the end of life. These activities associated with the whole life-cycle of the structures are grouped into the manufacturing phase, construction phase, use and maintenance phase and end of life phase. The LCA makes it possible to carry out an environmental impact assessment of a set of activities associated with the different stages of a structure's life-cycle and the global environmental impact by adding these phases. For all that, the LCA is an excellent tool to evaluate the environmental impact of structures. ISO 14040:2006 [22] provides guidance on carrying out the LCA, divided into four steps: (1) definition of goal and scope; (2) inventory analysis; (3) impact assessment; and (4) interpretation.

The first step defines all the specifications that will be considered in the LCA. This involves other features besides the definition of the goal and scope, such as the life-cycle inventory to be taken into account, the life-cycle assessment methodology considered, the functional unit and the assumptions and limitations that have been considered in the LCA. According to the guidance defined by ISO 14040:2006 [22], the characterization defines some assumptions and limitations of the LCA that condition the following life cycle inventory and life cycle assessment. Another important feature is the functional unit that represents the unit in which the assessment will be referred.

The inventory analysis is the collection of the data needed to define the inputs and outputs that represent the system studied. This data can be obtained in different ways: from direct measurements, literature, or other sources such as databases. The most common way to obtain data is from databases.

Once these first steps have been defined, the environmental impact assessment is used to evaluate the result of the inventory analysis to obtain a set of environmental indicators that represent the environmental profile of the product, process, or service. There are different methods of representing the environmental profile. These methods can be grouped into two different approaches: midpoint and endpoint assessments. The midpoint approach defines the environmental profile by means of a set of impact categories and the endpoint approach defines the environmental profile by means of a set of damage categories. There are three damage categories (human health, resource depletion and ecosystems) into which the impact categories are clustered. Therefore, although the midpoint approach provides a complete environmental profile, it is more difficult to interpret [23]. Conversely, the endpoint approach does not provide a detailed environmental profile like the midpoint approach but is easier to understand.

Finally, the information obtained must be interpreted. For this purpose, an analysis of the different stages of life-cycle of the bridge is carried out. In addition, a study of the environmental impact of a product, process, or service can be made to improve the environmental impact associated with its activities.

4. Case Study

For the purpose of this work, a bridge is selected to carry out the optimization-LCA. First, a cost-optimization of the bridge will be carried out and then a LCA of the cost-optimized bridge will be applied to obtain a complete environmental profile. In the next points, a precise description of the bridge will be presented and then the cost-optimization and the LCA will be described in detail for the bridge described.

4.1. Bridge Description

The bridge studied is a single span prestressed concrete precast bridge of 40 m. The section of the bridge is formed by two prestressed concrete precast isostatic beams with a U-shaped cross-section. The cross-section integrates a 12 m upper reinforced concrete slab. Note that the substructure is not included in the analysis since it depends on the ground characteristics and the orography. Figure 1 shows a general view of the bridge. The bridge is located in the eastern coastal area of Spain and the environmental ambient corresponds to XC-4 according to EN 206-1 [24]. Thus, corrosion is mainly caused by carbonation.

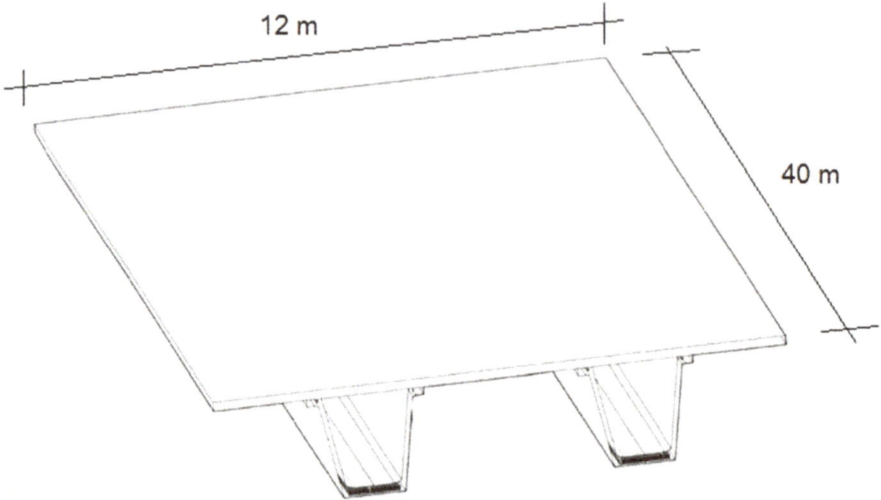

Figure 1. General view of the prestressed concrete precast bridge.

4.2. Optimization

In this section, the cost-optimization of the prestressed concrete precast bridge will be explained. This optimization process consists in the minimization of the cost C while some restrictions g_j are satisfied.

$$C = f(x_1, x_2, \ldots, x_n) \tag{1}$$

$$g_j(x_1, x_2, \ldots, x_n) \leq 0 \tag{2}$$

Note that $x_1, x_2, \ldots, x_n$ are the design variables used for the optimization. The objective function C expresses the cost of the bridge and the restrictions g_j are the serviceability limit states (SLS), the ultimate limit states (ULS), the durability limit states and the geometric and constructability constraints of the problem. There are 40 design variables, including eight variables that define the geometry of the section, two that define the concrete of the slab and the beam, four that define the prestressed steel and 26 that define the reinforcing steel. Furthermore, there are a set of parameters that have no influence on the optimization problem, such as the width, span and web inclination. Structural constraints have been considered according to the Spanish codes [25,26]. The ULSs verify if the ultimate resistance is greater than the ultimate load effect. Besides, the minimum amount of reinforcing steel for the stress requirements and the geometrical conditions are also considered. The SLSs examine different aspects. Cracking limit state requires compliance of the compression and tension cracks, as well as the decompression limit state in the area where the post-tensioned steel is located. Deflections are limited to 1/1000 of the free span length for the quasipermanent combination. In addition, the concrete and steel fatigue has been considered in this study. Table 1 summarizes of the ULSs and SLSs considered.

In this optimization, a hybrid memetic algorithm (MA) is applied. The MA is a population-based approach to stochastic optimization that combines the parallel search used by evolutionary algorithms with a local search of the solutions forming a population [27]. Regarding the local search used, a variable-depth neighbourhood search (VDNS) is used as a variant of the very large-scale neighbourhood search (VLSN) [28]. In this MA-VDNS, a set of 500 random solutions (*n*) is generated as the population. Then each of these solutions is improved by means of a VDNS search to reach a local optimum. To this end, the algorithm begins by changing only one variable and when ten consecutive movements have been performed without improvement (*no_imp*), there will be an increase in the number of variables (*var*) that are changed simultaneously, up to eight. Then, with this new improved population, a genetic algorithm is applied. The genetic algorithm develops the population, which is subjected to random movements (mutations and crossovers), preserving the better adapted solutions. The cost assessment takes into account a penalty cost; nevertheless, the VDNS does not consider the penalty cost (only feasible solutions are accepted) in order to avoid the early divergence of the algorithm. The VDNS is applied to the new generation up to 150 generations. Figure 2 shows a flow chart of the hybrid memetic algorithm.

Table 1. Ultimate and serviceability limit states.

Limit States
Flexure
Vertical shear
Longitudinal shear
Punching shear
Torsion
Torsion combined with flexure and shear
Fatigue
Crack width <0.2 mm
Compression and tension stress. Decompression in post-tensioned steel depth
Deflection for the quasipermanent combination <1/1000

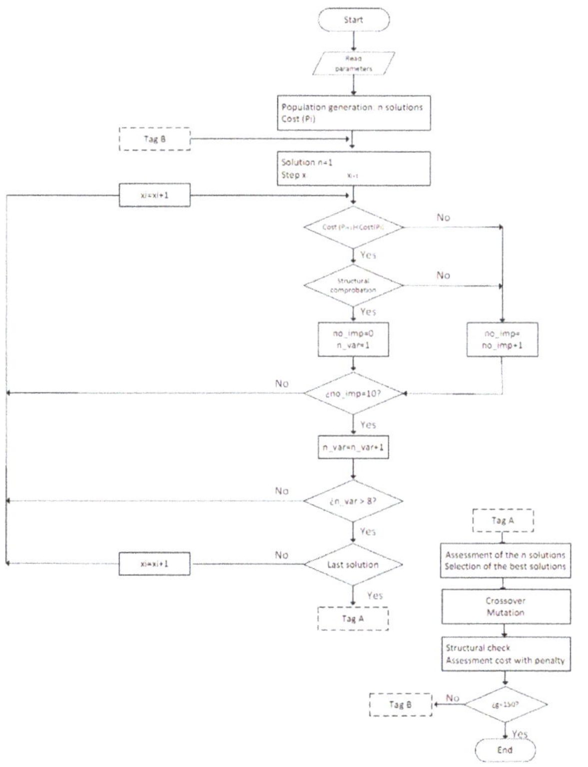

Figure 2. Hybrid memetic algorithm flow chart.

The solution obtained for the 40 m-long prestressed concrete precast bridge has a total cost of 108,274.45 €. The geometry of this bridge is shown in Figure 3. The amount of beam concrete used is 0.1117 m^3/m^2, with a strength of 35 MPa, while the amount of slab concrete used is 0.1797 m^3/m^2, with a strength of 40 MPa. Furthermore, the precast concrete beams require 6163 kg (12.52 kg/m^2) of reinforcing steel and 5184 kg (10.53 kg/m^2) of prestressed steel, while the concrete slab is defined by 11,772 kg (23.92 kg/m^2) of reinforcing steel.

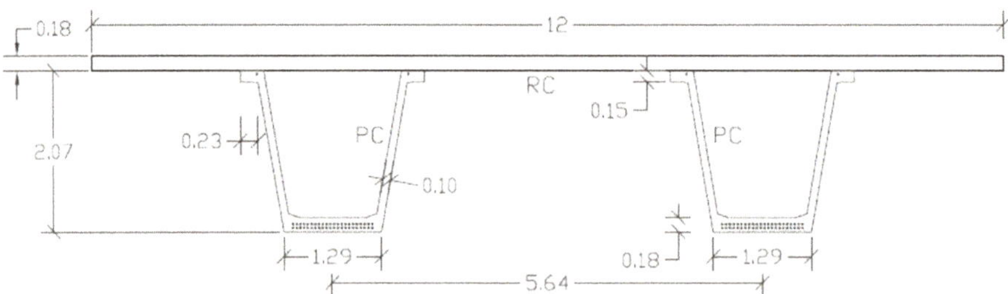

Figure 3. Geometry of optimized bridge.

4.3. Life-Cycle Assessment

In this section, the guidance defined by ISO 14040:2006 [22] will be applied to the bridge studied. For this purpose, the different steps will be particularized to the case of study, describing and taking into account the specific characteristics considered for this study. Figure 4 show a general view of the LCA process carried out.

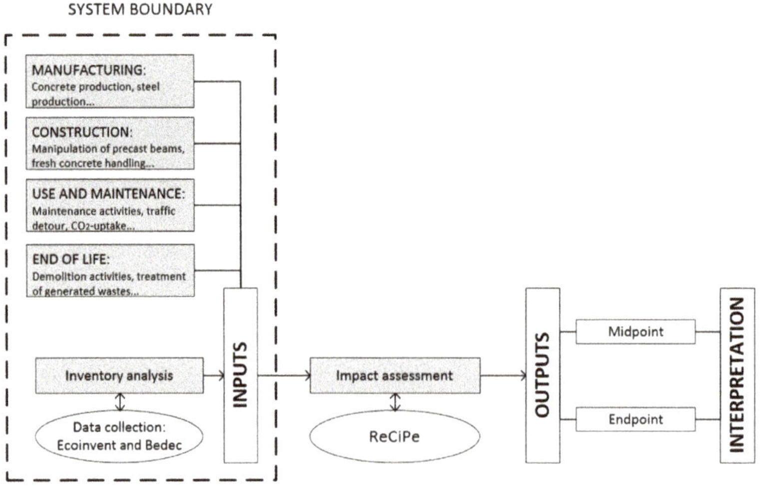

Figure 4. General scheme LCA process.

4.3.1. Goal and Scope

The LCA will be divided into the four main phases of the whole life-cycle of the bridge for a better understanding: (1) manufacturing; (2) construction; (3) use and maintenance; and (4) end of life. Each phase will be defined separately and thus each phase will be limited by its own system boundary. The functional unit will be 1 m of the length of the bridge. The final goal is to find the environmental impact of each phase and consequently the global environmental impact of the bridge by adding the environmental impacts of different phases.

Manufacturing

The manufacturing phase includes the upstream processes of the products used in the bridge and the associated transport, from the acquisition of raw materials to materials that are ready to be used in the construction of the bridge. The prestressed concrete precast bridge has three main components: beams of precast concrete, fresh concrete and steel. Therefore, first it is necessary delimit the activities associated with each product including the transport.

On one hand, the manufacture of the beams of precast concrete takes into account all the activities from the extraction of raw materials to the finishing of the beams in the precast plant, while the manufacture of the fresh concrete for the slab takes into account the activities from the extraction of raw material to the point when the concrete is ready to be used in the construction place. In both cases, the distance considered between the quarry and the precast plant or concrete plant is 50 km, the distance considered in the cement transportation is 20 km and the distance between the concrete plant and the construction site is 50 km. Furthermore, the dosage of concrete is taken into account to achieve the strength required. On the other hand, the manufacture of the steel takes into account all the activities from the acquisition of the raw material to the point when the steel is ready to be used in the precast plant or construction site. Considering that the bridge is built in Spain, the analysis takes the Spanish steel production characteristics. This implies that 67% of the steel is produced in an electric arc furnace and the remaining 33% is produced in a basic oxygen furnace. This ratio generates a recycling rate of steel of 71%. The distance considered between the steel production plant and the precast plant or construction site is 100 km. Table 2 shows the amount of material needed for the beam and slab and the dosage of the concrete in both cases.

Table 2. Amount of materials.

	Precast Concrete Beam	Concrete Slab
Strength (MPa)	35	40
Reinforcing steel (kg/m^2)	12.52	23.92
Prestressed steel (kg/m^2)	10.53	–
Concrete (m^3/m^2)	0.1117	0.1797
Cement (kg/m^3)	300	320
Gravel (kg/m^3)	848	829
Sand (kg/m^3)	1088	1102
Water (kg/m^3)	160	162
Superplasticizer (kg/m^3)	4	5

Construction

The construction phase includes all the materials and construction machinery necessary for the erection of the bridge. It includes the transportation and elevation of precast beams using special transport over 50 km. Furthermore, the bridge slab is considered to be cast in place. The construction machinery considered for the slab construction was obtained from the Bedec database [29]. The concrete machinery consumes 123.42 MJ of energy and emits 32.24 kg of CO_2 per m^3 of concrete. The distance travelled considered by the construction machinery is 50 km. In addition, the formwork is made by wood and can be reused 3 times.

Use and Maintenance

The maintenance and use phase includes everything that happens in the service life of the bridge. It takes some activities and processes (considering its own maintenance activities and the traffic detour due to the closure of the bridge) and the fixed CO_2. On one hand, the bridge needs one maintenance period of 2 days to satisfy with the regulations during its 120 years of service life. This maintenance activity considers that the concrete cover is replaced by a repair mortar. The maintenance action consists firstly of removing the concrete cover and providing a proper surface for the coating adhesion. Then, a bonding coat is applied between the old and new concrete. Finally, a repair mortar is placed to provide a new reinforcement corrosion protection [30]. Note that the study considers that the quality on-site work is adequate to guarantee that the bridge does not have durability problems during the service life. Besides, it is important to highlight that other maintenance activities to repair or replace equipment elements may take place. However, they are not evaluated in this study.

This study takes into account all the machinery necessary to repair the deterioration of the bridge including the transport to the bridge location and the increase in emissions generated due to the traffic detour [13,14]. The traffic detour is considered taking into account the average daily traffic of 8500 vehicles/day, where trucks comprise 10% of vehicles and a detour distance of 2.9 km. On the other hand, the fixation of the CO_2 by the concrete is a widely studied fact [31,32] that has been considered in the bridge studied.

End of Life

The end-of-life phase includes everything that happens after the service life of the bridge. All the activities and processes associated with this phase are related with the demolition of the bridge and the treatment of the generated wastes. On one hand, demolition activities for the destruction or dismantling of the bridge will be necessary. These demolition activities take into account all the machinery necessary for this purpose. On the other hand, the treatment of generated wastes takes into account a greater set of activities depending on the purpose of the processing. In this case, the bridge will be destroyed, after which all the wastes will be transported to a sorting plant where the concrete and steel will be separated. The concrete will be crushed and transported to a landfill and in this way, the complete carbonation of the concrete [32] and thus a higher fixation of CO_2 is assured. Seventy-one per cent of the steel will be recycled and in this way, the life-cycle of the bridge ends.

4.3.2. Inventory Analysis

The major part of the information of the products or processes used to define the activities of the whole life-cycle of the bridge is obtained from Ecoinvent database [17]. In the case of the information of the products or processes needed for the environmental impact assessment that do not exist in the Ecoinvent database, the data will be created by means of the data obtained from the literature or the Bedec database [29].

The Ecoinvent database is one of the most complete databases for the construction sector and has been created and grown thanks to the information obtained from different institutions. It was created in 2004 through the efforts of the several Swiss Federal Offices and research institutes. That implies that the major part of the information existing in the first versions of Ecoinvent was obtained from Swiss institutions but later, data from other countries were inserted. In this case, the bridge is located on the eastern coast of Spain. In the Ecoinvent database there is no information about this region and therefore it is necessary to consider information about the products or processes from other regions that do not coincide exactly with the products or processes used on the eastern coast of Spain. That means that there is inconsistency between the real data and the data from the Ecoinvent database. For this reason, uncertainty is applied to the Ecoinvent data. The uncertainty is divided into two parts: the first part concerns the type of product or process [33] and the second part concerns the differences between the real data and the data considered by means of the pedigree matrix [34].

4.3.3. Impact Assessment

There are many works in which the environmental impact assessment is carried out taking into account a small number of indicators, of which the CO_2 emissions are the most popular [35,36]. Despite the importance of the emission of CO_2, a complete impact assessment must consider a set of indicators that represent a complete environmental profile. That implies the use of environmental impact assessment methods. These methods can be separated depending on the approach used: midpoint or endpoint. On one hand, the midpoint approach defines the environmental profile by means of a set of impact categories. One of the most popular methods that take into account the midpoint approach is the CML. On the other hand, the endpoint approach defines the environmental profile considering only a small set of damage categories. One of the most frequently used methods that consider the endpoint approach is the Eco-indicator. Both approaches are necessary to carry out a complete environmental interpretation of the bridge. On one hand, the midpoint approach can provide a more accurate and complete environmental profile. On the other hand, the endpoint approach can be easier to interpret. For these reasons, the environmental impact assessment method used in this work is the ReCiPe method [18], whose main objective is to provide a combination of the Eco-indicator and CML, considering the midpoint and endpoint approaches.

4.3.4. Interpretation

The results are obtained considering the descriptions presented in the preceding sections. As stated above, the ReCiPe method will be used to carry out the environmental impact assessment of the bridge. For this purpose, by means of the midpoint approach, 18 impact categories will be shown with the associated uncertainty. In addition, the contribution of the different processes of the bridge life-cycle for the most popular impact categories will be represented. In the endpoint approach, the three damage categories are studied. Both approaches allow a higher level of interpretation.

Midpoint Approach

The midpoint approach of the ReCiPe method provides a complete environmental profile of each stage of the bridge life-cycle represented by 18 impact categories: agricultural land occupation (ALO), climate change (GWP), fossil depletion (FD), freshwater ecotoxicity (FEPT), freshwater eutrophication (FEP), human toxicity (HTP), ionizing radiation (IRP), marine ecotoxicity (MEPT), marine eutrophication (MEP), metal depletion (MD), natural land transformation (NLT), ozone depletion (OD), particulate matter formation (PMF), photochemical oxidant formation (POFP), terrestrial acidification (TAP), terrestrial ecotoxicity (TEPT), urban land occupation (ULO) and water depletion (WD). This large amount of information makes the results difficult to interpret. Although it is difficult to achieve a global assessment of the environmental impact of the bridge with the information obtained by means of the midpoint approach, it is very helpful to obtain more accurate knowledge of the impact of each category and the contribution of each process to the different impact categories.

As explained above, the data used for the environmental impact assessment do not correspond with the real data. This implies that the uncertainty associated with the different products or processes should be taken into account to obtain more realistic results. Table 3 shows the mean and coefficient of variance of each impact category for each bridge life-cycle phase. Although it is not possible to carry out a global assessment for each bridge life-cycle phase, it is possible to obtain information about the phase in which each impact category is the most significant and the variance of the information obtained. In this way, it can be observed that the manufacturing phase is the phase in which there are a higher number of impact categories with the highest contribution followed by the use and maintenance phase. The impact categories with the highest contribution in the manufacturing phase are ALO, GWP, FEPT, FEP, HTP, IRP, MEPT, MD, TETP, ULO and WD and the impact categories with the highest contributions to the use and maintenance phase are FD, MEP, NLT, ODP, PMFP, POFP and TAP. Neither the construction phase nor the end of life phase has impact categories with the highest contribution. All of this can be seen better in Figures 5 and 6, in which the bars represent the ratio of the contribution of each impact category to each life-cycle phase in relation to the highest contribution. In addition, Table 3 shows the variance of each result. In this way, although the GWP has the highest variance in the manufacturing phase, the manufacturing phase is the one in which more impact categories have the lowest variance, with a mean of 7.13%. The construction phase has the highest mean of variances (17.15%), followed by the end-of-life phase (13.16%) and the use and maintenance phase (10.58%). Furthermore, the impact category with the highest coefficient of variation is the ULO (17.28%) and the impact category with the lowest coefficient of variation is the ALO (8.04%).

Another type of information that can be obtained by the midpoint approach is the contribution of the different products or processes to each impact category. For illustrative purposes, only three of the most popular impact categories (GWP, OD and PMF) will be studied more exhaustively and will display the contribution of the different products or processes to each bridge life-cycle phase. Figures 7–10 show the contributions of the most important processes for each bridge life-cycle phase. Figure 7 corresponds to the manufacturing phase and it is possible to see that the most important associated processes are the cement production, steel production and transport. Cement production makes the highest contribution to the GWP, namely 46.49% of the total but in the PMF and OD categories, steel production has the higher ratio with percentages of 76.14 and 57.44% respectively. Furthermore, it can be seen that, although the GWP has a low percentage of other processes (6.07%), the cement production, steel production and transport represent a larger part of the environmental impact of this bridge life-cycle phase. Figure 8 corresponds to the construction phase and the processes that lead to practically all the environmental impacts are those due to the manipulation of fresh concrete and the transport and elevation of the precast beams. Figures 9 and 10 show the use and maintenance phase and end-of-life phase, in which the CO_2 fixed is taken into account. In the GWP impact category, it can be seen that there is a positive impact. On one hand, in the use and maintenance phase, the amount of CO_2 fixed is much lower than the CO_2 eq produced by the maintenance activities and the traffic detour because the concrete surface in contact with the environment represents a very low proportion of the total of amount of concrete in the bridge. The percentage of the CO_2 fixed is −3.84%, while the percentages of maintenance activities and traffic detour are 89.95% and 13.89%, respectively, adding a total of 100% due to that the global GWP impact in this phase is positive. The ratio of the contribution of the maintenance activities and traffic detour can be modified considerably in function of the features of the traffic diversion (distance, average daily traffic and percentage of trucks). On the other hand, in the end-of-life phase, the amount of CO_2 fixed is higher (−254.05%) than the CO_2 eq produced by the demolition activities (22.40%), the waste treatment (36.21%) and the associated transport (96.18%). The total contribution of the processes in the end-of-life phase is negative, adding a total of −100%. In the other impact categories (PMF and OD), the maintenance activities and transport make the major contribution to each bridge life-cycle.

Table 3. Midpoint approach.

Acronym	Reference Unit	Manufacturing		Construction		Use and Maintenance		EoL	
		m	cv (%)	m	cv (%)	m	cv (%)	m	cv (%)
ALO	$m^2 \times$ year	79.76	3.77%	2.59	7.46%	6.16	14.09%	1.73	6.84%
GWP	kg CO_2 eq	1838.55	16.86%	267.85	9.61%	1095.77	5.29%	−117.68	−6.97%
FD	kg oil eq	316.90	6.90%	51.48	17.52%	394.59	4.94%	11.00	16.57%
FEPT	kg 1,4-DB eq	38.15	2.93%	0.93	18.86%	8.53	26.70%	0.19	7.94%
FEP	kg P eq	0.82	4.19%	0.01	10.56%	0.08	14.00%	0.01	7.16%
HTP	kg 1,4-DB eq	1470.92	3.01%	22.58	16.26%	110.30	16.36%	5.77	7.80%
IRP	kg U235 eq	244.70	12.29%	18.96	10.35%	78.57	5.22%	10.22	7.14%
MEPT	kg 1,4-DB eq	37.90	2.92%	0.96	17.91%	7.65	26.08%	0.17	8.01%
MEP	kg N eq	0.29	8.79%	0.05	20.79%	0.49	2.90%	0.01	22.26%
MD	kg Fe eq	926.19	3.22%	5.34	17.35%	49.38	11.06%	0.77	22.90%
NLT	m^2	0.24	8.28%	0.05	18.78%	0.43	4.67%	0.01	24.03%
ODP	kg CFC-11 eq	0.00	8.59%	0.00	17.82%	0.00	4.61%	0.00	17.63%
PMFP	kg PM_{10} eq	3.84	5.67%	0.50	19.84%	4.33	3.12%	0.11	20.31%
POFP	kg NMVOC	5.76	9.12%	1.51	21.63%	14.03	2.77%	0.26	26.69%
TAP	kg SO_2 eq	5.30	8.90%	1.00	19.21%	8.40	3.12%	0.25	16.97%
TETP	kg 1,4-DB eq	0.45	4.60%	0.02	27.71%	0.06	12.68%	0.00	15.79%
ULO	$m^2 \times$ year	23.29	9.86%	3.50	29.32%	6.75	21.00%	0.17	8.93%
WD	m^3	8807.20	8.35%	219.49	7.63%	625.36	11.89%	146.17	6.83%

Figure 5. Impact categories of manufacturing and construction stage.

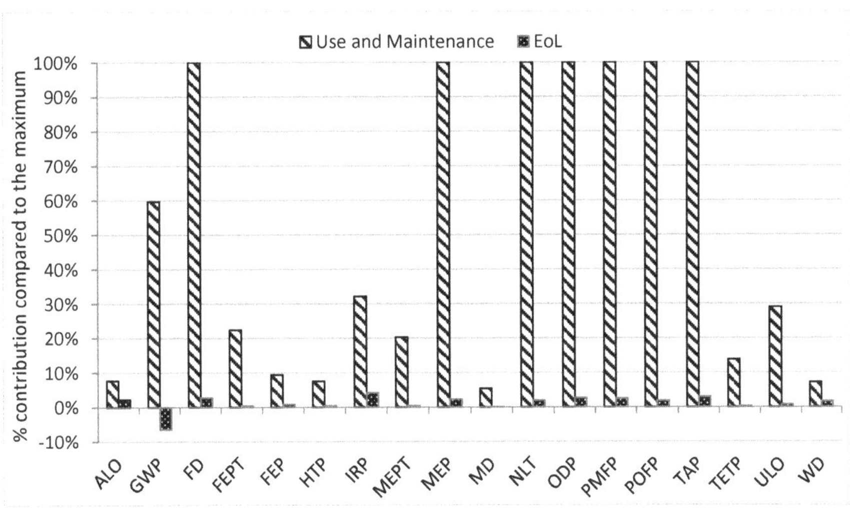

Figure 6. Impact categories of use and end of life stage.

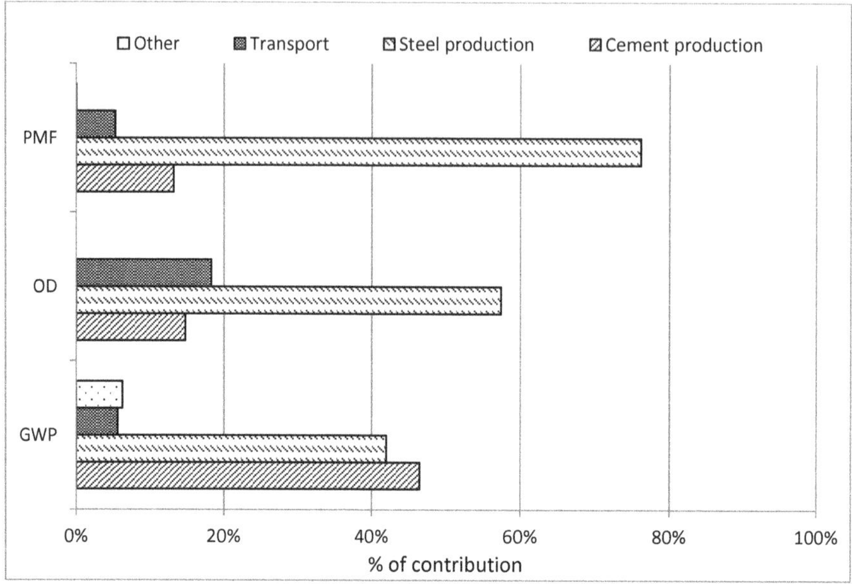

Figure 7. Manufacturing phase.

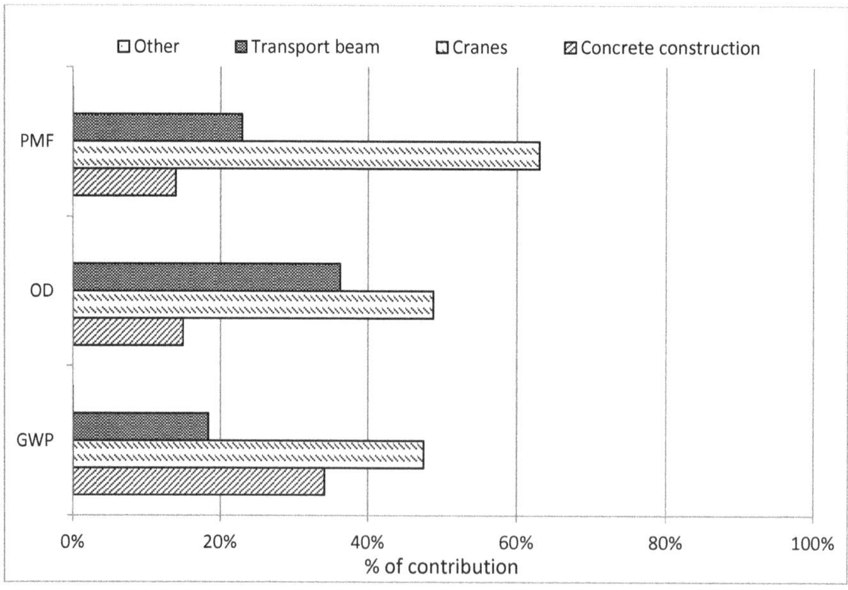

Figure 8. Construction phase.

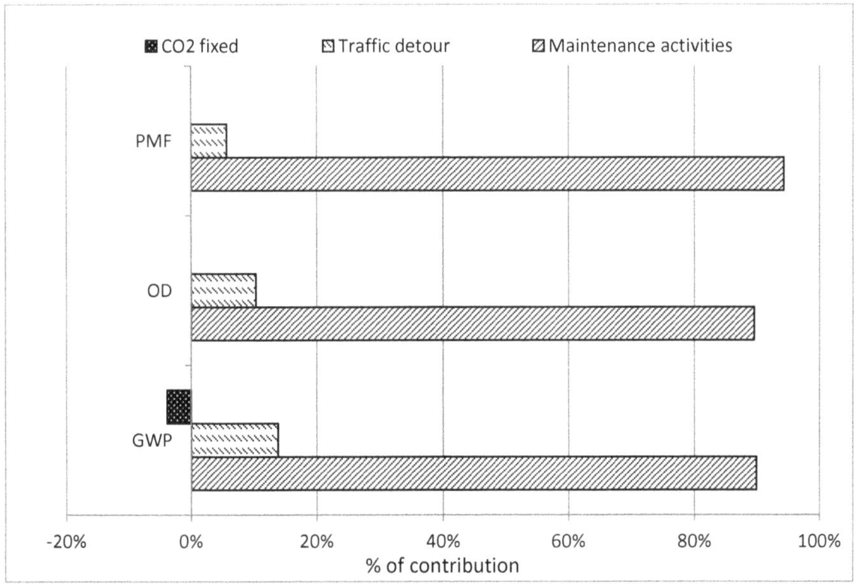

Figure 9. Use and maintenance phase.

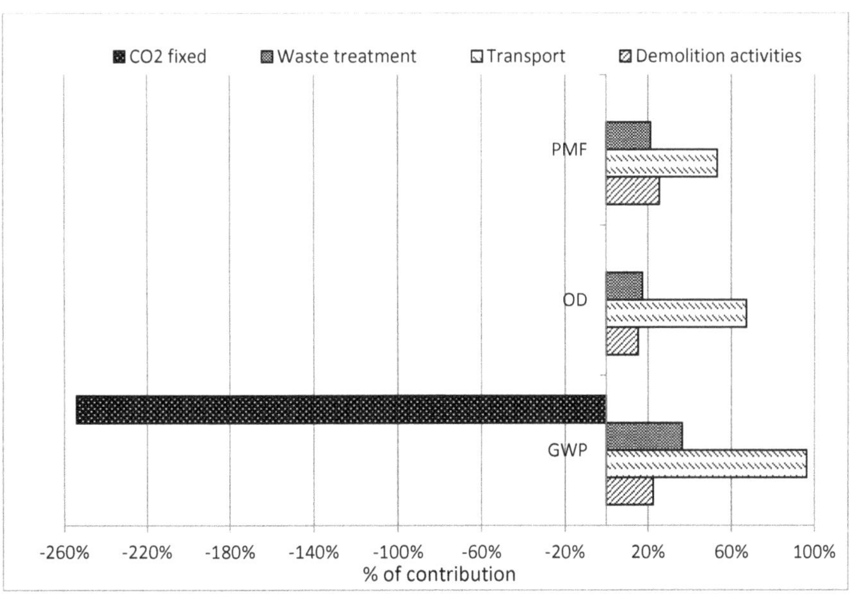

Figure 10. End-of-life phase.

Endpoint Approach

Despite the large amount of information obtained by means of the midpoint approach, it is very difficult to obtain a global environmental impact assessment. For this purpose, the endpoint approach is more useful. This approach provides only three damage categories (human health, resources and ecosystem), which are easier to interpret. Table 4 shows the mean and coefficient of variance of the three damage categories. Although the reference unit of the different damage categories remains different, carrying out the normalization and weighting of three categories is easier than doing so for 18 categories. In fact, ReCiPe allows the normalization of the three damage categories by converting the reference unit of each damage category into points. That makes it easier to interpret the global environment assessment of the bridge. Figure 11 shows the normalized value of each damage category of the whole life-cycle of the bridge and Figure 12 displays the contribution of each phase considering that the different damage categories have the same importance. On one hand, Figure 11 shows that human health is the most important damage category, followed by resources and ecosystem. On the other hand, in Figure 12 the contribution of different phases using the endpoint approach can be seen. The manufacturing phase is the phase with the highest contribution to the bridge life-cycle, followed by the use and maintenance phase and both the construction phase and the end-of-life phase make very low contributions compared to the other two phases.

Table 4. Endpoint approach.

Damage Category	Reference Unit	Manufacturing		Construction		Use and Maintenance		EoL	
		m	**cv (%)**	**m**	**cv (%)**	**m**	**cv (%)**	**m**	**cv (%)**
Human health	DALY	2.03×10^{-5}	11.69%	2.36×10^{-6}	11.68%	1.01×10^{-5}	4.89%	-8.86×10^{-7}	11.70%
Resource	$	1.19×10^{2}	4.01%	8.78×10^{0}	16.90%	6.91×10^{1}	5.60%	1.88×10^{0}	16.86%
Ecosystem	Species per year	4.58×10^{-3}	13.53%	5.18×10^{-4}	9.75%	2.75×10^{-3}	6.26%	-1.33×10^{-4}	7.64%

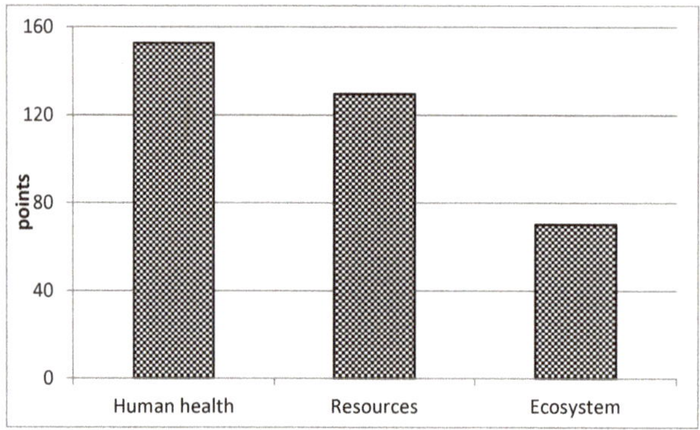

Figure 11. Damage categories.

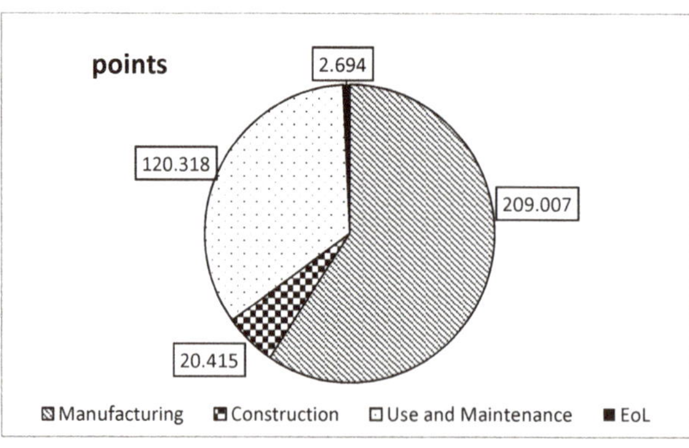

Figure 12. Contribution of bridge life-cycle phases.

5. Conclusions

Reduction of the environmental impact is a trend that must be taken into account due to the environmental problems that exist nowadays. In this respect, the construction sector has a large margin for improvement. The design of structures or buildings must consider the aspects of three pillars of sustainability. The assessment of the environmental impact during the whole life is a factor that must be taken into account in the design of structures or buildings. Although CO_2 emissions are not the only indicator to be considered in the environmental assessment, due to the relationship of this indicator with the cost, it is used to obtain a bridge with the lowest cost and a low environmental impact. Once this bridge has been obtained, a complete environmental assessment is carried out. For this purpose, a heuristic optimization by means of a hybrid memetic algorithm is used to obtain a cost-optimized prestressed concrete precast bridge and thus a low amount of associated CO_2. Then, the midpoint and endpoint approaches of the ReCiPe method are used to obtain a complete environmental profile of the bridge. These different approaches make it possible to obtain complementary data that provide different information. While the midpoint approach provides detailed information, the endpoint approach provides more concentrated information so it is possible to obtain only one score to assess all the environmental impacts.

Regarding the results of the midpoint approach, the manufacturing phase and use and maintenance phase are the phases with the higher environmental impact. With this knowledge, it is interesting to determine the processes that make the biggest contributions in these phases to try to reduce the environmental impact. Cement production and steel production are the processes with the highest environmental impact in the manufacturing phase, while the maintenance activities have the most environmental impact in the use and maintenance phase. Therefore, the midpoint approach indicates the process with the highest contribution in each impact category and in this way, it is possible to know which process to modify depending on the impact category to be improved. The midpoint approach provides detailed information but does not offer a single score that represents the global environmental impact of the bridge. For this purpose, the endpoint approach is used. As can be deduced, in the midpoint approach, the manufacturing phase and the use and maintenance phase are the ones with the higher environmental impact.

After studying both the midpoint and endpoint approaches, the results show the need for a complete environmental profile to evaluate the environmental impact of the bridge. The midpoint approach provides information that makes it possible to identify the processes in which improvements should be carried out to improve specific impact categories of the bridge but the endpoint approach provides a single score that is able to evaluate the global environmental impact of the bridge. Furthermore, although CO_2 emissions are an important indicator in the environmental impact assessment, in some cases it is not sufficient to obtain an accurate environmental evaluation and it is necessary to take into account all the other impact categories.

Acknowledgments: The authors acknowledge the financial support of the Spanish Ministry of Economy and Competitiveness, along with FEDER funding (Project: BIA2017-85098-R).

Author Contributions: This paper represents a result of teamwork. The authors jointly designed the research. Vicent Penadés-Plà drafted the manuscript. Tatiana García-Segura, José V. Martí and Víctor Yepes edited and improved the manuscript until all authors are satisfied with the final version.

Conflicts of Interest: The authors declare no conflict of interest.

References

1. United Nations. *World Commission on Environment and Development Our Common Future*; United Nations: New York, NY, USA, 1987.
2. Waas, T.; Hugé, J.; Block, T.; Wright, T.; Benitez-Capistros, F.; Verbruggen, A. Sustainability Assessment and Indicators: Tools in a Decision-Making Strategy for Sustainable Development. *Sustainability* **2014**, *6*, 5512–5534. [CrossRef]

3. Penadés-Plà, V.; García-Segura, T.; Martí, J.; Yepes, V. A review of multi-criteria decision-making methods applied to the sustainable bridge design. *Sustainability* **2016**, *8*, 1295. [CrossRef]

4. Zavadskas, E.K.; Antucheviciene, J.; Vilutiene, T.; Adeli, H. Sustainable decision making in civil engineering, construction and building technology. *Sustainability* **2018**, *10*, 14. [CrossRef]

5. Yepes, V.; Martí, J.V.; García-Segura, T. Cost and CO_2 emission optimization of precast-prestressed concrete U-beam road bridges by a hybrid glowworm swarm algorithm. *Autom. Constr.* **2015**, *49*, 123–134. [CrossRef]

6. Camp, C.V.; Assadollahi, A. CO_2 and cost optimization of reinforced concrete footings using a hybrid big bang-big crunch algorithm. *Struct. Multidiscip. Optim.* **2013**, *48*, 411–426. [CrossRef]

7. García-Segura, T.; Yepes, V.; Martí, J.V.; Alcalá, J. Optimization of concrete I-beams using a new hybrid glowworm swarm algorithm. *Lat. Am. J. Solids Struct.* **2014**, *11*, 1190–1205. [CrossRef]

8. García-Segura, T.; Yepes, V. Multiobjective optimization of post-tensioned concrete box-girder road bridges considering cost, CO_2 emissions, and safety. *Eng. Struct.* **2016**, *125*, 325–336. [CrossRef]

9. García-Segura, T.; Yepes, V.; Frangopol, D.M. Multi-objective design of post-tensioned concrete road bridges using artificial neural networks. *Struct. Multidiscip. Optim.* **2017**, *56*, 139–150. [CrossRef]

10. Laurent, A.; Olsen, S.I.; Hauschild, M.Z. Limitations of carbon footprint as indicator of environmental sustainability. *Environ. Sci. Technol.* **2012**, *46*, 4100–4108. [CrossRef] [PubMed]

11. Du, G.; Karoumi, R. Life cycle assessment of a railway bridge: Comparison of two superstructure designs. *Struct. Infrastruct. Eng.* **2012**, *9*, 1149–1160. [CrossRef]

12. Du, G.; Safi, M.; Pettersson, L.; Karoumi, R. Life cycle assessment as a decision support tool for bridge procurement: Environmental impact comparison among five bridge designs. *Int. J. Life Cycle Assess.* **2014**, *19*, 1948–1964. [CrossRef]

13. Hammervold, J.; Reenaas, M.; Brattebø, H. Environmental life cycle assessment of bridges. *J. Bridge Eng.* **2013**, *18*, 153–161. [CrossRef]

14. Pang, B.; Yang, P.; Wang, Y.; Kendall, A.; Xie, H.; Zhang, Y. Life cycle environmental impact assessment of a bridge with different strengthening schemes. *Int. J. Life Cycle Assess.* **2015**, *20*, 1300–1311. [CrossRef]

15. Zastrow, P.; Molina-Moreno, F.; García-Segura, T.; Martí, J.V.; Yepes, V. Life cycle assessment of cost-optimized buttress earth-retaining walls: A parametric study. *J. Clean. Prod.* **2017**, *140*, 1037–1048. [CrossRef]

16. Penadés-Plà, V.; Martí, J.V.; García-Segura, T.; Yepes, V. Life-cycle assessment: A comparison between two optimal post-tensioned concrete box-girder road bridges. *Sustainability* **2017**, *9*, 1864. [CrossRef]

17. Ecoinvent Center Ecoinvent v3.3. 2016. Available online: http://www.ecoinvent.org/database/older-versions/ecoinvent-33/ecoinvent-33.html (accessed on 15 January 2018).

18. Goedkoop, M.; Heijungs, R.; Huijbregts, M.; Schryver, A.D.; Struijs, J.; Van Zelm, R. *ReCiPe 2008. A Life Cycle Impact Assessment Which Comprises Harmonised Category Indicators at Midpoint and at the Endpoint Level*; Ministerie van Volkshuisvesting, Ruimtlejke Ordening en Milieubeheer: Den Haag, The Netherlands, 2009. Available online: https://www.researchgate.net/profile/Mark_Goedkoop/publication/230770853_Recipe_2008/links/09e4150dc068ff22e9000000.pdf (accessed on 15 January 2018).

19. Cohn, M.Z.; Dinovitzer, A.S. Application of structural optimization. *J. Struct. Eng.* **1994**, *120*, 617–650. [CrossRef]

20. Blum, C.; Puchinger, J.; Raidl, G.R.; Roli, A. Hybrid metaheuristics in combinatorial optimization: A survey. *Appl. Soft Comput.* **2011**, *11*, 4135–4151. [CrossRef]

21. Martí, J.V.; García-Segura, T.; Yepes, V. Structural design of precast-prestressed concrete U-beam road bridges based on embodied energy. *J. Clean. Prod.* **2016**, *120*, 231–240. [CrossRef]

22. International Organization for Standardization (ISO). *Environmental Management—Life Cycle Assessment—Principles and Framework*; ISO: Geneva, Switzerland, 2006.

23. Yi, S.; Kurisu, K.H.; Hanaki, K.; Hanaki, K. Life cycle impact assessment and interpretation of municipal solid waste management scenarios based on the midpoint and endpoint approaches. *Int. J. Life Cycle Assess.* **2011**, *16*, 652–668. [CrossRef]

24. European Committee. *European Committee for Standardization EN 206-1 Concrete—Part1: Specification, Performance, Production and Conformity*; European Committee: Brussels, Belgium, 2000.

25. Ministerio de Fomento. *EHE-08: Code on Structural Concrete*; Ministerio de Fomento: Madrid, Spain, 2008.

26. Ministerio de Fomento. *IAP-11: Code on the Actions for the Design of Road Bridges*; Ministerio de Fomento: Madrid, Spain, 2011.

27. Moscato, P. *On Evolution, Search, Optimization, Genetic Algorithms and Martial Arts—Towards Memetic Algorithms*; California Institute of Technology: Pasadena, CA, USA, 1989.

28. Dawkins, R. *The Selfish Gene*; Claarendon Press: Oxford, UK, 1976.

29. Catalonia Institute of Construction Technology. BEDEC PR/PCT ITEC Material Database 2016. Available online: https://www.itec.cat/nouBedec.c/bedec.aspx (accessed on 15 January 2018).

30. García-Segura, T.; Yepes, V.; Frangopol, D.M.; Yang, D.Y. Lifetime reliability-based optimization of post-tensioned box-girder bridges. *Eng. Struct.* **2017**, *145*, 381–391. [CrossRef]

31. Lagerblad, B. *Carbon Dioxide Uptake during Concrete Life Cycle—State of the Art*; Cement och Betong Institutet: Stockholm, Sweeden, 2005.

32. García-Segura, T.; Yepes, V.; Alcalá, J. Life cycle greenhouse gas emissions of blended cement concrete including carbonation and durability. *Int. J. Life Cycle Assess.* **2014**, *19*, 3–12. [CrossRef]

33. Frischknecht, R.; Jungbluth, N.; Althaus, H.-J.; Doka, G.; Dones, R.; Heck, T.; Hellweg, S.; Hischier, R.; Nemecek, T.; Rebitzer, G.; et al. The ecoinvent database: Overview and methodological framework. *Int. J. Life Cycle Assess.* **2005**, *10*, 3–9. [CrossRef]

34. Ciroth, A.; Muller, S.; Weidema, B.; Lesage, P. Empirically based uncertainty factors for the pedigree matrix in ecoinvent. *Int. J. Life Cycle Assess.* **2016**, *21*, 1339–1348. [CrossRef]

35. Itoh, Y.; Kitagawa, T. Using CO_2 emission quantities in bridge lifecycle analysis. *Eng. Struct.* **2003**, *25*, 565–577. [CrossRef]

36. Bouhaya, L.; Le Roy, R.; Feraille-Fresnet, A. Simplified environmental study on innovative bridge structure. *Environ. Sci. Technol.* **2009**, *43*, 2066–2071. [CrossRef] [PubMed]

 *sustainability*

Article

Can the SDGs Provide a Basis for Supply Chain Decisions in the Construction Sector?

Erica Russell [1,2,*], Jacquetta Lee [1] and Roland Clift [1]

[1] Centre for Environment and Sustainability, University of Surrey, Guildford GU2 7XH, UK; j.lee@surrey.ac.uk (J.L.); r.clift@surrey.ac.uk (R.C.)

[2] Previously also Carillion plc, Carillion House, 84 Salop Street, Wolverhampton WV3 0SR, UK

* Correspondence: e.f.russell@surrey.ac.uk; Tel.: +44-07713-177-236

Received: 30 November 2017; Accepted: 21 February 2018; Published: 28 February 2018

Abstract: The Construction sector is characterised by complex supply networks delivering unique end products over short time scales. Sustainability has increased in importance but continues to be difficult to implement in this sector; thus, new approaches and practices are needed. This paper reports an empirical investigation into the value of the UN Sustainable Development Goals (SDGs), especially Sustainable Consumption and Production (SDG12), when used as a framework for action by organisations to drive change towards sustainability in global supply networks. Through inductive research, two different and contrasting approaches to improving the sustainability of supply networks have been revealed. One approach focuses on the "bottom up" ethical approach typified by the Forest Stewardship Council (FSC) certification of timber products, and the other on "top-down" regulations exemplified by the UK Modern Slavery Act. In an industry noted for complex supply networks and characterised by adversarial relationships, the findings suggest that, in the long term, promoting shared values aligned with transparent, third-party monitoring will be more effective than imposing standards through legislation and regulation in supporting sustainable consumption and production.

Keywords: SDGs; construction; supply networks; sustainability; FSC; modern slavery

---

1. Introduction

1.1. Construction and a Sustainable Supply Network

Activity in the construction sector globally is expected to rise by 70% to 15 Trillion USD by 2025 [1], linked to global economic development associated with rising middle classes, increasing populations and a move to greater urbanisation. The sector generates around 50% of the waste by volume in developed countries, while buildings account for 19% of the world's energy-related CO_2 emissions. Construction is identified by the United Nations Environment Programme as a key area to address to mitigate climate change, a position echoed by the most recent IPCC Assessment Report [2,3].

In the UK, the construction industry has been dominated over the last 30 years by the development of a subcontracting culture, driven by "market forces" and leading to work allocated primarily through competitive tendering. The resulting construction networks have been characterised as "hollowed out conglomerates" [4]: "temporary multiple organisations", created to respond to bespoke client requirements requiring involvement of many value-adding organisations [5]. As a result, the sector forms a complex supply network including investors, developers, public bodies, architects/designers, contractors, manufacturers, raw materials suppliers and demolition experts (see Figure 1). Main contractors act as project managers for clients, drawing together all the skills, services and materials required to create a physical asset. It is their role to procure goods and services, albeit frequently to a pre-ordained plan, and they are increasingly reliant on subcontractors and suppliers; these purchases represent typically 75% of their turnover [6,7]. The resulting fragmentation has created a supply network in which relationships are highly competitive and frequently adversarial [8,9], leading to narrow profit margins for main contractors, a major cause of the recent demise of the company studied in this work (Carillion) [10]. Relationships between network actors remain primarily dyadic, i.e., between the client and main contractor or main contractor and Tier 1 supplier [11], and the structure provides a limited basis for the development of trust between network members or collaboration in a supply network that is rarely managed beyond the first tier [12–14].

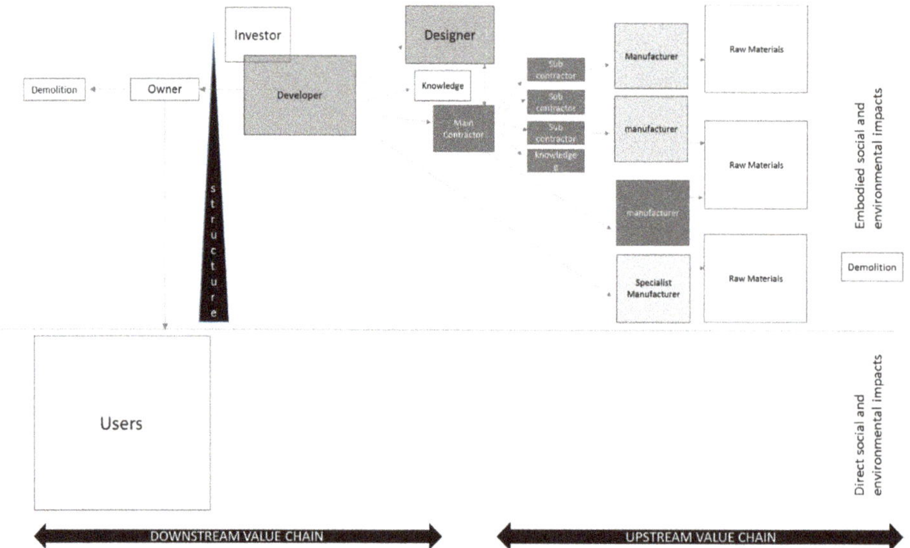

Figure 1. Simplified structure of a construction supply network [15]. Key: Increasing depth of shading in boxes demonstrates increased level of influence by Main Contractor. Increasing size of box indicates estimated increasing responsibility for impacts.

Despite these constraints, the UK Government has recognised that meeting sustainability targets in the construction sector depends on sustainable supply chain networks [16]. This creates a structural tension within the sector as the effective management of supply chains has become largely synonymous with collaborative forms of working [17]. Indeed, the sustainable supply chain management (SSCM) literature highlights collaboration and governance as powerful tools for facilitating sustainability initiatives [18]. For major construction contracts, changing business models have been shown to facilitate more collaborative working, including a move from "build" to "design and build" and "design, build, operate", the last an increasing feature of public–private partnerships. There has also been an associated increase in partnership working to win major contracts, e.g., involvement of multiple major contractors and manufacturers in the London Crossrail project or the formation of joint ventures (JV) such as CarillionAmey for MOD contracts. However, there is evidence that collaboration does not always result in beneficial outcomes [19,20]. It can also be difficult to achieve, as demonstrated in an industry-wide survey of 87 German firms by Brinkhoff and Thonemann which identified a 50% failure rate in collaborative supply chain relationships, the greatest issue being the difficulty of defining shared objectives [21].

There is a widespread view that sustainable consumption and production can only fully be achieved if the objectives of performance (economic, environmental and social) and principles of behaviour (quality) are shared, leading to common objectives, targets and requirements throughout the whole value network (e.g., [9,20,22,23]). This is not yet observed in the construction sector. For example, the industry annually procures over 380 million tonnes of resources [24] but inefficiencies within the network lead to 10–30% of waste materials being "unused product", including 800 ktons of shaped and sawn timber [25]. From a social perspective, the UK construction sector has been identified as a source of UK-based modern slave labour [26], with a significant proportion of the global 45.8 m people estimated to be subjected to modern slavery working within construction supply networks [27,28]. However, given the narrow profit margins noted above, the UK construction sector is typified by a focus on cost rather than value, with many companies seeking to transfer risk to others within the supply network. This acts as a strong inhibitor to undertaking initiatives directed at the environmental and social components of sustainability and heightens barriers to collaboration across the supply network.

Thus, although construction is an important sector for the promotion of sustainability in practice, it is characterised by serious structural problems. This paper reports on a heuristic study on the suitability of the Sustainable Development Goals (SDGs) in offering an effective collaborative framework to improve the social and environmental performance of the sector. In-depth interviews, surveys, and published sources have been examined, focusing on two different approaches implemented within one UK main contractor, Carillion plc (Wolverhampton, UK).

1.2. Sustainable Development Goals

The Sustainable Development Goals were adopted in 2015 [29] by the member states of the United Nations, in agreement with representatives of civil society and business, with the intention of guiding the global development agenda to 2030. They can be seen as an attempt to take the familiar representation of sustainability as "The art of living well within ecological limits" [30], complying with three sets of constraints—techno-economic efficiency, environmental compatibility and social equity [22]—and represent it in more tangible terms. In total, 17 different goals have been articulated, presumably divided in this way to aid understanding, acceptance and implementation. However, the SDGs are strongly interrelated, for example in ecological concerns such as the "nexus" of climate change, water availability and food supply, and in ethical concerns such as equity and justice. There are therefore questions, explored further in this paper, over the value of dividing the integrative concept of sustainability into so many apparently distinct goals.

Progress towards the SDGs to date has been slow [31]. The UN states that "Implementation and success … will be led by countries, with all stakeholders … expected to contribute to the … agenda" [32]. In the UK, in 2017, a review by the Environmental Audit Committee of the House of Commons [33] concluded that " … the Government seems more concerned with promoting the goals abroad, and has undertaken no substantive work to promote the Goals domestically or encourage businesses, the public sector and civil society to engage with the Goals and work towards meeting them". For commercial companies, the SDGs potentially represent a framework for articulating their social responsibilities and incorporating them in strategic planning. However, given the imprecise language and abstract level of many of the 169 targets, it is difficult to see how business could easily incorporate these into their practices. In addition, it is far from clear that the significance of the SDGs is sufficiently widely understood. For example, a recent survey [34] reported that more than 20% of European companies see "sustainability" as a source of competitive advantage but did not record to what extent they have truly grasped the concept as an imperative beyond commercial interest. The response most often reported is that companies are "developing products or services that will provide solutions in line with the Global Goals" [31], rather than attempting to embed the SDGs in planning and operations.

1.3. Responsible Consumption and Production

Notwithstanding the question over the value of subdividing sustainability goals into so many distinct categories, we focus in this paper on SDG12: Responsible Consumption and Production. This particular goal is selected because it is seen as one of the most central, relating closely to many of the other goals (many of which are subsidiary to this particular SDG), and yet is considered to be one of the five SDGs on which there has been least progress [30]. Furthermore, it requires attention to the whole supply network delivering products or services. Therefore, this goal provides a useful basis for exploring approaches to embed the SDGs into planning and practice.

Responsibility in supply chain management implies operating within ecological limits, as represented by the Planetary Boundaries defining the "safe operating space for humanity" [35,36], although there are serious difficulties in developing an operational approach to planning based on the Planetary Boundaries [37]. However, recognising that "sustainability" includes social equity, supply networks need to be more than just sequences of activities operating within ecological constraints; a supply network can be viewed as a set of relationships that convey benefits in both directions [23,38]. This view is exemplified by the fair trade movement, but has a more general significance in informing concepts of sustainable consumption [22]. Social responsibility in supply networks is a strong concern in the approaches examined here.

This paper reports an empirical investigation into approaches to network management in a UK-based multinational construction company, to explore how the SDGs—specifically No. 12: Responsible Consumption and Production (SDG12)—could be incorporated into planning and operations in a commercial company by contrasting "bottom-up" and "top-down" approaches to goal-setting. The results proved to be sufficiently informative to generate some conclusions with wider implications.

2. Selection and Exploration of Approaches to Embedding the SDGs

2.1. Methodology

This work forms part of a PhD programme, funded jointly by a public research agency (EPRSC) and what was until recently a leading private company (Carillion plc), directed at improving understanding of (un)sustainability in supply networks. The research has taken an inductive approach, aiming to develop a theory that is "grounded" in the data from which it has been derived [39,40].

It was important that the research was carried out in a main contractor, Carillion plc, recognised for its sustainability credentials. The company published its first "environmental" report in 1997 and won Price Waterhouse-Cooper's "Building Public Trust" award for Sustainability reporting in three consecutive years. The company has clearly articulated corporate values encompassing economic, environmental and social issues—"we care; we achieve together; we improve and we deliver"—and was already exploring how its existing approaches related to the UN Sustainable Development Goals. Its position as a main contractor made it the main node of the supply network. The company operated in the UK, Middle East and Canada, with an annual international procurement spend of 3.4 billion pounds sterling [41], working with over 8000 accredited first-tier suppliers and many thousands more in second and third tiers. Thus, involvement with Carillion offered an opportunity to examine how Responsible Consumption and Production can be incorporated into the operations of a commercial company.

During the initial phases of the PhD research in 2015, "purposeful sampling" [42] was undertaken to draw on the shared knowledge and experience of sustainable procurement within the company. Discussion with Carillion's Supply Chain Director and Sustainability Manager identified seven members of the supply chain team and two of the sustainability team likely to provide useful insights on interview. This provided a sample including both high and medium level decision makers, to provide a strategic overview of current procurement and sustainability within the company and the wider industry. Interviewees were also selected to offer a mix of job roles, from strategic to joint venture procurement, key project management, supplier accreditation and on-site sustainability monitoring (Table 1).

Table 1. Team and Job Role of Orientation Interview Participants.

Team	Role	Length of Interview
Supply Chain	Supplier Accreditation and Monitoring	1 h
Supply Chain	Supplier Accreditation and Management	1 h
Supply Chain	Managing Regional Strategy, supply chain procurement—multiple projects, client liaison	1 h
Supply Chain	Managing Regional Supply Chain Team-multiple projects, client liaison	1 h
Supply Chain	Managing Regional Supply Chain Team-multiple projects, client liaison	1 h
Supply Chain	Managing Procurement—Joint Venture	1.15 h
Supply Chain	Leading team for large public-sector project, delivery, client liaison	1 h
Sustainability	Corporate Sustainability—policy, strategy and reporting	45 min
Sustainability	Business Unit Sustainability Strategy—monitoring, reporting, leading project sustainability	1 h

At this initial stage of the research, the "orientation interviews" used a semi-structured format to prompt responses but to also allow the interviewer to explore themes or comments in more depth (see Appendix A for prompt questions). Open coding of the interview transcripts identified two sustainability issues where participants noted proactive, multi-tier engagement with the supply network: responsible sourcing of timber through Forest Stewardship Council (FSC) certification and Modern Slavery. Consideration of these two issues by the paper's authors confirmed that they represented contrasting approaches to supply network sustainability (see Table 2). They were therefore selected as suitable cases for further exploration.

Table 2. Comparison of Forest Stewardship Council (FSC) and Modern Slavery (MS) supply network approaches identified during "orientation interviews".

	FSC	Modern Slavery
Status	Optional	Mandatory
Lead	bottom-up	top-down
Time	20 years of experience/mature process	2 years since implementation of UK act. Process still developing
Corporate Drivers for Action	Initially NGO pressure and consumer concern on product providers [1]	NGO pressure driving legislation
Personal Drivers for Supply Chain team	High decision makers with values aligned to FSC social and environmental aims	Meeting legalisation, alignment with general values
Network Collaboration	longer term collaboration has allowed development of relationships and trust within network	short development time resulting in collaboration primarily with peers
Implementation	FSC policy, Chain of Custody	MS Policy, and Audit

[1] EU timber import regulation introduced in 2010 [43].

To develop understanding of how the company approached FSC certification and Modern Slavery, multiple data collection methods were adopted to allow triangulation of results. Further interviews were added with individuals whose importance was highlighted during the semi-structured interviews:

two supply chain team members responsible for procurement of timber with FSC certification. The researcher was also allowed to observe internal meetings on Modern Slavery by the Carillion Sustainable Procurement Steering Group. To understand the perspective of Carillion procurement teams on FSC certification and Modern Slavery, these issues were included in an on-line survey sent to all 94 supply chain team members and the sustainability manager. A response rate of 72% provided 68 completed surveys. The demographics of the respondents were 75% male, with 6.3% of respondents being aged 15–24, 32.8% aged 25–44 and 60.9% being aged 45–64. Respondents were also classified by decision making level, verified by the Supply Chain Director. High Level decision makers accounted for 18.8% of respondents, medium level 57.8% and 23.4% at the limited decision-making level. Questions were rated by participants using the Likert scale to provide more nuanced answers and these were analysed using SPSS. Team members were also given the opportunity to comment. Transcribed interviews, meeting notes, other observations and materials were coded in MAXQDA software, building on the initial open coding and allowing the emergence of themes and concepts. In the final, selective coding phase [44] illustrative quotes were derived. Additionally, "company" information was supplemented and cross referenced with industry literature, Carillion annual sustainability reports and other materials available publicly online. Carillion staff provided feedback and identified errors and/or omissions. The resulting "bottom-up" and "top-down" cases were then analysed (Section 3) using the methodology for cross-case analysis proposed by Khan and Van Wynsberghe [45] for making deductions from a small number of cases.

2.2. "Bottom-up" Goal Setting: Forest Stewardship Certification

The Forest Stewardship Council (FSC) Chain of Custody represents an attempt to frame objectives, targets and requirements that can be shared throughout the supply network for forest products. The FSC originated in the early 1990s through discussions between the World Wildlife Fund (WWF, now Worldwide Fund for Nature) and several major UK DIY chains, concerned about the impact their procurement of wood was having on rainforests and the risks this entailed [46]. The group, WWF95+, wanted an industry-wide approach to ensure that the timber they purchased could be "guaranteed" as ethically sourced. This desire was compounded by the failure of governments at the Rio Earth Summit to reach an agreement to stop deforestation. From the outset, FSC took a non-governmental approach, harnessing commercial drivers to effect change. FSC has striven for, and increasingly achieved, a membership-based approach to forest management. It now provides all interested groups a voice at their membership forum and is overseen by a board elected by the membership. Decisions on forest management include indigenous people through local consultation. Local workers are prioritized for employment, which must include training in safety and use of equipment, and must be paid a "decent" salary [47]. Audits are carried out to ensure that the principles behind the approach are met. The social and environmental benefits and the associated reduction in risks resulting from this local, ethical approach to forest management are then propagated through the supply network, with the "Chain of Custody" assured by recording each step in the process. Evidence from WWF indicates that many smaller producers have benefited financially from FSC whilst larger scale producers have derived less advantage [48]. Since its inception in 1993, FSC has grown to be a significant market mechanism to promote responsible forest management, now covering 180 million hectares of forest in 112 countries.

By 1998, WWF95+ had grown to 86 participants including Tarmac (from which Carillion (Wolverhampton, UK) emerged in 1999). Tarmac was the first construction company to be engaged in the WWF95+ group and was seen by WWF as an important agent to bring benefits to both construction and forest industries. Carillion, driven primarily by corporate environmental values and reputational risk management, published a Sustainable Timber Policy, ratified by the Carillion board, to purchase only timber and wood-derived products with FSC Chain of Custody certification or, where this could not be achieved, to use sources that were independently verified as legal and sustainable. Requiring procurement teams to source only certified sustainable timber represents a major commitment, and it is clear from interview comments that this continued irrespective of client demand: "no client ever requested FSC apart from occasionally" (high-level decision maker). The principle was communicated to clients, sub-contractors and other suppliers.

Carillion accepted that, without a certified standard, it could not guarantee it was not colluding with or procuring timber from illegal logging operations; FSC provided a way for Carillion to ensure that its supply network did not contribute to illegal and destructive deforestation and thereby to avoid potential reputational risk. However, it was also made clear that Carillion's aim reached beyond its own network, "to promote demand and improve competitive pricing for FSC timber within the construction industry as a whole" [49]. Indeed, this was seen most recently in Oman, when Carillion's joinery workshop was the first in the country to have FSC Chain of Custody certification [41]. Carillion continued to support FSC and its incorporation into its procurement and operational processes. In 2013, Carillion updated its Timber Policy as part of a wider corporate 2020 sustainability strategy.

Operating the FSC "Chain of Custody" demanded additional commitment. This included guidance and training material for Carillion procurement and operations staff and ongoing, intensive engagement with suppliers and subcontractors bringing timber onto site as part of contract works packages. As noted by a high-level decision maker, "whilst it remains outside the industry norm you just have to continuously communicate it. People still don't really know what they are buying. They don't know how to maintain chain of custody". The company set up internal systems to manage the monthly reporting of timber usage and report this annually in its independently audited Annual Sustainability Report. Nevertheless, even after twenty years of commitment to the FSC Chain of Custody, Carillion continued to experience several major challenges in requesting FSC timber as its primary option and propagating this principle through the supply network. These included influencing or controlling sub-tiers in the supply network where there is a lack of contractual status, ensuring record keeping met Chain of Custody requirements and, linked to this, committing staff and management time to resolving reporting anomalies such as "they'd used more FSC timber on the job than the whole of the UK in a year" (high-level decision maker) [49].

Despite these concerns, this "bottom-up" ethical approach to timber management delivered some successes. By 2009, timber with no certification represented only 7.9% of Carillion's total purchases and continued to reduce, with annual fluctuations, to 5% during 2015 [50,51]. Carillion was committed to 100% purchases of certified sustainable timber and wood by 2020. Increased availability of FSC or a similar Chain of Custody organisation, the Programme for the Endorsement of Forest Certification (PEFC), has enabled more wholesalers and suppliers to offer Chain of Custody products. Increased internal monitoring and awareness supported and were supported by this change. Carillion continued to place an emphasis on the role of FSC in preventing deforestation [41]. For example, the 2017 Carillion supply chain team survey identified that 67.7% believed that the most important reason for the company supporting responsible sourcing of timber was its alignment with Carillion's environmental and social values, ensuring that forests remained "alive for future generations" [52]. Minimisation of reputational risk due to unethical sourcing was also identified as a major benefit of FSC Chain of Custody, rated as highly important or important by 65% of respondents. Whilst still facing challenges in procuring FSC timber, the company worked to identify and support improvement in its supply network, focusing on "temporary" wood products where there is less awareness of FSC materials amongst subcontractors. Carillion promoted FSC certification to its upstream suppliers, but engagement was more limited in the sector's downstream value chain: whilst certified timber does gain credits within building standards such as BREEAM, few clients directly specify FSC or other responsibly sourced timber materials. Furthermore, unlike companies supplying the consumer market, there is no direct communication between main contractors in the construction sector and the end users of the structures, most of whom will be unaware of timber sources.

2.3. Top-down Goal Setting: Modern Slavery

The issue of forced labour, employed directly or more remotely through supply networks, has for some time been addressed through voluntary codes of conduct such as that promoted by the International Labour Organisation (ILO) or the UN Global Compact. However, the UK Government, in 2015, passed the Modern Slavery Act [53], requiring all UK companies to address the issue of modern slavery in their own businesses and their supply networks. Modern slavery is considered to be delineated by bonded labour, poor wages, working and living conditions, intimidation and violence or human trafficking. Companies with a turnover of more than £36 m must demonstrate the action they are taking and publicly report this on an annual basis. The act affects all UK-based companies and is also generally accepted to apply to all non-UK based subsidiaries. In setting this legislation, the government has imposed values and specified the process by which all UK companies must engage with this issue, in marked contrast to the way FSC certification has developed more organically.

The UK Government initially suggested that there could be between 10,000 and 13,000 people enduring modern slavery in the UK [54]. The worldwide figure has been estimated at 45 million [27], with construction identified as a major area of concern because of its high reliance on flexible, temporary labour and highly diverse global supply networks.

Globally it is estimated that 7% of the world's workforce is employed in the construction sector [27]. The complexity of labour issues in construction is compounded by the large number of different materials used (see Section 2.2), with up to 10,000 different component parts being required for the construction and use phase of buildings [55]. Companies primarily manage labour issues as part of their product supply networks but the high numbers of products and components, often originating from unknown global sources, makes it difficult to ensure transparency in employment practices. Even in relatively short supply networks, such as within the UK, mapping labour practices can become complex. Complexity can make the different forms of modern slavery, which are frequently informal and transient in nature, hard to detect and therefore persistent [56–58].

Carillion first used assessment tools in 1999 to review the environmental performance of suppliers. From this work, they identified that only 50% of suppliers broadly met requirements. As a result, Carillion began to address the social aspects of its suppliers' services and products by engaging with suppliers to promote sustainable sourcing of products and materials, "with high risk suppliers being encouraged to change practices rather than being delisted" [49]. The company made it clear that ensuring human rights was a key company principle and that they had "an ongoing commitment to improve the living and working conditions ... not just for direct employees but also for our subcontractor teams" [49]. The Modern Slavery Act 2015 went further, demanding greater transparency across the supply network, with clear evidence of company engagement. Along with many of its peers, Carillion included questions on their supplier registration system relating to human rights, asking for confirmation that companies had employment practices in line with the ILO or UN Global Compact on human rights, i.e., that they ensured fair wages and freedom of association, with no forced labour. They also asked companies if they engaged in responsible sourcing within their own supply networks.

A senior-level working group within Carillion reviewed existing company approaches and risks and, as a result, accepted that for many smaller suppliers, the Act and the concept of modern slavery represented a little-known issue. This was reiterated by the supply chain team with one member stating, of suppliers, "There is limited knowledge out there and even less on how it will be implemented" (medium-level decision maker). A large part of the company's efforts was therefore directed at engagement and awareness raising. At a company level this was achieved by direct communication with Tier 1 suppliers, changes to the supplier registration process, information and awareness raising via Carillion's own website and Carillion's own externally facing supply chain teams and operational staff. Questions on the internal supplier registration system were expanded to include the term "modern slavery" and, to support smaller companies, and in 2016 Carillion's Labour Standards Charter was developed [59] which suppliers could sign and adopt if they did not have their own processes in place. However, Carillion also identified that slavery was an industry-wide concern which, whilst highlighted by legislation, strongly resonated with the values of their peers and would benefit from collaborative efforts. In 2012, to meet gaps in sub-sector specific sustainability skills, Carillion, along with industry peers, helped create and co-fund an on-line skills platform, the Supply Chain Sustainability School (SCSS). By 2015, the free-to-use "school" had thousands of industry members. To overcome the knowledge gap on Modern Slavery, a group from the SCSS encouraged and facilitated main contractors, clients and major manufacturers to work collaboratively to create new slavery guidance directed specifically at the construction industry. Skills modules on Modern Slavery and the Act, along with video materials and written information, were developed and promoted by main contractors and clients to organisations in their supply networks [60]. However, progress of awareness across the network remains slow, with 21% of Carillion supply chain team identifying, in 2017, that most or many of the Tier 1 suppliers they worked with did not know about the Modern Slavery Act [52].

For companies operating in the UK, the Modern Slavery Act aligns with general societal norms and values and builds on existing legal requirements; and yet slavery can still occur. Detection becomes even more difficult when company operations or supply networks span countries or regions where different operating principles are accepted and where government engagement may be less developed. As a specific example, Carillion identified its highest risk area as its Middle East and North Africa (MENA) businesses, particularly in relation to worker welfare standards. Carillion had operated for more than 40 years in the Middle East, a region that has been experiencing an immense building boom. Organisations such as the Business and Human Rights Resource Centre have identified major human rights abuses in the region, such as migrant workers being subjected to high recruitment fees, non-payment of wages and restricted mobility. A separate Carillion-MENA Welfare Steering Group was therefore established.

In 2009, Carillion established a business in Qatar to provide construction, infrastructure and facilities management services; it grew to employ approximately 1100 people directly, with a further 6000 employed through subcontractors [6]. Carillion entered into a commitment that employees would be paid in accordance with Qatar Labour law but, in addition, that employees would also receive flights home, holiday pay, health insurance and accommodation and food. They set standards for accommodation that landlords had to meet prior to contracting, and required accommodation to be audited to ensure the standards continued to be met. Carillion put in place processes to ensure that employees have freedom of association, routes to express grievances and work to the same Health and Safety standards as in the UK, replicating the "Don't Walk By" culture used on all UK construction sites. However, in 2014, they were publicly accused of having subcontractor labour on site who had been forced to surrender their passports and were living in poor accommodation and receiving only a small part of the promised wages [61]. In response, Carillion implemented a similar approach for workers employed through sub-contractors. In one of the most contentious areas, that of recruitment, Carillion worked with "preferred suppliers" who had been reviewed for financial, ethical and professional conduct. It also carried out spot checks and terminated contracts with companies that charged excessive fees or had been unethical in their approach. A company like Carillion does not have direct control over its subcontractors; however, they are expected, as a minimum, to comply with Qatari labour laws. Carillion proactively reviewed and monitored the employment practices and accommodation of its suppliers and their subcontractors; only those that met Carillion's standards were included in the preferred supplier list. During 2016, the Carillion Board members visited two accommodation sites in the Middle-East as part of the audit process. When the Business and Human Rights Resource Centre approached the top 100 construction companies working in Qatar and UAE requesting feedback on their approach to worker's rights, Carillion was one of the first companies to respond but only 22 companies in total replied [62]. Qatar has seen an increased drive by global construction contractors to remove modern slavery from their supply networks but media reports, mainly fuelled by the reports of campaigning NGOs, continue to focus on poor worker conditions. In December 2016 Qatar abolished its "kafala" system of worker recruitment and, whilst Amnesty International notes "it leaves the same basic system intact", this does demonstrate that the state is responding to pressure and supporting change [63].

3. Observations and Comparisons

Cross-case comparison provides the opportunity to learn from different cases and to gather critical evidence to modify policy [45]. This type of comparison, considering general and distinctive characteristics between a small number of cases, has already been utilised to support policy recommendations on the role of supply chain collaboration in affordable housing development [64] and to examine sustainable supply chain management in the textile sector [65]. Using a case-oriented process, the FSC and Modern Slavery approaches outlined in Sections 2.2 and 2.3, were analysed to look for patterns of similarities and differences. The themes identified have been collated in Table 3, illustrating the main benefits and challenges within each theme. The observations reveal how the SDGs align with the top-down and bottom-up approaches.

Table 3. Benefits and challenges associated with top-down and bottom-up approaches.

Theme		Modern Slavery (Top-Down Goal Setting)		FSC (Bottom-Up Goal Setting)
Defining 'What is right'	B	Commitment by Government to a legal 'solution' defines the 'ethical' position for the supply network.	C	Requires commitment, to buy FSC timber and create market demand, which can be difficult in a business-to-business sector.
			B	Negotiated' agreement across the supply chain—engagement with personal and corporate values
Collaboration	B	In the UK the legal requirement has created a level playing field and engendered collaboration between construction industry organisations/companies. This has resulted in shared costs.	B	Demands collaboration along the supply network.
			C	Supply network collaborators may have different goals (i.e., improved living conditions, reducing loss of rainforest, minimising cost).
Relationships	C	It is difficult to get beyond Tiers 1 & 2 especially in global networks; modern slavery most likely to occur in tiers 4, 5 and beyond.	B	Considers social, environmental and economic issues making it attractive for local communities to engage and support.
	C	Tentative relationships with NGOs	B	Strong supportive engagement of NGOs offering critical assessments and validation.
			B	Positive benefits to downstream SMEs engaged in process.
			B	Senior procurement staff are engaged with downstream end suppliers (FSC)
			B	Reduces the likelihood of modern slavery as it can remove exploitative drivers e.g., illegal logging
Control	C	Modern slavery is driven by issues outside the control of corporate organisations i.e., inequalities, legal protection of vulnerable workers in some countries	B	Operates as a non-governmental process, unrestricted by national borders.
Ability to Deliver	C	Demands for 'no slavery in the supply network' are strained by time pressured delivery requirements.	C	Documenting Chain of Custody is critical to maintain credibility but increases costs and is complex to manage
	C	Modern slavery is frequently linked to 'illegal' labour and exploitation—policies, charters and audits struggle to reach lower Tiers		
Transparency	C	Reporting by major companies but currently weak driver across rest of supply chain	B	Detailed and transparent reporting

Key: B: Benefit to Contractor. C: Challenge to Contractor.

Both FSC and Carillion plc have committed to support the SDGs [51,66]. Indeed, the FSC Chain of Custody approach provides a well-established collaborative supply network of the type identified by the UN as necessary to effect change. However, this work confirmed that those engaged in the FSC timber value chain support the SDG "Sustainable Consumption and Production"—indeed, the Forest Stewardship Council have identified it as one of the goals to which they aspire—but showed that they do not see it as a primary focus. FSC believes that another goal, SDG 15—"Life on Land"—is most relevant to their work, specifically target 15.2 "progress towards sustainable forest management". The Council has furthermore identified that, for those within the supply network, FSC accreditation provides a tool that supports 11 SDG goals and 35 targets (Appendix B, Table A1) [66]. Carillion had also reviewed the SDGs and undertook a major materiality survey, asking staff, clients and other stakeholders to identify the goals they felt Carillion could effectively support. Stakeholders identified five goals, all of which focused on social equality or business innovation, but of these only Goals 5 (Gender Equality) and 8 (Decent Work and Affordable Growth) aligned with FSC's goal selection. In 2017, Carillion published its Annual Sustainability Report noting support for nine SDGs. The additional goals recognized the importance of impacts within its supply network (SDG12) and on environmental issues (especially SDG 15: Life on Land) and also SDG 11 (Sustainable Cities and Communities) [51]. Such variation in goal alignment would suggest that the position and role of an organisation within the supply network influences its view of how it can effect change and thus which SDGs are most relevant. This view is supported by Schmidt et al. [67] who note that, whilst stakeholders advocate "monolithic" outcomes across the supply chain, very different issues are salient for different companies, which therefore set different goals depending on their position within the supply network [68]. This highlights the difficulty of aligning goals and targets between different parts of the value network, even where strong relationships already exist.

The Modern Slavery approach demonstrates less well-developed collaboration within the supply network: collaboration may reach beyond the first tier of contractors but lacks the clarity and consistency provided by a Chain of Custody process. This study has confirmed how construction companies, such as Carillion, with strong social and ethical stances will implement policies, undertake audits, and work collaboratively with employees and local groups in an attempt to prevent slavery. However, other stakeholders in the supply system have identified that they have a role to play in supporting Decent work and affordable growth (SDG8) but did not link this with SDG12, the performance of the whole supply network. Unlike FSC, there is currently no "bottom-up" approach; it appears that an "imposed" principle, such as the prevention of modern slavery, is extremely difficult to deliver throughout the supply network. This becomes increasingly problematic where peer organisations, and public bodies are not engaged with the issue. A recent report by Segall and Labowitz [69] concluded that breaking the "cycle of abuse" needs stronger legal enforcement; in particular, much greater regulation of the recruitment process and collaboration between all actors in the supply network is needed, to include countries where migrant workers are recruited.

4. Operationalising the SDGs—Value Driven Approaches

Improvements in the sustainability of supply networks are generally limited to incremental improvements; of doing "less bad". Although the SDGs have been presented as a way to enable more radical reductions in unsustainability, they do not challenge the incremental approach and as yet have not offered targets based on operating boundaries. The targets and indicators that do currently exist offer limited guidance on how commercial actions can be effectively aligned with the SDGs or the Planetary Boundaries [37]. Currently the only explicitly "business focused" indicator for SDG12 is Indicator 12.6.1—"Number of companies publishing sustainability reports" [70]—which is focussed on process rather than outcome. Such weakly articulated aspirations do little to discourage the current business focus on "how a company can contribute" rather than "how it will deliver" [71].

Therefore, there is a role for trade associations and commercial organisations to develop sectoral approaches and guidelines, loosely analogous to the Product Category Rules (PCRs) used in product labelling and based, like PCRs, on shared goals and aspirations. Some large corporations, who are identified as adopting leading sustainable practises, appear to have the necessary "shared, organisation-wide long-term vision" and exhibit "core values and cultures and a sense of purpose beyond the economic bottom line" [72]. In a traditional view of supply networks this is demonstrated by the "corporation" acting as the driving force for implementing CSR through selective commercial pressures on organisations in its upstream supply network. It operates as a hub for stakeholder engagement, on the implicit assumption that the principles of the corporation have precedence over those of other organisations within the network [23]. This is echoed by Jorgensen and Knudsen [73] who note that larger buyers, acting as change agents, exert pressure on their supplier tiers to comply with their environmental and social requirements. It has been argued [23,74,75] that, as the corporation develops these relationships, it moves beyond an adversarial negotiating stance to one of co-operation as it seeks to propagate its own principles through the supply network. This would appear to be the current position with modern slavery: the corporation, driven by legislation and supported by NGOs, tries to require specific social values to be upheld throughout the supply network.

By contrast, the approach followed by FSC has led to development of a different dynamic in the supply network. Through the creation of the FSC "brand" and rigorous certification structure, forest owners have built an equally strong position in the supply network, with network power being found at both raw material and "retail" ends of the network [76]. Both dominant groups value the importance of their reputation and, in the approach presented, both Carillion and FSC shared a similar set of "values". Both organisations had strong self-interest to co-operate but in doing so were also working towards goals with longer-term benefits, beyond those which both parties originally expected. These included generating wealth for local communities, especially local SMEs, and providing buyers with a high level of confidence in the fair treatment of people working in the supply network.

5. Conclusions

The FSC multi-stakeholder approach highlights the practical value of shared goals and principles as the basis for long-term supply network relationships and collaboration. NGO oversight and certification creates transparency and ensures compliance even by actors in the supply network whose commitment to the goals may be weaker. FSC appears to operate most effectively at a sector level rather than just a single supply network: what started as a "bottom-up" approach has developed into a shared position of network "power". Interestingly, whilst aligned goals support a shared vision, the complementarity between the roles of FSC and Carillion in the supply network could ensure the achievement of sustainable outcomes: FSC represents those directly involved in forestry, working to overcome environmental and social issues associated with illegal logging, whilst Carillion could offer the economic driver to deliver change. We argue that this complementarity promotes change towards sustainability but makes alignment with a single unifying SDG unrealistic.

Whilst the SDGs do set "slavery" within the context of wider sustainability goals, they may still be seen as merely rebadging earlier failed agreements: 178 nations are signatories to the ILO Forced Labour Convention of 1930 [77] but this has not eliminated slavery from supply networks. SDGs do not provide a new practical framework for successful delivery of fair labour. Where moves to eliminate modern slavery from the construction sector have been successful, they have exhibited some of the characteristics demonstrated in the FSC Chain of Custody approach; i.e., engagement with peers and NGOs that expands stakeholder collaboration and the creation of a more transparent corporate approach.

Thus, the difference in effectiveness between the two approaches reviewed in this work shows that, without alignment of principles and goals or shared vision throughout all tiers of the supply network, it is hard to motivate actors in the supply network to engage so that progress in reducing unsustainability is limited. Principles and goals are, however, individual, and vary between cultures, industrial sectors, organisations of different sizes, etc. They develop within an organisation through the complex interaction of information, experience and surrounding behaviours. The potential is high for divergent and non-complementary behaviours within individual SDGs between different actors in the supply network, resulting from their different values and priorities. Hence, it is possible that the SDGs will struggle to be universally adopted without complementary behaviours within the supply network.

Whilst this research has considered the role of supply networks within the construction sector, we suggest that the scenarios explored will have resonance in other sectors where brand dominance and consumer pressure is limited. The UN Sustainable Development Goals have undoubtedly succeeded in raising awareness amongst a broad range of actors and stakeholders of the issues grouped under the heading of "sustainability", but this research suggests that they may be less successful in providing companies with a practical decision-making framework, especially in the context of complex global supply networks. Much more work is needed to make the Goals operational. The top-down and bottom-up goal-setting approaches examined in this paper reflect two different ways to embed sustainability in an industrial sector. Based on the exploration reported here, we suggest that the bottom-up approach is ultimately more likely to be successful because it promotes alignment of goals and/or principles between the different actors in the supply network, so that all actors can gain benefit from the relationship and have the flexibility to focus on the goals that are most relevant to them. Given the demise of Carillion, after the research reported here, there is no possibility to continue this particular investigation. However, a longitudinal study is really needed to explore and compare the effectiveness of different approaches in embedding more sustainable practices in companies in this and other sectors.

Acknowledgments: Carillion plc was and EPSRC is a funder of the University of Surrey Doctoral Practitioner student Erica Russell. The work of Jacquetta Lee is funded by the University of Surrey. Roland Clift's contribution is unfunded.

Author Contributions: The paper was jointly conceived by all authors through discussion and written communication. The initial framing of the paper was suggested by Roland Clift and the analysis of different approaches has been drawn from PhD thesis material by Erica Russell based on work within Carillion plc. Jacquetta Lee led the results and observations sections with contributions from all authors.

Conflicts of Interest: The authors declare no conflict of interest. Carillion plc played no role in the design, collection, analysis or interpretation of the data. They did read the manuscript prior to submission and noted several factual inaccuracies which were amended.

Appendix A. Semi Structured Orientation Interview Questions

Note: The original survey included additional prompts for several of the questions based on corporate procedure. Some of these details are confidential and have not been included in this Appendix.

Question to be asked by Interviewer to prompt discussion

1. Please could you outline your role and how this fits within the supply chain (SC) team.
2. How do you select suppliers and monitor supplier performance?
3. What typically is the relationship/communication routes that the SC team have with suppliers?
4. How do you see Carillion's supply chain? (those involved in face to face meetings to be shown the three basic models, Figures A1–A3).
5. How far down the chain do you think Carillion have direct or indirect influence currently?
6. When you report KPIs for Carillion, how far down the chain do you report?
7. What do you think suppliers understand about sustainability? (Does it matter? to whom)
8. When, as part of tendering process, is Sustainability flagged as an important criterion?
9. If you talk to suppliers what do you say are the key sustainability goals that Carillion are looking to achieve through their work.
10. How do you keep up to date with the company's sustainability objectives/goals?
11. If suppliers don't know about sustainability where do you suggest they go if they want help?

12. Can suppliers respond to requests for more innovative approaches/more sustainable approaches? (prompt: Examples of success)
13. What do you think are the big barriers/issues that need to be turned into opportunities?

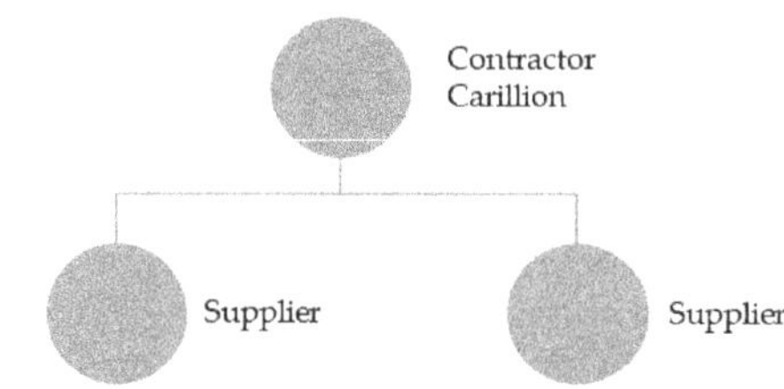

Figure A1. Upstream Tier 1 Supplier Model.

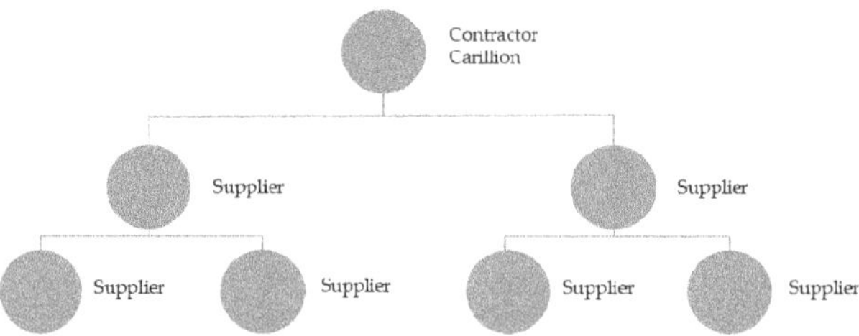

Figure A2. Upstream Multiple Supplier Model.

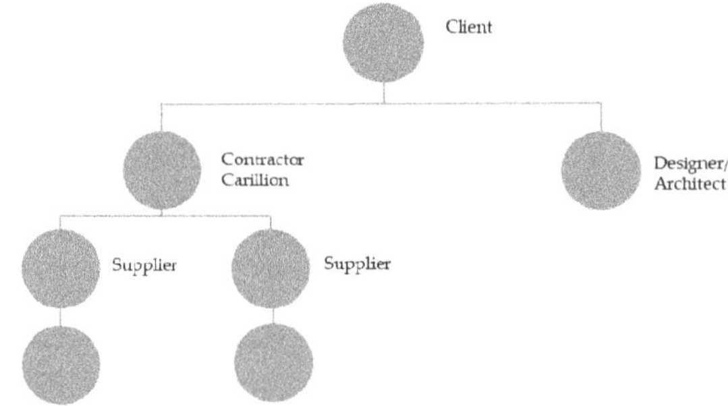

Figure A3. Upstream and Downstream Supply Network Model.

Appendix B. Sustainable Development Goals Supported by FSC

Table A1. Sustainable Development Goals Supported by FSC.

SDGs Supported	SDG Targets Supported
Primary Goal	
15. Life on Land	**Main Target:** 15.2 By 2020, promote the implementation of sustainable management of all types of forests, halt deforestation, restore degraded forests and substantially increase afforestation and reforestation globally
	Secondary Targets:
	15.1, 15.3, 15.4, 15.5, 15.7, 15.8, 15.c

Table A1. *Cont.*

SDGs Supported	SDG Targets Supported
Additional Goals	
1. No Poverty	1.5
2. Zero Hunger	2.4
5. Gender Equality	5.5, 5.a
6. Clean water and sanitation	6.4, 6.5, 6.6, 6.7
7. Affordable and clean energy	7.2
8. Decent work and economic growth	8.4, 8.5, 8.7, 8.8,

References

1. Oxford Economics. *Global Construction 2025*; Oxford Economics: Oxford, UK, 2013.
2. Independent Police Complaints Council (IPCC). *Climate Change 2014: Mitigation of Climate Change. Contribution of Working Group III to the Fifth Assessment Report of the Intergovernmental Panel on Climate Change*; Edenhofer, O., Pichs-Madruga, R., Sokona, Y., Farahani, E., Kadner, S., Seyboth, K., Adler, A., Baum, I., Brunner, S., Eickemeier, P., et al., Eds.; Cambridge University Press: Cambridge, UK; New York, NY, USA, 2014.
3. United Nations Environment Programme (UNEP). *Buildings and Climate Change: Summary for Decision Makers*; United Nations Environmental Programme—Sustainable Buildings and Climate Initiative: Paris, France, 2009.
4. Green, S. The evolution of corporate social responsibility in construction. In *Corporate Social Responsibility in the Construction Industry*; Taylor and Francis: London, UK, 2008.
5. Cherns, A.; Bryant, D. Studying the client's role in construction management. *Constr. Manag. Econ.* **1984**, *2*, 177–184. [CrossRef]
6. Carillion. A Better Tomorrow: Sustainability Report 2015. Wolverhampton: Carillion plc. 2016. Available online: http://sustainability2015.carillionplc.com/assets/files/carillion-sr2015-full.pdf (accessed on 2 February 2017). (Archived by WebCite® at http://www.webcitation.org/6xBZOaUQ0).
7. Scholman, H.S.A. *Uitbesteding Door Hoofdaannemers (Subcontracting by Main Contractors)*; Economisch Instituut Voor de Bouwnijverheid: Amsterdam, The Netherlands, 1997.
8. Korczynski, M. The low-trust route to economic development: Inter-firm relations in the UK engineering construction industry in the 1980s and 1990s. *J. Manag. Stud.* **1996**, *33*, 787–808. [CrossRef]
9. Akitoye, A.; Mcintosh, G.; Fitzgerald, E. A survey of supply chain collaboration and management in the UK construction industry. *Eur. J. Purch. Supply Manag.* **2000**, *6*, 159–168. [CrossRef]
10. The Economist. Where Did Carillion Go Wrong? Available online: https://www.economist.com/news/britain/21735047-mistakes-caused-mega-contractors-demise-are-common-outsourcing-industry-where (accessed on 18 January 2018). (Archived by WebCite® at http://www.webcitation.org/6xBXBdihE).
11. King, A.P.; Pitt, M.C. Supply chain management: A main contractors perspective. In *Construction Supply Chain Management: Concepts and Case Studies*; Pryke, S., Ed.; Wiley-Blackwell: Oxford, UK, 2009; pp. 189–192.
12. Saad, M.; Jones, M.; James, P. A review of the progress towards the adoption of supply chain management (SCM) relationships in construction. *Eur. J. Purch. Supply Manag.* **2002**, *8*, 173–183. [CrossRef]
13. Briscoe, G.; Dainty, A. Construction supply chain integration: An elusive goal? *Supply Chain Manag. Int. J.* **2005**, *10*, 319–326. [CrossRef]
14. Skitmore, M.; Smyth, H. Marketing and pricing Strategy. In *Construction Supply Chain Management Concepts and Case Studies*; Pryke, S., Ed.; Wiley Blackwell: Oxford, UK, 2009.
15. Russell, E.F. Investigation into the Use of Main Contractor Category Management to Improve Sustainability within the Construction Supply Network. Ph.D. Thesis, University of Surrey, Guildford, UK, 2017.
16. Bank for International Settlements (BIS). *Industrial Strategy: Industry and Government in Partnership: Construction 2025*; HMSO: London, UK, 2013.

17. Fawcett, S.E.; Magnum, G.M. The rhetoric and reality of supply chain integration. *Int. J. Phys. Distrib. Logist. Manag.* **2002**, *32*, 339–361. [CrossRef]

18. Vurro, C.; Russo, A.; Perrrini, F. Shaping sustainable value chains: Network determinants of supply chain governance models. *J. Bus. Ethics* **2009**, *90*, 607–621. [CrossRef]

19. Fernie, S.; Tennant, S. The non-adoption of supply chain management. *Constr. Manag. Econ.* **2013**, *31*, 1038–1058. [CrossRef]

20. Nystrom, J. Partnering: Definition, Theory and Evaluation. Ph.D. Thesis, Royal Institute of Technology (KTH), Stockholm, Sweden, 2007.

21. Brinkhoff, A.; Thonemann, U.W. Perfekte Projekte in der Lieferkette. *Harv. Bus. Manag.* **2007**, *7*, 6–9.

22. Clift, R.; Sim, S.; Sinclair, P. Sustainable consumption and production: Quality, luxury and supply chain equity. In *Treatise in Sustainability Science and Engineering*; Jawahir, I.S., Sikhdar, S., Huang, Y., Eds.; Springer: Heidelberg, Germany, 2013; pp. 291–309.

23. Spence, L.; Bourlakis, M. The evolution from corporate social responsibility to supply chain responsibility: The case of Waitrose. *Supply Chain Manag.* **2009**, *14*, 291–302. [CrossRef]

24. Hobbs, G. *Construction and Resources Roadmap*; BRE Report Prepared for DEFRA's Business Waste and Resource Efficiency Programme (BREW); BRE: Watford, UK, 2008; Available online: https://www.bre.co. uk/filelibrary/pdf/rpts/waste/Roadmap_final.pdf (accessed on 28 December 2017).

25. WRAP. Current Practices and Future Potential in Modern Methods of Construction, Ref WAS003-001. 2007. Available online: http://www.wrap.org.uk/sites/files/wrap/Modern%20Methods%20of%20Construction% 20Full.pdf (accessed on 24 January 2018). (Archived by WebCite® at http://www.webcitation.org/ 6xBXNwYDo).

26. HM Government. UK Annual Report on Modern Slavery. 2017. Available online: https: //www.gov.uk/government/uploads/system/uploads/attachment_data/file/652366/2017_uk_ annual_report_on_modern_slavery.pdf (accessed on 27 December 2017). (Archived by WebCite® at http://www.webcitation.org/6xBXZdy7I).

27. Walk Free Foundation. *The Global Slavery Index 2016*; The Minderoo Foundation Pty Ltd.: Dalkeith, Australia, 2016.

28. ILO. Global Estimates of Modern Slavery: Forced Labour and Marriage. 19 September 2017. Available online: http://www.ilo.org/wcmsp5/groups/public/---dgreports/---dcomm/documents/ publication/wcms_575479.pdf (accessed on 21 January 2017). (Archived by WebCite® at http://www. webcitation.org/6xBXiAzXF).

29. United Nations. *Transforming Our World: The 2030 Agenda for Sustainable Development. A/RES/70/1*; United Nations: New York, NY, USA, 2015.

30. Jackson, T. Keeping out the giraffes. In *Long Horizons*; Tickell, A., Ed.; British Council: London, UK, 2010; p. 20.

31. Globescan. *Evaluating Progress towards the Sustainable Development Goals. Globescan/Sustainability Survey*; Globescan: Toronto, ON, Canada, 2017.

32. UN. United Nations Sustainable Development Goals, Sustainable Development Agenda. 2015. Available online: http://www.un.org/sustainabledevelopment/development-agenda/ (accessed on 9 January 2018). (Archived by WebCite® at http://www.webcitation.org/6xBXp5g2H).

33. Environmental Audit Committee. *Sustainable Development Goals in the UK*; House of Commons: London, UK, 2017.

34. Ethical Corp. *Sustainability in Europe—Top Trends*; Ethical Corporation: London, UK, 2017.

35. Rockström, J.; Steffen, W.; Noone, K.; Persson, A.; Chapin, F.S., III; Lambin, E.F.; Lenton, T.M.; Scheffer, M.; Folke, C.; Schellnhuber, H.J.; et al. A safe operating space for humanity. *Nature* **2009**, *461*, 472–475.

36. Steffen, W.; Richardson, K.; Rockström, J.; Cornell, S.E.; Fetzer, I.; Bennett, E.M.; Biggs, R.; Carpenter, S.R.; Vries, W.D.; de Wit, C.A.; et al. Planetary boundaries: Guiding human development on a changing planet. *Science* **2015**, *347*. [CrossRef] [PubMed]

37. Clift, R.; Sim, S.; King, H.; Chenoweth, J.L.; Christie, I.; Clavreul, J.; Mueller, C.; Posthuma, L.; Boulay, A.M.; Chaplin-Kramer, R.; et al. The challenges of applying planetary boundaries as a basis for strategic decision-making in companies with global supply chains. *Sustainability* **2017**, *9*, 279. [CrossRef]

38. Clift, R. Metrics for Supply Chain Sustainability. *Clean Technol. Environ. Policy* **2003**, *5*, 240–247. [CrossRef]

39. Glasser, B.G.; Strauss, A.L. *The Discovery of Grounded Theory: Strategies for Qualitative Research*; Aldine Publishing Company: Hawthorne, NY, USA, 2008.

40. Gilbert, N.; Stoneman, P. *Researching Social Life*, 4th ed.; Sage: London, UK, 2016.

41. Carillion. Sustainability Report 2014: Our Business. Wolverhampton: Carillion plc. 2015. Available online: http://sustainability2014.carillionplc.com/assets/files/carillion-sr2014-full.pdf (accessed on 7 May 2017). (Archived by WebCite® at http://www.webcitation.org/6xBZ8AZCh).

42. Sandelowski, M. Qualitative analysis: What it is and how to begin. *Res. Nurs. Health* **1995**, *18*, 371–375. [CrossRef] [PubMed]

43. European Union. Council Regulation (EU) No. 995/2010 Laying down the obligations of operators who place timber and timber products on the market. *Off. J. Eur. Union* **2010**, *53*, 23–34. Available online: http://eur-lex.europa.eu/legal-content/EN/TXT/PDF/?uri=OJ:L:2010:295:FULL&from=EN (accessed on 12 September 2017).

44. Strauss, A. *Qualitative Analysis for Social Scientists*; Cambridge University Press: Cambridge, UK, 1987.

45. Khan, S.; VanWynsberghe, R. Cultivating the Under-Mined: Cross-Case Analysis as Knowledge Mobilization. Forum Qualitative Sozialforschung/Forum: Qualitative Social Research, [S.l.]. 2008, Volume 9. Available online: http://www.qualitative-research.net/index.php/fqs/article/view/334/729 (accessed on 18 February 2018).

46. Murphy, D.F.; Bendell, J. Do-it yourself or do-it together? The implementation of sustainable timber purchasing policies by DIY retailers in the UK. In *Greener Purchasing: Opportunities and Innovations*; Russel, T., Ed.; Greenleaf Publishing: Sheffield, UK, 1998; pp. 118–134.

47. FSC. The 10 FSC Principles. 2017. Available online: http://www.fsc-uk.org/en-uk/about-fsc/what-is-fsc/fsc-principles (accessed on 3 May 2017). (Archived by WebCite® at http://www.webcitation.org/6xBXwVgke).

48. Breukink, G.; Levin, J.; Mo, K. Profitability and Sustainability in Responsible Forestry: Economic Impacts of FSC Certification on Forest Operators, WWF. 2015. Available online: http://d2ouvy59p0dg6k.cloudfront.net/downloads/profitability_and_sustainability_in_responsible_forestry_main_report_final.pdf (accessed on 13 September 2017).

49. Carillion. *We Are Making Choices: Carillion's Environment, Community and Social Report 1999–2000*; Carillion plc: Wolverhampton, UK, 2000.

50. GFTN. GFTN-UK Forest Product Reporting Summary for 2015: Carillion plc. 2016. Available online: https://carillionplc-uploads-shared.s3-eu-west-1.amazonaws.com/wp-content/uploads/2016/08/1332BJ-carillion-gftn-uk-2015-published-timber-report-original.pdf (accessed on 3 May 2017). (Archived by WebCite® at http://www.webcitation.org/6xBY8pzrJ).

51. Carillion. How We're Making Tomorrow a Better Place: Carillion Sustainability Report 2016. Wolverhampton: Carillion plc. 2017. Available online: http://sustainability2016.carillionplc.com/assets/files/carillion-sr2016-full.pdf (accessed on 6 June 2017). (Archived by WebCite® at http://www.webcitation.org/6xBaRFVYP).

52. Carillion. *Supply Chain Team Survey*; Carillion: Wolverhampton, UK, 2017; unpublish.

53. UK Government. UK Modern Slavery Act. 2015. Available online: http://www.legislation.gov.uk/ukpga/2015/30/contents/enacted (accessed on 3 May 2017). (Archived by WebCite® at http://www.webcitation.org/6xBYqVD4m).

54. Silverman, B. Modern Slavery: An application of Multiple Systems Estimation, UK Government, London. 2014. Available online: https://www.gov.uk/government/uploads/system/uploads/attachment_data/file/386841/Modern_Slavery_an_application_of_MSE_revised.pdf (accessed on 3 May 2017). (Archived by WebCite® at http://www.webcitation.org/6xBYHOaez).

55. United Nations Environment Programme (UNEP). Greening the Building Supply Chain. United Nations Environment Programme. In *Division of Technology, Industry and Economics*; Sustainable Buildings and Climate Initiative: Paris, France, 2014.

56. Allain, J.; Crane, A.; LeBaron, G.; Behbahani, L. *Forced Labour's Business Models and Supply Chains*; Joseph Rowntree Foundation & Queens University: Belfast, UK, 2013.

57. Gold, S.; Trautrims, A.; Trodd, Z. Modern slavery challenges to supply chain management. *Supply Chain Manag. Int. J.* **2015**, *20*, 485–494. [CrossRef]

58. New, S. Modern slavery and the supply chain: The limits of corporate social responsibility? *Supply Chain Manag. Int. J.* **2015**, *20*, 697–707. [CrossRef]

59. Carillion. Labour Standards Charter. 2017. Available online: https://carillionplc-uploads-shared.s3-eu-west-1.amazonaws.com/wp-content/uploads/2017/01/0919FQ-cai0209_labour-standards-charter_jan2017-original.pdf (accessed on 5 May 2017). (Archived by WebCite® at http://www.webcitation.org/6xBYgCHY9).

60. Action Sustainability, Supply Chain Sustainability School: Modern Slavery. 2016. Available online: https://www.supplychainschool.co.uk/default/modern-slavery.aspx (accessed on 7 May 2017). (Archived by WebCite® at http://www.webcitation.org/6xBYQAJ3f).

61. Lloyd-Roberts, S. Qatar 2022: Construction Firms Accused Amid Building Boom, BBC Newsnight. 8 December 2014. Available online: http://www.bbc.co.uk/news/business-30295183 (accessed on 9 November 2017). (Archived by WebCite® at http://www.webcitation.org/6xBamEK3H).

62. Basset Hound Rescue Southern California (BHRSC). *A Wall of Silence: The Construction Sector's Response to Migrant Rights in Qatar and UEA*; Business and Human Rights Resource Centre: London, UK, 2016.

63. Booth, R.; Kelly, A. Migrant Workers in Qatar Still at Risk Despite Reforms, Warns Amnesty, The Guardian. 13 December 2016. Available online: https://www.theguardian.com/global-development/2016/dec/13/migrant-workers-in-qatar-still-at-risk-despite-reforms-warns-amnesty (accessed on 9 November 2017). (Archived by WebCite® at http://www.webcitation.org/6xBaeDovb).

64. Cai, X.; Tsai, C.; Wu, W. Are they neck and neck in the affordable housing policies? A cross case comparison of three metropolitan cities in China. *Sustainability* **2017**, *9*, 542. [CrossRef]

65. Oelze, N. Sustainable supply chain management implementation–Enablers and barriers in the textile industry. *Sustainability* **2017**, *9*, 1435. [CrossRef]

66. FSC. FSC®: A Tool to Implement the Sustainable Development Goals. 2016. Available online: https://ic.fsc.org/en/web-page-/fsc-contributions-to-achieving-the-sustainable-development-goals (accessed on 6 May 2017). (Archived by WebCite® at http://www.webcitation.org/6xBaz1uYG).

67. Schmidt, C.; Foerstl, K.; Schaltenbrand, B. The supply chain position paradox: Green practices and firm performance. *J. Supply Chain Manag.* **2017**, *53*, 3–25. [CrossRef]

68. Gualandris, J.; Klassen, R.D.; S Vachon, S.; Kalchschmidt, M. Sustainable evaluation and verification in supply chains: Aligning and leveraging accountability to stakeholders. *J. Oper. Manag.* **2015**, *38*, 1–13. [CrossRef]

69. Segall, D.; Labowitz, S. *Making Workers Pay: Recruitment of the Migrant Labor Force in the Gulf Construction Industry*; NYU Stern Centre for Business and Human Rights: New York, NY, USA, 2017.

70. UNSD. Global Indicator Framework for the Sustainable Development Goals and Targets of the 2030 Agenda for Sustainable Development, Global Indicator Framework A RES 71 313 Annex, Statistical Commission Pertaining to the 2030 Agenda for Sustainable Development. United Nations, 2017. Available online: https://unstats.un.org/sdgs/indicators/Global%20Indicator%20Framework_A.RES.71.313%20Annex.pdf (accessed on 8 November 2017). (Archived by WebCite® at http://www.webcitation.org/6xBbEFjMi).

71. University of Cambridge Institute for Sustainability Leadership (CISL). *Towards a Sustainable Economy: The Commercial Imperative for Business to Deliver the UN Sustainable Development Goals*; Vrettos, A., Ed.; CISL; University of Cambridge: Cambridge, UK, 2017; p. 5.

72. Carter, C.R.; Rogers, D.S. A framework of sustainable supply chain management: Moving toward new theory. *Int. J. Phys. Distrib. Logist. Manag.* **2008**, *38*, 360–387. [CrossRef]

73. Jorgensesn, A.L.; Knudsen, J.S. Sustainable competitiveness in global value chains: How do small Danish firms behave. *Corp. Gov. Int. J. Bus. Soc.* **2006**, *6*, 449–462. [CrossRef]

74. Spekman, R.; Kamauff, J.; Myhr, N. An empirical investigation into supply chain management: A perspective on partnerships. *Int. J. Phys. Distrib. Logist. Manag.* **1998**, *34*, 414–433. [CrossRef]

75. Duffy, R.; Fearne, A. Partnerships and alliances in UK supermarket supply networks. In *Food Supply Cain Management*; Bourlakis, M., Weightman, P., Eds.; Blackwell: Oxford, UK, 2004; pp. 136–152.

76. Roberts, S. Supply Chain Specific? Understanding the Patchy Success of Ethical Sourcing Initiatives. *J. Bus. Ethics* **2003**, *44*, 159–170. [CrossRef]
77. ILO Forced Labour Convention (No29), Geneva, Switzerland, 1930. Available online: http://www.ilo.org/wcmsp5/groups/public/@asia/@robangkok/documents/genericdocument/wcms_346435.pdf (accessed on 28 December 2017). (Archived by WebCite®at http://www.webcitation.org/6xBYYUDum).

Article

Effect of Crack on Durability of RC Material under the Chloride Aggressive Environment

Yongchun Cheng, Yuwei Zhang, Guojin Tan * and Yubo Jiao

College of Transportation, Jilin University, Changchun 130025, Jilin, China; chengyc@jlu.edu.cn (Y.C.);
ywzhang15@mails.jlu.edu.cn (Y.Z.); jiaoyb@jlu.edu.cn (Y.J.)
* Correspondence: tgj@jlu.edu.cn; Tel.: +86-0431-8509-5446

Received: 5 January 2018; Accepted: 4 February 2018; Published: 7 February 2018

Abstract: Exposed to aggressive environments, the rebar in reinforced concrete (RC) bridges will be corroded gradually. Durability of RC material mostly depends on the rebar corrosion behavior. In this research, influences of crack on rebar corrosion were investigated. Firstly, RC specimens with different crack number, width and spacing were prepared and the rebar corrosion was conducted through an accelerated chloride penetration method. Then, corrosion current densities of rebar were calculated from electrochemical test methods including liner polarization (LP), Tafel potentiodynamic polarization (TPP) and electrochemical impedance spectroscopy (EIS) measurements. Finally, the discussion was presented about a more reasonable electrochemical testing method for rebar corrosion in RC material. Besides, the significant influence factor among crack width, number and spacing was evaluated based on both One-way analysis of variance (One-way ANOVA) and Turkey's honest significant difference (Turkey's HSD) test. The results revealed that a more reasonable way to obtain corrosion current densities of rebar is combining EIS measurement with TPP measurement. Crack number shows the most significant effect on corrosion behavior of rebar, while crack spacing possesses the least one.

Keywords: rebar corrosion; cracked RC material; aggressive environment; durability; electrochemical test methods

---

1. Introduction

Reinforced concrete (RC) is one of the most widely used materials for bridge structures. The deteriorated durability of RC structures causes structural safety problems and expensive maintenance costs. Besides, in the process of bridge reinforcement, environmental problems including noise pollution, dust pollution and consumption of natural resources will occur. Meanwhile, the corrosion of rebar in concrete is crucial to the durability of RC structures, especially in chloride aggressive environments [1,2]. Rebar in concrete are not prone to corrosion within a short time in sound concrete due to the high and stable alkalinity of concrete pore solution. However, cracks will be formed due to material shrinkage, thermal gradients and repeated mechanical loading causing a higher osmotic pressure, which often leads to a higher chloride diffusion coefficient in concrete [3,4]. Besides, cracks provide a convenient channel for water ingression, harmful ions and oxygen towards the internal concrete and the surface of rebar, which could result in accelerating corrosion of rebar. Therefore, it is necessary to investigate corrosion behavior of rebar under the effect of crack in chloride aggressive environment [5–8].

A large number of laboratorial studies have been conducted focusing on durability of concrete and rebar corrosion under the actions of chloride penetration and crack. Wang et al. [9] introduced feedback controlled splitting tests to generate crack width-controlled concrete specimens and studied the relationship between crack characteristics and concrete permeability. Ye et al. [10] established a model of chloride penetration into cracked concrete subject to drying-wetting cycles based on Fick's second law. Marsavina et al. [11] made artificial cracks on concrete specimens by a thin copper to perform a concrete chloride penetration test. They demonstrated that the chloride diffusivity in cracked concrete is much stronger than sound concrete. Du et al. [12] and Liu et al. [13] simulated chloride diffusivity in cracked concrete using multi-component ionic transport models and found that the geometry of crack affects chloride transport. Papakonstantinou and Shinozuka [14] established a probabilistic model for rebar corrosion in reinforced concrete structures of large dimensions considering crack effects to simulate the complex phenomena involved in a detailed and simple way, appropriate for implementation on large-scale, real structures. Cao et al. [15] found that oxygen is a crucial factor influencing rebar corrosion propagation process during the full RC structures' service life. Zhu et al. [16] investigated the influence of load-induced cracking behavior on the process of chloride penetration into concrete. Pacheco et al. [17] analyzed bending cracks in RC samples by measuring the electrical resistance across the crack and illustrated that cracks in concrete represent fast routes for chloride penetration, which can result in rebar corrosion. Besides, Šavija and Schlangen [18] claimed that chloride induced corrosion of rebar is one of the most important mechanisms causing deterioration of RC structures and the requirements for their premature repair or replacement. Pedrosa and Andrade [19] indicated that rebar corrosion causes several damage types that influence the structure service life. They analyzed the effect of a range of distinct corrosion rates on the crack growth rate, and they established an empirical model to describe the relation between crack width and the corrosion rate applied for the first stage of cracking. Based on the existed researches, there are different types of cracks caused by temperature, loading and rebar corrosion deteriorating the durability of RC structures, and each crack type has its own distribution inside concrete cover. Thus, performing a laboratorial study on comprehensive effects of crack parameters (such as crack number, width and spacing) on corrosion behavior of rebar could better illustrate the effect of crack on the durability of RC structures.

Nowadays, electrochemical measurements are popular to analyze the corrosion behavior of rebar. The reason is that rebar corrosion is an electrochemical process. An oxidation reaction occurs ($Fe \rightarrow Fe^{2+} + 2e^-$) at the anodic area of rebar. The electrons given by anodic area are consumed at the cathodic area ($O_2 + 2H_2O + 4e^- \rightarrow 4OH^-$). Pore solution as a conducting medium for the transportation of electrons and ions ensure the corrosion process to proceed [20]. The electrochemical nature of corrosion means that electrochemical techniques can be used to monitor the corrosion behavior such as corrosion rate or corrosion current density of rebar in concrete. The commonly used electrochemical techniques include liner polarization (LP), Tafel potentiodynamic polarization (TPP) and electrochemical impedance spectroscopy (EIS) measurements [21]. These techniques are widely used in the studies of carbon steel and alloy steel corrosion [22]. There are a large volume of published researches describing the corrosion behavior of rebar by immersing them into chloride solution [23] or cement extract solution [24,25]. Meanwhile, some researches focus on the corrosion behavior of rebar in real concrete material. Andrade et al. [26] analyzed influence of environmental factors and cement chemistry on corrosion behavior of rebars in concrete by using EIS measurement, and they indicated that redox activity caused by harmful ions in the rebar's oxides layer greatly influences the electrochemical behavior of rebars in the passivity potential domain. In addition, Andrade et al. [27] indicated that different geometrical dispositions of the electrodes used in EIS measurement may affect the test results much. Wang et al. [21] analyzed corrosion rate of rebar in concrete under cyclic freeze-thaw and chloride salt action using LP and TPP measurements. They illustrated that TPP measurement is rapid and easy to operate, read corrosion current directly and provides sufficiently accurate results. Gerengi et al. [28] used EIS measurements to investigate corrosion behavior of rebar in reinforced concrete exposed to sulphuric acid, and they claimed that EIS measurement is one of the most widely used techniques in recent years. Andrade and Alonso [29] applied a non-destructive electrochemical test method for the estimation in large size concrete structures of the instantaneous corrosion current density and discussed the accuracy and applicability of this measurement used in real RC material. In summary, most of the literatures applied one of the electrochemical measurements.

However, combining multiple electrochemical measurements may show a more accurate way to undertake the study on rebar corrosion.

In this paper, corrosion current density (i_{corr}) of rebar was treated as the index for estimating the effect of crack on durability of RC material. Firstly, i_{corr} values of rebar were tested by TPP, LP and EIS measurements. Subsequently, a more reasonable electrochemical testing method was recommended for rebar in RC material. Finally, the effect of crack width, number and spacing on durability of RC material was analyzed by statistical analysis methods.

2. Materials and Methods

2.1. Materials and Mixture

Q235 rebars (equivalent to SS400 and A36, i.e., yield strength is 235 MPa) with diameter 7 mm were cut into 110 mm long. Rebars were polished with 340# to 2000# grit silicon carbide emery paper in order to remove the passivation layers and guarantee no pit corrosion on their surfaces. Subsequently, according to a national standard [30], ethanol and acetone treatments were used to degrease surfaces of rebar, so the electrochemical properties of each rebar were sensitive to the effect of chloride ions. One end of each rebar was welded to a copper wire.

PO 42.5 type ordinary Portland cement was used in this study conforming to the requirements of the national standard [31]. Crushed stones with diameters ranging from 2.36 mm to 20.00 mm and natural sands with fineness modulus of 2.7 were adopted as coarse and fine aggregates, respectively. The mixture proportions of concrete used in this study are listed in Table 1. Slump of the mixture was tested to be 40 mm, which means the concrete mixture has favorable cohesiveness and meets well with the requirement of the national standard [32].

Table 1. Mixture proportions of concrete.

Materials	Nominal Proportions (kg/m^3)
Cement	433
Water	195
Fine aggregate	567
Coarse aggregate	1205

2.2. Specimen Preparation

RC samples used in this study were composed of concrete (cracked or sound) and rebar. Samples were made with artificial cracks by means of positing and removal of required thin copper sheets with different number, width and spacing inside samples. Figure 1 shows the dimensions of the specimen and relative position among concrete, crack and rebar. The length of each crack is 60 mm and the depth is 19 mm. The length direction of each crack is perpendicular to the length direction of rebar. The combinations of crack width, number and spacing in samples are given in Table 2.

Table 2. Design of specimens.

Symbol	Crack Width (mm)	Crack Number	Crack Spacing (mm)	Sketch of Specimen
N	0	0	-	
A1	0.05	1	-	
A2	0.1	1	-	
A3	0.2	1	-	
B1	0.1	2	15	
B2	0.1	2	25	
C1	0.1	3	15	
C2	0.1	3	25	

Cylinders of φ100 mm × 50 mm were casted in plastic modules and compacted by vibrating table. They were allowed to cure at the condition of 20 °C and 95% relative humidity. Copper sheets were removed from specimens after a 4-h curing period. Specimens were removed from moulds after a 24-h curing period. Besides, the weld and exposed parts of rebar must be covered with a layer of epoxy resin to protect rebar from corrosion during the curing time. After that, all specimens were cured under the normal curing condition (20 °C and 95% relative humidity) for 28 days before further experiments.

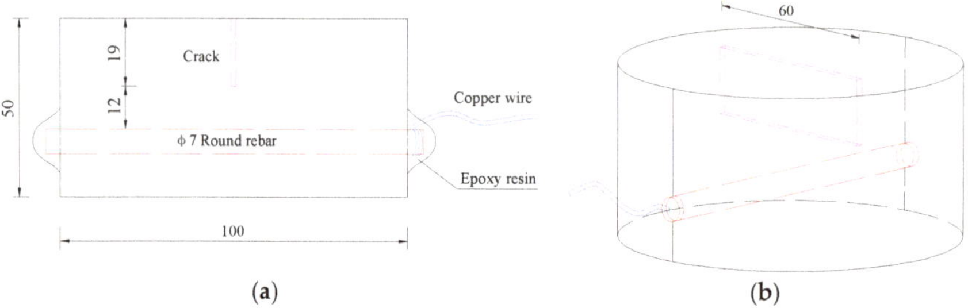

Figure 1. Specimen dimension and rebar arrangement: (**a**) Front view of the sample; (**b**) Stereoscopic view of the sample.

2.3. Experimental Methods

2.3.1. Scheme Design

After curing, in order to investigate the influence of crack on corrosion behavior of rebar in concrete, an experimental procedure was determined and shown in Figure 2. Firstly, RC samples must be vacuum-saturated by vacuum water saturation instrument to guarantee they were well conductive. Secondly, electrochemical properties of rebar in concrete were tested by LP, TPP and EIS measurements. Subsequently, the accelerated chloride penetration experiment was conducted for 96 h so that chloride ions can rapidly penetrate into concrete along with the depth direction of crack and corrode the rebar. Then, the vacuum-saturated treatment will be conducted in a second time before testing electrochemical properties of rebar again. Finally, electrochemical test measurements were performed again to obtain electrochemical properties of rebar after the accelerated chloride penetration experiment. The rationality of electrochemical test procedure and test times will be discussed in the Section 2.3.4.

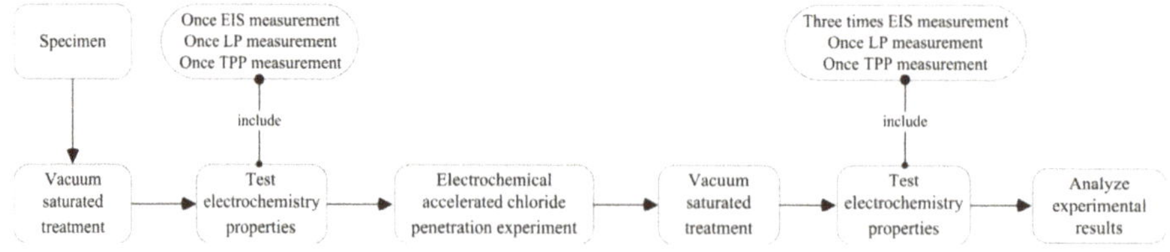

Figure 2. Experimental procedure.

2.3.2. Vacuum-Saturated Treatment

According to a national standard [30], the vacuum-saturated treatment must be performed before the accelerated chloride penetration experiment in order to ensure chloride ions could penetrate into concrete steadily. Also, it is necessary before testing electrochemical properties of rebar to guarantee RC samples were conductive. The vacuum-saturated procedure consisted of conditioning RC samples into a vacuum desiccator and applying a vacuum pressure of 5 kPa for 3 h. Subsequently, the desiccator was filled with deionized water until all samples were immersed, and the pressure was maintained for another one hour. Finally, samples were continuously immersed for 18 h without vacuum pressure.

2.3.3. Accelerated Chloride Penetration Experiment

In general, it will take a long time to observe the corrosion of rebar due to the slow corrosion process of chloride ions by immersing RC samples into chloride solution. However, the accelerated chloride penetration experiment used in this study showed a more effective way to accelerate the corrosion process of rebar affected by chloride ions, which could simulate the natural aggressive environment well. The experimental setup is displayed in Figure 3. Each specimen was placed between two acrylic cells defined as cathode cell and anode cell. The volume of each cell is 270 cm^3. Cathode cell was filled with 3.5 wt % NaCl solution, and anode cell was filled with 0.3 mol/L NaOH solution. Two stainless steel-based plates as cathode electrode and anode electrode were placed into these two cells. The cracked surface of specimen was placed into the cathode cell, and the other side was placed into the anode cell. Then, these two cells were connected to a DC regulated power supply, and the applied voltage was 20.0 V. This designed accelerated chloride penetration system was electrified for 96 h.

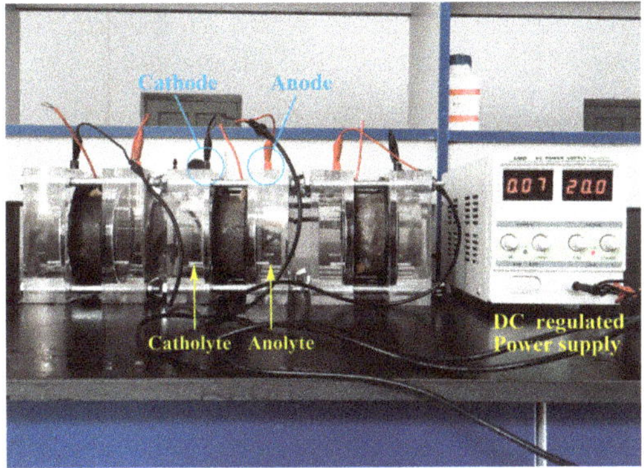

Figure 3. Experimental setup for accelerated chloride penetration.

2.3.4. Corrosion Density Test for Bar Based on Electrochemical Measurements

Nowadays, electrochemical measurements including TPP, LP and EIS measurements are popularly used in the studies of steel corrosion under different conditions. These three measurements were employed in this study to evaluate the electrochemical properties of rebar under an accelerated chloride penetration. Corrosion current density (i_{corr}) is used as the index to evaluate the corrosion behavior of rebar.

The electrochemical measurements were carried out by CS350H electrochemical workstation in this study based on the three-electrode system shown in Figure 4. In this system, the rebar embedded in concrete was treated as the working electrode (WE) with a polarization area of 21.98 cm^2, a saturated calomel electrode (SCE) with potassium chloride salt bridge placed in a Luggin capillary was used as the referenced electrode (RE) and a stainless steel-based plate with 2 mm × 100 mm × 150 mm was used as the counter electrode (CE).

Corrosion density is used to evaluate the corrosion behavior of rebar, and it can be obtained from TPP measurement directly or calculated by Stern-Geary equation

$$i_{corr} = \frac{B}{R_p} \tag{1}$$

where i_{corr} is the corrosion density ($\mu A/cm^2$); B is the Stern-Geary coefficient (mV/Decade); R_p is polarization resistance ($\Omega \cdot cm^2$). The value of B can be calculated from TPP measurement or estimated to fall within the range from 25 mV to 52 mV. Song [33] demonstrated the estimated range value of B is applicable only in a uniform corrosion system at its corrosion potential, whereas the RC structure may be subjected to a non-uniform corrosion. The typical value (25 mV~52 mV) of B is not acceptable in a RC corrosion system. Therefore, the value of B should be calculated from TPP curves more accurately. Meanwhile, R_p can be obtained from LP and EIS measurements. The details of these three measurements are as follows.

TPP measurement is an electrochemical test method to characterize corrosion properties of metal materials based on the three-electrode system. TPP curves can be obtained by monitoring the potential used steady fixed levels of current at specific potential. In TPP curves, the current density I (A/cm^2) and polarization potential E (V) under potentiodynamic polarization of rebar always conform to the Butler–Volmer equation:

$$I = i_{corr}\left\{\exp\left[\frac{2.303(E - E_{corr})}{b_a}\right] - \exp\left[\frac{2.303(E_{corr} - E)}{b_c}\right]\right\} \tag{2}$$

where E_{corr} is the corrosion potential (V vs. SCE), respectively. b_a and b_c are the anodic Tafel slope and cathodic Tafel slope, respectively (mV/Decade). Values of these electrochemical parameters can be obtained by a curve-fitting approach named Tafel extrapolation method. Schematic illustration parameters in TPP curve are shown in Figure 5.

The Stern–Geary coefficient B related to b_a and b_c can be calculated by

$$B = \frac{b_a \times b_c}{2.303 \times (b_a + b_c)} \tag{3}$$

After obtaining a stable open circuit potential (OCP), TPP measurement could be carried out from -250 mV to 250 mV vs. OCP, the scanning rate of which is 1 mV/s.

Besides, LP measurement is another electrochemical method to obtain resistances of metal materials accurately. Polarization curves in LP measurement can be obtained as the same method in TPP measurement. As shown in Figure 6, the value of R_p is defined as the slope of potential to current density, can be obtained as follows:

$$R_p = \left.\frac{\Delta E}{\Delta I}\right|_{I \to 0}^{E \to E_{corr}} \tag{4}$$

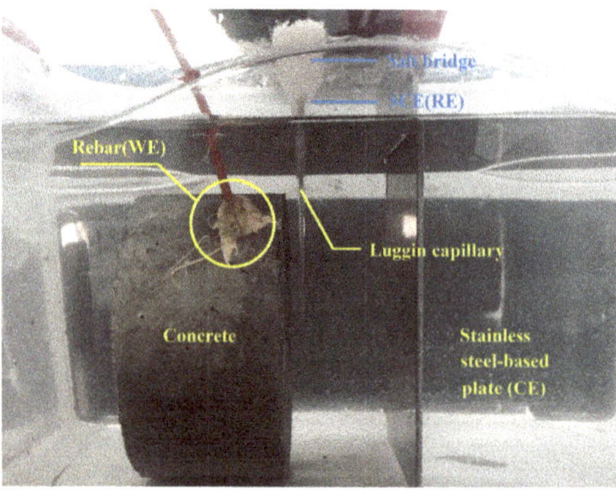

Figure 4. Experimental setup for electrochemical measurements (a three-electrode system).

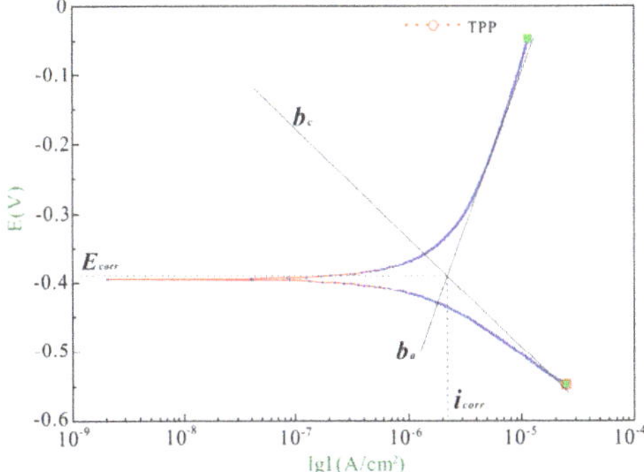

Figure 5. Schematic illustration of the Tafel potentiodynamic polarization (TPP) curve with Tafel electrode slopes.

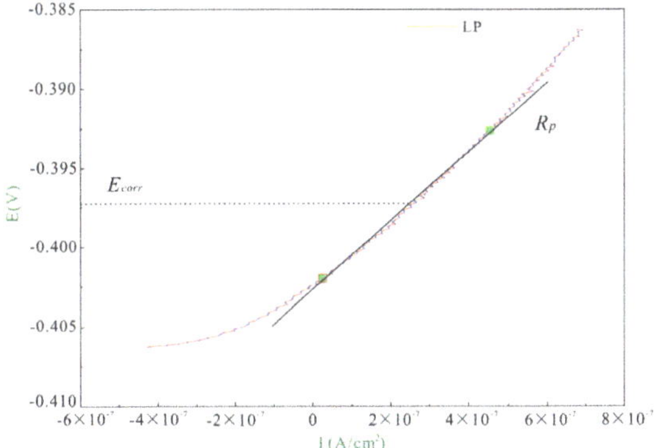

Figure 6. Schematic illustration of polarization resistance.

In this study, LP measurements were carried out from -10 mV to 10 mV vs. OCP with the scanning rate of 0.167 mV/s.

In addition, EIS measurement based on alternating currents (AC) used to analyze corrosion mechanisms of steel has been well documented for many years [34]. Applied AC amplitude signal was equal to 10 mV over the frequency range from 0.01 Hz to 100,000 Hz. To better understand the electrochemical phenomena, an equivalent circuit was used to simulate the electrochemical behavior of rebar. The equivalent circuit is illustrated in Figure 7, which shows the solution resistance (R_{p1}), concrete resistance (R_{p2}), polarization resistance of rebar (R_{p3}), external concrete capacitance (CPE_1) and double layer capacitance on the surface of rebar (CPE_2).

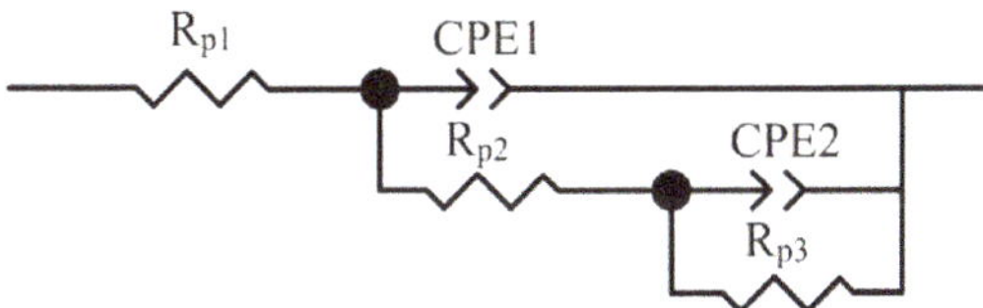

Figure 7. Equivalent circuit for reinforced concrete (RC) samples.

The immittance Z ($\Omega \cdot cm^2$) of this equivalent circuit is given by

$$Z = R_{p1} + \cfrac{1}{T_1(j\omega)^{P_1} + \cfrac{1}{R_{p2} + \cfrac{1}{T_2(j\omega)^{P_2} + \frac{1}{R_{p3}}}}} \tag{5}$$

in which,

$$T_n(j\omega)^{P_n} = T_n \omega^{P_n} \left[\cos(\frac{P_n \pi}{2}) + j\sin(\frac{P_n \pi}{2}) \right] (n = 1, 2) \tag{6}$$

where T_n and p_n are parameters of CPE ($\mu F \cdot cm^2$); ω is the frequency of applied AC (Hz); $j = \sqrt{-1}$. The obtained impedance data could be analyzed by Z view program to fit with the immittance equation of this equivalent circuit.

The voltage applied in EIS measurements is much smaller than those of LP and TPP measurements, which means EIS measurements display a smaller disturbance and better reproducibility of electrochemical systems than the others [35]. Therefore, EIS measurements can be performed continuously, while the others cannot because the error in a continuous test procedure can be very large. The disturbances of these three measurements are ranked from small to large: EIS measurements, LP measurements, TPP measurements. Therefore, the test sequence should be the same as the disturbance. Besides, EIS, LP and TPP measurements should be conducted only once on each specimen before an accelerated chloride penetration experiment. Because the electrochemical property of each rebar in concrete is very stable after 28-day curing, and the test results are treated as unique initial values to evaluate the differences of corrosion behavior of rebar. Furthermore, EIS, LP and TPP measurements should be conducted three times, one time and one time, respectively, after accelerated chloride penetration experiments in order to obtain more statistical random sampling test results and quantitatively analyze the effects of crack on electrochemical properties of rebar.

2.4. Statistical Analysis Methods

After obtaining i_{corr} of rebar, One-way analysis of variance (ANOVA) combined with Turkey's honest significant difference (Turkey's HSD) test was performed to analyze the effect of crack on rebar corrosion and determine the most significant factor that influences durability of RC material. One-way ANOVA was used in this study to determine whether crack width, number or spacing is the most significant factor, and Turkey's HSD test was performed for further judging the accuracy of the One-way ANOVA results. The details of these two methods are as follows:

2.4.1. One-Way ANOVA

The most commonly used statistical method to analyze the contribution of each level in one factor of experimental results is One-way ANOVA and F-tests method, which can take advantage of sums of squares to separate the overall variance in the response into variances caused by measurement error and processing parameters [36].

In One-way ANOVA, the sum of square SS_T can be calculated by

$$SS_T = \sum_{i=1}^{m}\sum_{j=1}^{n} y_{ij}^2 - \frac{1}{N}(\sum_{i=1}^{m}\sum_{j=1}^{n} y_{ji})^2 \tag{7}$$

where m, n are the number of levels in each factor and the number of experimental results in each level, respectively. y_{ij} is the jth experimental result of level i. N is the number of all experimental results, which is equal to $m \times n$.

The level square of deviance SS_i can be calculated by

$$SS_i = \frac{1}{n}\sum_{i=1}^{m}\left(\sum_{j=1}^{n} y_{ij}\right)^2 - \frac{1}{N}(\sum_{i=1}^{m}\sum_{j=1}^{n} y_{ji})^2 \tag{8}$$

The error square of deviance SS_e can be obtained by

$$SS_e = SS_T - SS_i \tag{9}$$

The estimate of variance is given by

$$MS_i = SS_i / d_{f_i} \tag{10}$$

$$MS_e = SS_e / d_{f_e} \tag{11}$$

where MS_i and MS_e are the estimate of variance for level i and error, respectively. d_{f_i} and d_{f_e} are corresponding degrees of freedom, which can be obtained by

$$d_{f_i} = m - 1 \tag{12}$$

$$d_{f_e} = N - 1 \tag{13}$$

F-value is given by

$$F = MS_i / MS_e \tag{14}$$

In this method, whether the influence factor is significant or not depends on F-value. The larger F-value is, the higher the influence factor is.

2.4.2. Turkey's HSD Test

Meanwhile, Turkey's HSD is a statistical test procedure. It is a post-hoc and single-step multiple comparison procedure, which can be performed combining with analysis of variance method to determine means that are significantly different from each other or not [37,38].

In Turkey's HSD test, HSD value is the critical value to judge whether the influence of factor is significant or not. HSD critical value can be calculated by

$$HSD_a = q_a(m, d_{f_e}) \times \sqrt{\frac{MS_e}{n}} \tag{15}$$

where $q_a(m, d_{f_e})$ is studentized range; α is significant level; m is number of levels in each factor; d_{f_e}, MS_e and n are number of degrees of freedom, estimate of variance of error, number of test results in each level, respectively. In this study, the range values of means between each two levels in specific factor will be calculated and compared with HSD_α. If the range value of means is larger than HSD_α, the two levels are said to have significant effects on rebar corrosion.

3. Results and Discussion

3.1. Recommendation of Reasonable Electrochemical Test Method

As an example, i_{corr} values of sample B2 calculated from 3 measurements are analyzed to illustrate the reasonable electrochemical test method for rebar in the RC corrosion system.

In TPP measurements, anodic and cathodic potentiodynamic polarization curves for rebar in sample B2 are shown in Figure 8. The curve tested before the chloride penetration is defined as initial curve, while the curve tested after the chloride penetration is defined as final curve.

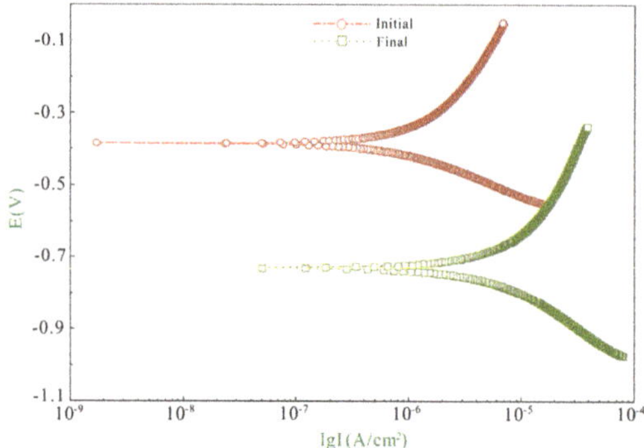

Figure 8. TPP curves of rebar in sample B2.

The anodic and cathodic curves are extrapolated up to their intersection at a point where corrosion current density and corrosion potential can be obtained, as shown in Figure 5. The anodic Tafel slopes (b_a), cathodic Tafel slopes (b_c), Stern-Geary coefficient (B), corrosion current density (i_{corr}) and corrosion potential (E_{corr} vs. SCE) obtained from TPP curves are listed in Table 3.

It can be concluded i_{corr} increases and E_{corr} (vs. SCE) decreases under the action of chloride penetration. Meanwhile, both b_a and b_c increase, which illustrates the anodic and cathodic reactions are all accelerated and affected by the environmental loading.

Table 3. Electrochemical parameters from TPP curves of rebar in sample B2.

	b_a (mV/Decade)	b_c (mV/Decade)	B (mV/Decade)	i_{corr} (μA/cm^2)	E_{corr} (V)
Initial	382.11	160.64	49.17	1.32	−0.38
Final	583.29	282.93	82.72	10.55	−0.73

In LP measurements, the current density can be also calculated by Equation (1). In Equation (1), B is obtained from TPP curves and R_p is calculated from line polarization curves by Equation (4). The calculated R_p and i_{corr} from LP measurements are listed in Table 4.

Table 4. Electrochemical parameters from liner polarization (LP) curves of rebar in sample B2.

	B (mV/Decade)	R_p (Ω·cm^2)	i_{corr} (μA/cm^2)
Initial	49.17	34,384	1.43
Final	82.72	7673	10.78

It can be seen that R_p is reduced while i_{corr} increases under the action of chloride penetration, which infers that rebar is continuously corroded.

In addition, the corrosion current density can be also obtained by EIS measurements. Figure 9 depicts the impedance spectra of rebar in sample B2. The curves before and after the chloride penetration experiment are defined as one initial curve and three final curves, respectively. From Figure 9, it can be seen that the radius of the capacitive loop shrinks after the chloride penetration experiment decreases, which indicates the corrosion resistance of rebar is declined. The equivalent circuit is fitted with impedance spectra by Equation (5), and the results are shown in Table 5.

In Table 5, i_{corr} is calculated by Equation (1), where B is obtained from TPP curves, and R_p is equal to R_{p3}. From Table 5, it can be concluded that the impedance spectra of rebar in sample B2 obtained from three continuous tests is reproducible. Besides, both R_{p2} and R_{p3} decrease after the chloride penetration, which indicates that material properties of concrete were degraded and the rebar was corroded gradually under the process of chloride penetration.

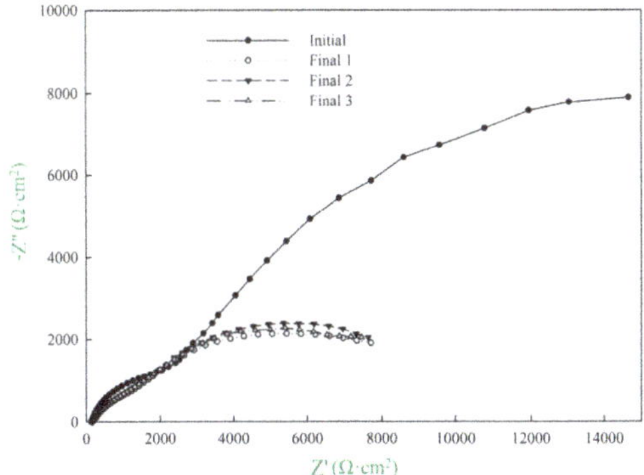

Figure 9. Impedance spectra of rebar in sample B2.

Table 5. Fitting parameters from electrochemical impedance spectroscopy (EIS) of rebar in sample B2.

	R_{p1} ($\Omega \cdot cm^2$)	CPE_1-T ($\mu F \cdot cm^2$)	CPE_1-P ($\mu F \cdot cm^2$)	R_{p2} ($\Omega \cdot cm^2$)	CPE_2-T ($\mu F \cdot cm^2$)	CPE_2-P ($\mu F \cdot cm^2$)	R_{p3} ($\Omega \cdot cm^2$)	i_{corr} ($\mu A/cm^2$)
Initial	132.5	3.52×10^{-5}	0.76	2720	2.40×10^{-4}	0.69	30,077	1.63
Final 1	154.2	9.07×10^{-5}	0.62	2256	2.25×10^{-4}	0.68	7540	10.97
Final 2	156.1	8.99×10^{-5}	0.62	2197	2.35×10^{-4}	0.64	7609	10.87
Final 3	154.6	8.54×10^{-5}	0.62	2094	2.21×10^{-4}	0.62	7554	10.95

The corrosion current densities obtained from 3 measurements are compared and shown in Figure 10. It can be seen the corrosion current densities from EIS measurements are larger than those of TPP and LP measurements. The reason is that R_p in TPP and LP measurements contain not only the resistance of rebar but also the resistance of concrete in the test conductive circuit, while R_p from EIS measurements is exactly the resistance of rebar. Thus, R_p from EIS measurements is smaller than those from TPP and LP measurements. According to Equation (1), i_{corr} calculated from EIS measurements is greater than the others. Inductively, the reasonable way to calculate i_{corr} of rebar is EIS measurement combined with TPP measurement, namely, R_p and B should be calculated by EIS measurement and TPP measurement. After that, i_{corr} can be calculated by Stern-Geary equation. In this way, the corrosion behavior of rebar can be judged more accurately compared with only a single measurement.

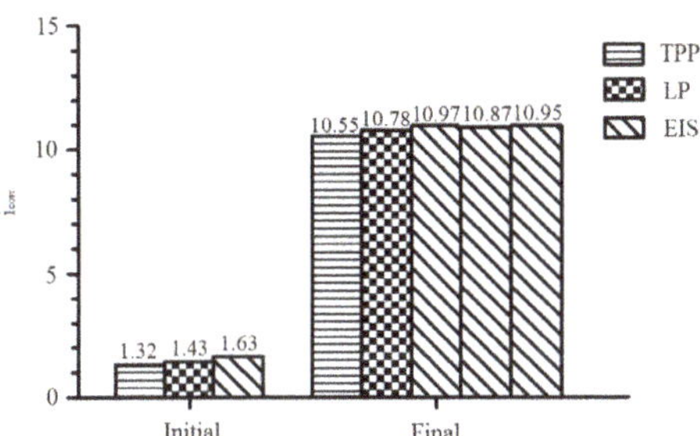

Figure 10. Comparison of corrosion current densities obtained from 3 measurements.

3.2. The Effect of Crack on Corrosion Behavior of Rebar

i_{corr} of rebar in each specimen could be obtained by the same test method of B2 discussed above. The variation of corrosion current densities is calculated by

$$i_{corr}^n = i_F{}^n - i_I{}^0 (n = 1, 2, 3) \tag{16}$$

where i_{corr}^n, $i_F{}^n$ and $i_I{}^0$ are the nth variation of i_{corr}, the nth corrosion current density and initial corrosion current density, respectively. The results are listed in Table 6. The average value for 3 variations of i_{corr} and corresponding standard deviations are also calculated. The results can be used to analyze effects of crack width, number and spacing on corrosion behavior of rebar under the aggressive environment. Comparative evaluations were conducted based on One-way ANOVA and Turkey's HSD test.

Table 6. Variation of corrosion current density in each sample.

Symbol	Crack Width (mm)	Crack Number	Crack Spacing (mm)	i_{corr}^1 (μA/cm^2)	i_{corr}^2 (μA/cm^2)	i_{corr}^3 (μA/cm^2)	Average Value (μA/cm^2)	Standard Deviation (μA/cm^2)
N	0	0	-	4.00	3.88	3.63	3.84	0.19
A1	0.05	1	-	7.81	7.76	7.68	7.75	0.07
A2	0.1	1	-	8.06	8.12	8.04	8.07	0.04
A3	0.2	1	-	8.92	8.86	8.88	8.89	0.03
B1	0.1	2	15	9.48	9.36	9.44	9.43	0.06
B2	0.1	2	25	9.24	9.32	9.34	9.30	0.05
C1	0.1	3	15	12.86	12.75	12.78	12.80	0.06
C2	0.1	3	25	12.44	12.54	12.45	12.48	0.06

In One-way ANOVA, corrosion current densities of rebar in sample N, A1, A2 and A3 are used to analyze the effect of crack width on rebar corrosion. One-way ANOVA results for crack width are listed in Table 7.

Table 7. One-way ANOVA for crack width.

Source of Variance	Square of Deviance	Degree of Freedom	Estimate of Variance	F-Value	$F_{0.01}$ (3,8)	Significance
Crack width	45.618	3	15.206	1427.80	7.59	**
Error	0.085	8	0.011			
Total	45.703					

** Correlation is significant at the 0.01 level.

Besides, corrosion current densities of rebar in sample N, A2, B1 and C1 are used to analyze the effect of crack number on rebar corrosion. Results of One-way ANOVA for crack number are listed in Table 8.

Table 8. One-way ANOVA for crack number.

Source of Variance	Square of Deviance	Degree of Freedom	Estimate of Variance	F-Value	$F_{0.01}$ (3,8)	Significance
Crack number	123.733	3	41.244	3721.29	7.59	**
Error	0.089	8	0.011			
Total	123.822					

** Correlation is significant at the 0.01 level.

In addition, corrosion current densities of rebar in sample B1 and B2 are used to analyze the effect of crack spacing on rebar corrosion. Results of One-way ANOVA for crack spacing are listed in Table 9.

Table 9. One-way ANOVA for crack spacing.

Source of Variance	Square of Deviance	Degree of Freedom	Estimate of Variance	F-Value	$F_{0.05}$ (1,4)	$F_{0.1}$ (1,4)	Significance
Crack spacing	0.024	1	0.024	7.367	7.71	4.54	*
Error	0.013	4	0.003				
Total	0.037						

* Correlation is significant at the 0.1 level.

From Tables 7–9, it can be concluded that F-values of crack width and crack number are larger than $F_{0.01}$ (3,8), which means the probability is 99% that crack width and crack number present a significant influence of rebar corrosion. Meanwhile, F-value of crack number is larger than that of crack width, which means the influence of crack number is larger than crack width. Besides, F-value of crack spacing is between $F_{0.1}$ (1,4) and $F_{0.05}$ (1,4), which means that the probability is 90% that crack spacing is the significant effect on rebar corrosion.

According to the conclusions discussed above, it is more likely that crack width and crack number are the statistically significant effects of rebar corrosion. However, it doesn't mean that there is a larger magnitude between these two factors [39]. Therefore, Turkey's HSD test as a post-hoc multiple comparisons method should be conducted to determine the significant effects of three influence factors on rebar corrosion.

Similar to One-way ANOVA, means of corrosion current densities of rebar in sample N, A1, A2 and A3 are performed Turkey's HSD test to determine if crack width is the significant effect on rebar corrosion. Results of Turkey's HSD test are listed in Table 10.

Table 10. Turkey's honest significant difference (HSD) test results for crack width.

Source of Range Value	Range Value	$q_{0.05}$ (4,8)	$q_{0.01}$ (4,8)	MS_e	$HSD_{0.05}$	$HSD_{0.01}$	Significance
$K_{0.05}$–K_0	3.91						**
$K_{0.1}$–$K_{0.05}$	0.32						-
$K_{0.2}$–$K_{0.1}$	0.82	7.35	11.5	0.011	0.445	0.70	**
$K_{0.1}$–K_0	4.23						**
$K_{0.2}$–K_0	5.05						**
$K_{0.2}$–$K_{0.05}$	1.14						**

** Correlation is significant at 0.01 level. - Correlation is not significant.

In Table 10, most range values between each two levels ($K_{0.05}$–K_0 etc.) in crack width factor is larger than $HSD_{0.01}$, which means the influence between each two levels is significant at 0.01 level. It is also proved that crack width has a statistically significant influence on rebar corrosion.

Besides, means of corrosion current densities of rebar in sample N, A2, B1 and C1 are adopted to perform Turkey's HSD test to determine if crack number is the significant effect on rebar corrosion. Results are listed in Table 11.

Table 11. Turkey's HSD test results for crack number.

Source of Range Value	Range Value	$q_{0.05}$ (4,8)	$q_{0.01}$ (4,8)	MS_e	$HSD_{0.05}$	$HSD_{0.01}$	Significance
K_1–K_0	4.23						**
K_2–K_1	1.36						**
K_3–K_2	3.37	7.35	11.5	0.011	0.445	0.70	**
K_2–K_0	5.59						**
K_3–K_0	8.96						**
K_3–K_1	4.73						**

** Correlation is significant at 0.01 level.

In Table 11, all range values between each two levels (K_1–K_0 etc.) in crack number factor is larger than $HSD_{0.01}$, which means the influence between each two levels is significant at 0.01 level. It is also proved that crack number has a statistically significant influence on rebar corrosion.

In addition, means of corrosion current densities of rebar in sample B1 and B2 are adopted to perform Turkey's HSD test to determine if crack spacing is the significant effect on rebar corrosion. Results are listed in Table 12.

Table 12. Turkey's HSD test results for crack spacing.

Source of Range Value	Range Value	$q_{0.05}$ (2,4)	$q_{0.01}$ (2,4)	MS_e	$HSD_{0.05}$	$HSD_{0.01}$	Significance
K_{15}–K_{25}	0.13	9.8	22.3	0.003	0.31	0.70	-

- Correlation is not significant.

In Table 12, the range value of means between 15 mm spacing level and 25 mm spacing level in crack spacing factor is not significantly different, which can be concluded that crack spacing is not a significant effect on rebar corrosion. Moreover, the results of Turkey's HSD test reveal that crack width and crack number possess a greater effect on rebar corrosion. In summary, the influence degrees of three factors including crack width, number and spacing on rebar corrosion can be ranked as crack number, crack width and crack spacing from the greatest to the least.

Crack leads to the greater capillarity and osmotic pressure, resulting in a serious corrosion of rebar in the vicinity of cracks. Meanwhile, the larger crack width causes a more oxygen and adverse ion aqueous solution to diffuse into concrete. It is said that rebar corrosion rate maybe depend on total crack width. To some extent, the total crack width would increase with increasing of crack number. Therefore, crack number presents the most significant effect on corrosion of rebar.

4. Conclusions

Electrochemical tests were carried out for evaluating the corrosion behavior of rebar (reflecting the durability of RC material) under the effect of crack. The major work of this study includes the recommendation of the reasonable electrochemical test method for rebar in RC material and quantitative analysis of the effect of crack number, width and spacing on rebar corrosion. The following conclusions were drawn:

(1) Due to the EIS excluding the polarization resistance (R_p) error caused by concrete and Stern-Geary coefficient (B) from TPP reflecting the non-uniform corrosion of rebar in RC material, a more accurate electrochemical test method combining EIS with TPP measurements was recommended for rebar corrosion behavior in RC material.

(2) The influences of crack parameters (i.e., crack width, number and spacing) on durability of RC material were analyzed based on One-way ANOVA and Turkey's HSD test. Results revealed that crack number presents the most significant effect, while crack spacing possesses the least one. As for wondering that if the influence on rebar corrosion caused by total crack width of multiple cracks is equal to that caused by an individual crack width, this also has become the topic which the authors further deliberated from now on.

Acknowledgments: The authors express their appreciation for the financial supports of National Natural Science Foundation of China (Nos. 51478203, 51408258).

Author Contributions: Yongchun Cheng and Yuwei Zhang designed the experiments and wrote the paper; Yuwei Zhang and Guojin Tan performed the experiments and analyzed the experimental results; Yubo Jiao wrote a part of the paper.

Conflicts of Interest: The authors declare no conflict of interest.

References

1. Kwon, S.J. Current Trends of Durability Design and Government Support in South Korea: Chloride Attack. *Sustainability* **2017**, *9*, 417. [CrossRef]
2. Yang, K.H.; Singh, J.; Lee, B.Y.; Kwon, S.J. Simple technique for tracking chloride penetration in concrete based on the crack shape and width under steady-state conditions. *Sustainability* **2017**, *9*, 282. [CrossRef]
3. Wang, H.L.; Dai, J.G.; Sun, X.Y.; Zhang, X.L. Characteristics of concrete cracks and their influence on chloride penetration. *Constr. Build. Mater.* **2016**, *107*, 216–225. [CrossRef]
4. Ji, Y.; Hu, Y.; Zhang, L. Laboratory studies on influence of transverse cracking on chloride-induced corrosion rate in concrete. *Cem. Concr. Compos.* **2016**, *69*, 28–37. [CrossRef]
5. Zhu, X.; Zi, G.; Cao, Z.; Cheng, X. Combined effect of carbonation and chloride ingress in concrete. *Constr. Build. Mater.* **2016**, *110*, 369–380. [CrossRef]
6. Lambert, P.; Page, C.L.; Vassie, P.R.W. Investigations of reinforcement corrosion. 2. Electrochemical monitoring of steel in chloride-contaminated concrete. *Mater. Struct.* **1991**, *24*, 351–358. [CrossRef]
7. Shaheen, F.; Pradhan, B. Effect of chloride and conjoint chloride–sulfate ions on corrosion of reinforcing steel in electrolytic concrete powder solution (ECPS). *Constr. Build. Mater.* **2015**, *101*, 99–112. [CrossRef]
8. Tennakoon, C.; Shayan, A.; Sanjayan, J.G.; Xu, A. Chloride ingress and steel corrosion in geopolymer concrete based on long term tests. *Mater. Des.* **2017**, *116*, 287–299. [CrossRef]
9. Wang, K.; Jansen, D.C.; Shah, S.P.; Karr, A.F. Permeability study of cracked concrete. *Cem. Concr. Res.* **1997**, *27*, 381–393. [CrossRef]
10. Ye, H.; Jin, N.; Jin, X.; Fu, C. Model of chloride penetration into cracked concrete subject to drying–wetting cycles. *Constr. Build. Mater.* **2012**, *36*, 259–269. [CrossRef]
11. Marsavina, L.; Audenaert, K.; Schutter, G.D.; Faur, N.; Marsavina, D. Experimental and numerical determination of the chloride penetration in cracked concrete. *Constr. Build. Mater.* **2009**, *23*, 264–274. [CrossRef]
12. Du, X.; Jin, L.; Ma, G. A meso-scale numerical method for the simulation of chloride diffusivity in concrete. *Finite Elem. Anal. Des.* **2014**, *85*, 87–100. [CrossRef]
13. Liu, Q.F.; Yang, J.; Xia, J.; Easterbrook, D.; Li, L.Y. A numerical study on chloride migration in cracked concrete using multi-component ionic transport models. *Comput. Mater. Sci.* **2015**, *99*, 396–416. [CrossRef]
14. Papakonstantinou, K.G.; Shinozuka, M. Probabilistic model for steel corrosion in reinforced concrete structures of large dimensions considering crack effects. *Eng. Struct.* **2013**, *57*, 306–326. [CrossRef]
15. Cao, C.; Cheung, M.M.S.; Chan, B.Y.B. Modelling of interaction between corrosion-induced concrete cover crack and steel corrosion rate. *Corros. Sci.* **2013**, *69*, 97–109. [CrossRef]
16. Zhu, W.; François, R.; Fang, Q.; Zhang, D. Influence of long-term chloride diffusion in concrete and the resulting corrosion of reinforcement on the serviceability of RC beams. *Cem. Concr. Compos.* **2016**, *71*, 144–152. [CrossRef]
17. Pacheco, J.; Šavija, B.; Schlangen, E.; Polder, R.B. Assessment of cracks in reinforced concrete by means of electrical resistance and image analysis. *Constr. Build. Mater.* **2014**, *65*, 417–426. [CrossRef]
18. Šavija, B.; Schlangen, E. Chloride ingress in cracked concrete—A literature review. In Proceedings of the International PhD Conference on Concrete Durability, Madrid, Spain, 19 November 2010; Andrade, C., Gulikers, J., Eds.; RILEM Publications: Madrid, Spain, 2012; pp. 133–142.
19. Pedrosa, F.; Andrade, C. Corrosion induced cracking: Effect of different corrosion rates on crack width evolution. *Constr. Build. Mater.* **2016**, *133*, 525–533. [CrossRef]
20. Zhou, Y.; Gencturk, B.; Willam, K.; Attar, A. Carbonation-Induced and Chloride-Induced Corrosion in Reinforced Concrete Structures. *J. Mater. Civ. Eng.* **2014**, *27*, 46–70. [CrossRef]
21. Wang, Z.; Zeng, Q.; Wang, L.; Yao, Y.; Li, K. Corrosion of rebar in concrete under cyclic freeze–thaw and Chloride salt action. *Constr. Build. Mater.* **2014**, *53*, 40–47. [CrossRef]
22. Andrade, C.; Garcés, P.; Martínez, I. Galvanic currents and corrosion rates of reinforcements measured in cells simulating different pitting areas caused by chloride attack in sodium hydroxide. *Corros. Sci.* **2008**, *50*, 2959–2964. [CrossRef]
23. Qi, X.; Mao, H.; Yang, Y. Corrosion behavior of nitrogen alloyed martensitic stainless steel in chloride containing solutions. *Corros. Sci.* **2017**, *120*, 90–98. [CrossRef]

24. Liu, G.; Zhang, Y.; Ni, Z.; Huang, R. Corrosion behavior of steel submitted to chloride and sulphate ions in simulated concrete pore solution. *Constr. Build. Mater.* **2016**, *115*, 1–5. [CrossRef]
25. Liu, M.; Cheng, X.; Li, X.; Zhou, C.; Tan, H. Effect of carbonation on the electrochemical behavior of corrosion resistance low alloy steel rebars in cement extract solution. *Constr. Build. Mater.* **2016**, *130*, 193–201. [CrossRef]
26. Andrade, C.; Keddam, M.; Nóvoa, X.R.; Pérez, M.C.; Rangel, C.M.; Takenouti, H. Electrochemical behaviour of steel rebars in concrete: Influence of environmental factors and cement chemistry. *Electrochim. Acta* **2001**, *46*, 3905–3912. [CrossRef]
27. Andrade, C.; Soler, L.; Alonso, C.; Nóvoa, X.R.; Keddam, M. The importance of geometrical considerations in the measurement of steel corrosion in concrete by means of AC impedance. *Corros. Sci.* **1995**, *37*, 2013–2023. [CrossRef]
28. Gerengi, H.; Kocak, Y.; Jazdzewska, A.; Kurtay, M.; Durgun, H. Electrochemical investigations on the corrosion behaviour of reinforcing steel in diatomite- and zeolite-containing concrete exposed to sulphuric acid. *Constr. Build. Mater.* **2013**, *49*, 471–477. [CrossRef]
29. Andrade, C.; Alonso, C. Test methods for on-site corrosion rate measurement of steel reinforcement in concrete by means of the polarization resistance method. *Mater. Struct.* **2004**, *37*, 623–643. [CrossRef]
30. Ministry of Housing and Urban-Rural Development of the People's Republic of China. *Standard for Methods of Long-Temper Formance and Durability of Ordinary Concrete*; Ministry of Housing and Urban-Rural Development of the People's Republic of China: Beijing, China, 2009. (In Chinese).
31. General Administration of Quality Supervision, Inspection and Quarantine of the People's Republic of China. *Common Portland Cement*; General Administration of Quality Supervision, Inspection and Quarantine of the People's Republic of China: Beijing, China, 2007. (In Chinese).
32. Ministry of Housing and Urban-Rural Development of the People's Republic of China. *Standard for Quality Control of Concrete*; Ministry of Housing and Urban-Rural Development of the People's Republic of China: Beijing, China, 2011. (In Chinese).
33. Song, G. Theoretical analysis of the measurement of polarisation resistance in reinforced concrete. *Cem. Concr. Compos.* **2000**, *22*, 407–415. [CrossRef]
34. Song, H.W.; Saraswathy, V. Corrosion Monitoring of Reinforced Concrete Structures-A Review. *Int. J. Electrochem. Sci.* **2007**, *2*, 1–28.
35. Martínez, I.; Andrade, C. Polarization resistance measurements of bars embedded in concrete with different chloride concentrations: EIS and DC comparison. *Mater. Corros.* **2015**, *62*, 932–942. [CrossRef]
36. Keleştemur, O.; Yildiz, S.; Gökçer, B.; Arici, E. Statistical analysis for freeze–thaw resistance of cement mortars containing marble dust and glass fiber. *Mater. Des.* **2014**, *60*, 548–555. [CrossRef]
37. Brown, A. A new software for carrying out one-way ANOVA post hoc tests. *Comput. Methods Program Biomed.* **2005**, *79*, 89–95. [CrossRef] [PubMed]
38. Stoline, M.R. The Status of Multiple Comparisons: Simultaneous Estimation of all Pairwise Comparisons in One-Way ANOVA Designs. *Am. Stat.* **1981**, *35*, 134–141.
39. Montgomery, D.C. *Design and Analysis of Experiments*, 8th ed.; Wiley and Sons: Hoboken, NJ, USA, 2012.

Article

Comprehensive Sustainability Evaluation of High-Speed Railway (HSR) Construction Projects Based on Unascertained Measure and Analytic Hierarchy Process

Yongzhi Chang, Yang Yang and Suocheng Dong *

Institute of Geographic Sciences and Natural Resources Research, The Chinese Academy of Sciences, Beijing 100101, China; changyz@igsnrr.ac.cn (Y.C.); yangy.15b@igsnrr.ac.cn (Y.Y.)
* Correspondence: dongsc@igsnrr.ac.cn; Tel.: +86-10-6488-9430

Received: 4 January 2018; Accepted: 26 January 2018; Published: 5 February 2018

Abstract: This paper aims to evaluate the sustainability of high-speed railway (HSR) construction projects in a comprehensive manner. To this end, the author established an index system, involving 4 primary indices, 9 secondary indices, and 32 tertiary indices. The analytic hierarchy process (AHP) and the unascertained measure were introduced to calculate the weights of these indices. Then, the index system was applied to evaluate the sustainability of the China's Harbin-Dalian Passenger Dedicated Line (PDL). The results show that the Harbin-Dalian PDL project achieved good results in terms of process, economic benefit, impact, and sustainability, and will bring long-term benefits in the fields of tourism, economy, and transport capacity, as well as many other fields. In spite of its good overall sustainability, the project needs to further increase its economic benefits and reduce its negative environmental impact. For this purpose, it is necessary to adopt the management mode of "separation between network and transportation" and apply noise prevention measures like noise barriers, tunnels, and overhead viaducts. This research lays a solid basis for the sustainability evaluation of HSR construction projects, and simplifies the modelling process for designers of HSR.

Keywords: high-speed railway (HSR); sustainable development; unascertained measure; analytic hierarchy process (AHP); Harbin-Dalian Passenger Dedicated Line (PDL)

1. History of HSR

High-speed railway (HSR) is considered to be one of the most important breakthroughs in passenger transport technology made in the 20th century. Being a safe, rapid, reliable, comfortable, and convenient mode of transport, HSR has injected new vitality into the existing railway transport system and became a symbol of modern society [1–6].

Opened in 1964, the Tōkaidō Shinkansen marked the dawn of the high-speed era, kicking off half a century of high-speed construction and research. At the sight of Japan's success, developed countries quickly followed suit.

The most famous project is the *Train à Grande Vitesse* (TGV) in France. TGV Sud-Est, the first line of the project, was funded by the French government in 1976, and opened to the public between Paris and Lyon on 27 September 1981. The trains operate on the line at a maximum speed of 270 km·h^{-1} [7]. The commercial success of the first TGV led to an expansion of the network to different parts of France. The statistics of the International Union of Railways (UIC) show that the TGV had 2142 km of HSR in operation, 634 km under construction, and 1786 km being planned by the end of April 2017 [8].

The success of the TGV motivated other European countries to construct HSR within and across their borders, forming a high-speed network across the continent. These countries include Germany

(1988), Italy (1988), Spain (1992), Belgium (1997), the United Kingdom (2003), and the Netherlands (2009). After the turn of the century, the high-speed boom spread eastwards to Asia. Different forms of HSR were built in Korea (2004) and China (2008).

Over the past 50 years, the global HSR network has transported about 15 billion passengers, about twice the world population. According to UIC statistics, 34,000 km of HSR had entered service by the end of 2016 [9], an increase of 14.86% compared to 2015 (Figure 1); the mileage rocketed up to 37,300 km by the end of April 2017, with another 15,900 km under construction [10]. Asia and Europe accounted for 98.07% of all HSR in service and under construction around the world (Figure 2).

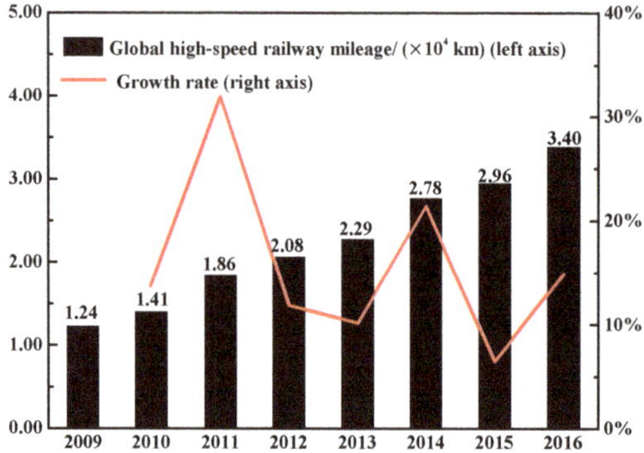

Figure 1. Global total mileage of HSR till 2016.

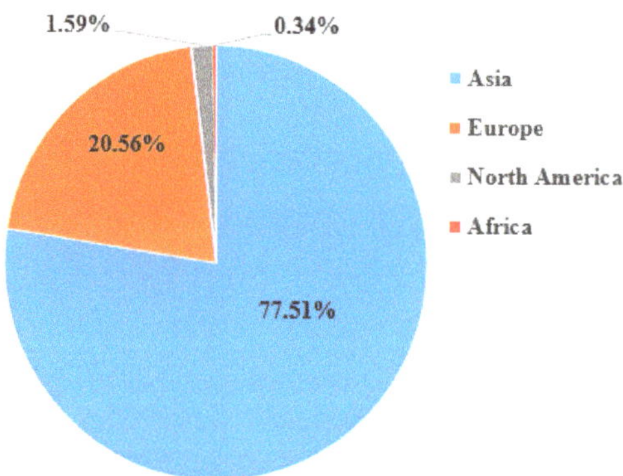

Figure 2. The global percentage of in-service under construction mileage of HSR in Asia, Europe, North America, and Africa in April 2017.

HSR construction is in full swing across China. In 2008, the Beijing–Tianjin Intercity Railway became the first line in the country to accommodate trains travelling at maximum speeds above 300 km. A boom of HSR construction ensued. By the end of April 2017, China had built the world's longest high-speed network, consisting of 84 high-speed lines (segments). In total, there were 23,900 km of route in service, and 10,700 km under construction. The two mileages take up 57.87% and 97.72% of the global total, respectively [8].

With only a couple of years, China has caught up with the first movers in HSR and become the focal point of HSR development. The HSR has been recognized as a first product of Chinese manufacturing [11–15]. Guided by the "going global" strategy, China is actively building HSR linking up the countries along the overland and the maritime silk roads. Once completed, these railways will boost the development of infrastructure and regional economy in all the countries radiated by the "the Belt and Road" [16].

2. Requirements on HSR

HSR is a major booster of socioeconomic development and the cornerstone of many industries. However, the environmental impact of HSR should never be overlooked. Despite being more environmental friendly than traditional transport modes (e.g., highway), HSR has led to some environmental and resource problems due to the largescale construction and leapfrogging development. To solve the problems, the planners, constructors, and operators of HSR must strike a balance between railway, population, economy, environment, and resources, so that the HSR develops in a sustainable manner.

Sustainable development is the ultimate goal of transportation. The key to sustainable development of transportation industry lies in its coordination with society, economy, environment, and resources [17–22]. In this research, the sustainable development of HSR is investigated from both internal and external perspectives that are required to satisfy the operating framework in Figure 3 [23,24].

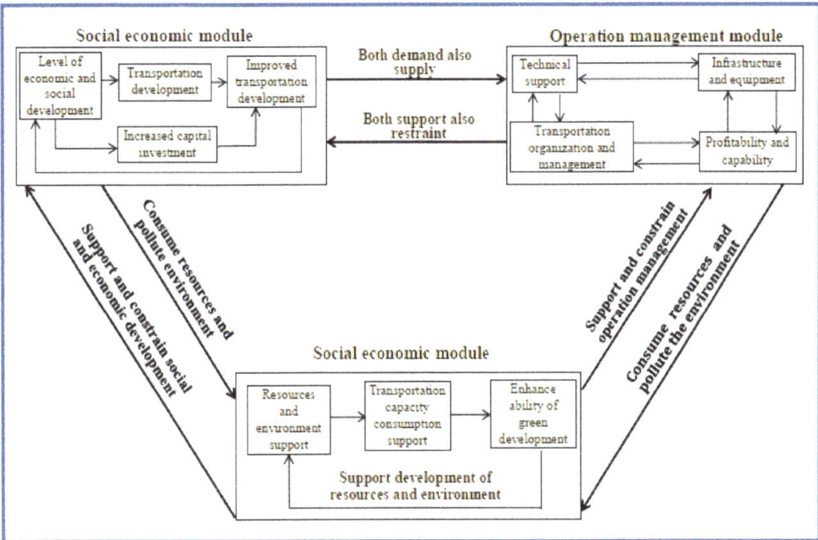

Figure 3. Operating framework for the sustainable development of HSR [23,24].

The sustainable development of HSR should meet three conditions. First, HSR must adapt to the socioeconomic situation. The development of HSR both depends on social and economic factors and promotes socioeconomic development. Second, HSR development must be coordinated internally. The internal elements include infrastructure, transport equipment, scheduling, service provision, and software-hardware integration. Third, HSR should pursue green development and ecological harmony. For this purpose, the land and non-renewable resources ought to be utilized rationally, and environmental pollution and traffic accidents must be avoided.

Facing the above requirements, it is urgent to rationalize the planning and construction of HSR projects and realize the sustainable development of the HSR network. One of the viable options is to evaluate every aspect of HSR development and weigh the pros and cons of existing and impending projects. The comprehensive evaluation is critical to the healthy development of HSR [25].

3. Literature Review

The sustainability evaluation of HSR construction projects started late at home and abroad. The existing research emphasizes theoretical analysis over empirical evidence and lacks sufficient evaluation contents, systematic indices and feasible methods. The rationality, normality, and feasibility of the evaluation are yet to be improved.

The current evaluation methods include analytical hierarchy process (AHP) [26–28], neural network [29], fuzzy mathematics [27,30,31], multi-criteria decision analysis [32,33], fault tree analysis [34], visco-elastic model [35], transfer function [36], finite element [37], etc. Most of these approaches tackle a single aspect of sustainable development. Only a few support comprehensive evaluation, namely, neural network and fuzzy mathematics. The problem is these few methods perform poorly in index selection and weight determination.

The sustainability evaluation of HSR construction projects involves much uncertainty and covertness. In particular, many attributes are uncertain in the decision-making process. Hence, the evaluation should cover both quantitative and qualitative indices. The unascertained measure theory is an ideal way to deal with uncertain information and achieve a comprehensive evaluation. Nevertheless, it is difficult to determine the weight of a complex index system that relies solely on the unascertained measure theory.

To overcome the defect, it is necessary to introduce the AHP to the evaluation process. Taking the object as a system, the AHP makes decisions through decomposition, comparative judgment, and totalization. By this method, the factors of a complex system are divided into interconnected, orderly layers; the importance of each layer relative to the other layer is quantified, and the weight of each layer is determined mathematically. In this way, the decision is made according to objective reality.

In view of the above facts and the features of HSR construction projects, this paper establishes a complete index system for sustainability evaluation. The system has such four primary indices as process, economic benefits, effect, and sustainability. Then, the unascertained measure was adopted to create the sustainability evaluation model for HSR construction projects. Then, the model was applied to the sustainability evaluation of the Harbin-Dalian Passenger Dedicated Line (PDL). This research lays a solid basis for sustainability evaluation of HSR construction projects and simplifies the modelling process for designers of HSR.

4. Establishment of Evaluation Index System

The index system is essential to the accuracy and reliability of the sustainability evaluation. Based on the literature review, the author decided to carry out a questionnaire survey before setting up the index system. The questionnaire was prepared after interviewing 30 HSR experts working in national and local governments, construction companies, design institutions, railway management departments, advisory bodies, and research institutes. Prior to the survey, the preliminary questionnaire was validated with selected samples and optimized into a formal questionnaire. After the survey, a field investigation was carried out to extract common factors through the factor analysis.

Considering the correlation between the factors, the author constructed the index system for the sustainability evaluation of HSR construction projects. Four layers of indices were introduced, namely the target layer (main index), criteria layer (primary indices), sub-criteria layer (secondary indices), and solution layer (tertiary indices). The solution layer consists of 32 indices that explain the factors on the sub-criteria layer (Table 1). The goal is to describe the elements in the tertiary indices and obtain the values of the evaluation indices in light of the actual situation of the object. As mentioned above, the index system has such four primary indices as process, economic benefits, effect, and sustainability.

Table 1. The evaluation index system for the sustainable development of HSR construction project [23,27].

Total Index	First Index	Secondary Index	Third Index
Index system of sustainable development on HSR construction project X	process evaluation F_1	process evaluation of pre-bid decision-making S_1	necessity of project construction T_1
			scientific of decision-making program T_2
			survey and design T_3
		construction process evaluation S_2	safety management T_4
			main technical indices T_5
			construction quality T_6
			investment control T_7
			time limit T_8
		operation process evaluation S_3	preparation of operational effect T_9
			operation management level T_{10}
	evaluation of economic benefits F_2	financial benefit evaluation S_4	payback period of investment T_{11}
			debt service coverage ratio (DSCR) T_{12}
		national economic benefits evaluation S_5	contribution to GDP T_{13}
			freight cost reduction benefits T_{14}
			economic internal rate of return (IRR) T_{15}
			economic net present value (NPV) T_{16}
	effect evaluation F_3	social effect evaluation S_6	drive labor and employment T_{17}
			appreciation rate of land and house T_{18}
			outage rate of cars and planes along the line T_{19}
			increased transport capacity T_{20}
			average travel time T_{21}
			per Capita GNP along the line T_{22}
			compensate degree of benefit damage group T_{23}
			boost national prestige T_{24}
		environmental effect evaluation S_7	ecological environmental effect T_{25}
			project pollution control T_{26}
	sustainability evaluation F_4	evaluation of internal sustainability factors S_8	project management mechanism T_{27}
			growth rate of traffic volume T_{28}
			technical factors T_{29}
		evaluation of external sustainability factors S_9	benign circulation factors T_{30}
			project satisfaction T_{31}
			capital factors T_{32}

The process evaluation index covers pre-bid decision-making, construction, and operation. These indices evaluate the actual project situation in the early stage, including work performance, construction quality, and operation efficiency. Based on the evaluation results, one can learn from the successes and failures, enhance the project quality, and elevate the management level.

The economic benefits include financial benefit and national economy. As its name suggests, the former index measures the financial benefit of the project. To evaluate the index, the income and expenses were acquired from the project's financial statements, and the implementation of the national tax system was taken as a reference. According to *Railway construction project economic evaluation methods and parameters (3nd edition)* (Ministry of Housing and Urban-Rural Development of the PRC, National Development and Reform Commission of the PRC, Ministry of Railways of the PRC, 2012) [38] and literature [39], the financial benefit of the project was further split into profitability and debt paying ability. The latter index, national economy, is fundamental to the sustainable development of HSR construction projects. The index reflects the net contribution of project to the national economy, and discloses the economic effect of project cost and investment.

The social effect evaluation exists in the forms of social effect and environmental effect. From the sociological angle, this paper explores the social effect from four dimensions (Table 2). The environmental effect refers to the chemicals, wastes, noises, and electromagnetic radiations released during project construction and operations in the natural environment (e.g., water, air and soil). The evaluation results make it possible to minimize the adverse effects on the life of local residents and ecological system.

Table 2. The index system of social effect evaluation.

Index	Contains the Content	Content Description
Social Effect Evaluation	effect on residents' lives	Per Capita GNP along the line Average travel time
	effect on the interest group	drive labor and employment appreciation rate of land and house compensate degree of benefit damage group
	effect on traffic structure	outage rate of cars and planes along the line increased transport capacity
	effect on the national image	boost national prestige

The sustainability evaluation is prospective in nature. For sustainable development, the relevant internal and external factors are evaluated in scientific means, along with the realization and influence of the project. In light of the evaluation results, one can give advice on how to promote sustainable development from the perspectives of technology, economy, society, and environment.

5. Unascertained Measure Theory

Uncertainty exists extensively in objective and subjective worlds; the uncertain information includes two basic meanings: randomness and ambiguity [40,41]. In 1836, Mill [42] proposed firstly the concept of "uncertainty". Stochastic problem was first proposed by the Soviet mathematician Kolmogorov in 1933, and he established probability theory and the axiomatic method [43]. In 1965, the concept of fuzzy information and fuzzy set theory was created by American scholar Zaden, who developed the study field of uncertainty [44]. Chinese scholar Deng founded the grey system theory in 1982 [45]. In 1991, Wang built a universal grey set on the basis of grey system theory; it involved a lot uncertain information [46]. Academician Wang, a famous Chinese scientist, put forward various types of information concepts, namely, unascertained information in 1990, which are different from random information and fuzzy information [47]. At present, there is a unified understanding of fuzzy information, random information, and gray information in uncertain information, but there is no uniform definition for unascertained information. However, theorists basically agree that unascertained information is the subjective uncertainty of decision makers because they lack information to determine the real state and quantity of the object. This creates subjective and cognitive uncertainty for decision makers due to the lack of objective information. It is fundamentally different from concepts like randomness (which only deals with what happens in the future), fuzziness (which reflects the nature of a certain object that does not have a clear definition or an evaluation target), and grayness.

Thanks to the concerted efforts of scholars like Álvaro [48] and Wu [49], so far, a systematic theory and method have been developed for unascertained information, which was first proposed by Academician Wang [47].

6. Establishment of Unascertained Measure Model

Set $x_1, x_2, \cdots, x_n$ as evaluation objects of news sensitivity, set universe $U = \{x_1, x_2, \cdots, x_n\}$. The evaluation $x_i \in U$ $(i = 1, 2, \cdots, n)$ has m first indices $I_1, I_2, \cdots, I_m$ and $\bar{I} = \{I_1, I_2, \cdots, I_m\}$. For $I_i \in \bar{I}$ has k secondary evaluation indices $I_{i1}, I_{i2}, \cdots, I_{ik}$, and $\bar{I_i} = \{I_{i1}, I_{i2}, \cdots, I_{ik}\}$. Therefore, x_{ij} can be expressed as k dimensional vector $x_{ij} = \{x_{ij1}, x_{ij2}, \cdots, x_{ijk}\}$, x_{ijr} means the value of the secondary indices of I_j, which is x_i's first index. Each x_{ijr} has p evaluate grades $c_1, c_2, \cdots, c_p$, the evaluation space is $C = \{c_1, c_2, \cdots, c_p\}$ [40,41].

6.1. Single-Index Measure

6.1.1. Single-Index Measure Matrix

Set $\mu_{ijrq} = \mu\left(x_{ijr} \in c_q\right)$ to express the degree that x_{ijr} belongs to c_q, which is the qth evaluation class (rating). μ must meet the conditions as follows:

$$0 \leq \mu\left(x_{ijrq} \in c_q\right) \leq 1, i = 1, 2, \cdots, n; \ j = 1, 2, \cdots, m; \ r = 1, 2, \cdots, k; q = 1, 2, \cdots, p \tag{1}$$

$$u\left(x_{ijr} \in C\right) = 1, i = 1, 2, \cdots, n; \ j = 1, 2, \cdots, m; \ r = 1, 2, \cdots, k \tag{2}$$

$$\mu\left(x_{ijr} \in \bigcup_{l=1}^{q} c_l\right) = \sum_{l=1}^{q} \mu\left(x_{ijr} \in c_l\right) \ q = 1, 2, \cdots, p \tag{3}$$

Define Equation (2) as the normalization and Equation (3) as the additivity. That which meets the three equations above is unascertained measurement. The matrix that follows is a single index measure matrix [50].

$$\left(\mu_{ijrq}\right)_{k \times p} = \begin{bmatrix} \mu_{ij11} & \mu_{ij12} & \cdots & \mu_{ij1p} \\ \mu_{ij21} & \mu_{ij22} & \cdots & \mu_{ij2p} \\ \cdots & \cdots & \ddots & \cdots \\ \mu_{ijk1} & \mu_{ijk2} & \cdots & \mu_{ijkp} \end{bmatrix} \ i = 1, 2, \cdots, n; j = 1, 2, \cdots, m \tag{4}$$

6.1.2. Distinction Weight of Single-Index

Using the concept of information entropy to define the peak of index I_{ijr}.

$$V_{ijr} = 1 + \frac{1}{\ln p} \sum_{q=1}^{p} \mu_{ijrq} \ln \mu_{ijrq} \tag{5}$$

p in Equation (5) represents the number of the evaluate ratings, μ_{ijrq} is the measure of single index, and the value of V_{ijr} expresses the degree that I_{ijr} different to each evaluation class. The distinction weight is as follows:

$$\omega_{ijr} = \frac{V_{ijr}}{\sum\limits_{r=1}^{k} V_{ijr}} \ i = 1, 2, \cdots, n; j = 1, 2, \cdots, m; r = 1, 2, \cdots, k \tag{6}$$

$\sum\limits_{r=1}^{k} \omega_{ijr} = 1, 0 \leq \omega_{ijr} \leq 1$, ω_{ijr} is the classification weights of I_{jr}. $\omega_{ij} = \left(\omega_{ij1}, \omega_{ij2}, \cdots, \omega_{ijk}\right)$ is the classification weight vector of secondary grade index [51].

6.2. First Grade Index Measure

Set $\mu_{iq} = \mu\left(x_i \in c_q\right)$ expresses the degree that sample x_i belongs to c_r, which is the rth evaluation class (rating).

$$\mu_{iq} = \sum_{j=1}^{m} \omega_{ij} \mu_{ijq}; i = 1, 2, \cdots, n; q = 1, 2, \cdots, p \tag{7}$$

Due to $0 \leq \mu_{iq} \leq 1$, and $\sum\limits_{q=1}^{p} \mu_{iq} = \sum\limits_{q=1}^{p} \sum\limits_{j=1}^{m} \omega_{ij}\mu_{ijq} = \sum\limits_{j=1}^{m} \omega_{ij} \sum\limits_{q=1}^{p} \mu_{ijq} = \sum\limits_{j=1}^{m} \omega_{ij} = 1$, μ_{iq} is the unascertained measure. Define $\left(\mu_{i1}, \mu_{i2}, \cdots, \mu_{ip}\right)$ as measure evaluation vector of x_i's composite indicator. The matrix $\left(\mu_{iq}\right)_{n \times p}$ is measure matrix of comprehensive index [52].

$$
\left(\mu_{iq}\right)_{n \times p} =
\begin{bmatrix}
\mu_{11} & \mu_{12} & \cdots & \mu_{1p} \\
\mu_{21} & \mu_{22} & \cdots & \mu_{2p} \\
\cdots & \cdots & \ddots & \cdots \\
\mu_{n1} & \mu_{n2} & \cdots & \mu_{np}
\end{bmatrix}
\tag{8}
$$

6.3. Determination of First Grade Index Weight by AHP

AHP was proposed by Saaty, an American operational research expert, in the 1970s [53]. It is a method of combining qualitative and quantitative, systematized and hierarchical qualities. It is a process of modeling and quantifying decision makers' decision thinking processes for complex systems. By using AHP, decision makers decompose the complex problems into several levels and factors, and make simple comparisons and calculations among the factors, so that they can get the weight of different plans and provide the basis for the best plan selection. In AHP, in order to make the judgment quantified, the key is to quantitatively describe the relative superiority of any two schemes to a certain criterion. For a single criterion, the comparison between the two plans can always demonstrate the advantages and disadvantages. AHP adopts the 1–9 scale method to give the quantitative scale for the evaluation of different situations. This scale is adopted in matrices to look for relative criteria's weights and to compare the alternatives linked to every criterion. Table 3 summarizes the basic ratio scale. All final weighted coefficients are shown in matrices. Alternatives and criteria can be ranked based on the overall aggregated weights in matrices. The alternative with the highest overall weight would be the most preferable [28,54].

Table 3. Saaty's scale for AHP pairwise comparisons [28,54].

Weight	Description
1	equal importance
3	moderately more important
5	strongly more important
7	very strongly more important
9	dominant importance
2, 4, 6, 8	representing the intermediate values of the above adjacent judgments
Reciprocals	anti-comparison

Based on this first index's judgment matrix, the weights of every first grade index can be calculated by the geometric calculation method of mean.

$$
\overline{\omega_i} = \sqrt[n]{\prod_{j=1}^{n} a_{ij}} \; (i = 1, 2, \cdots, n; j = 1, 2, \cdots, n)
\tag{9}
$$

Then, by employing normalized processing, using the following equation:

$$
\omega_i = \frac{\overline{\omega_i}}{\sum\limits_{i=1}^{n} \overline{\omega_i}}
\tag{10}
$$

The weight vector of first index is obtained: $\omega = \left(\omega_1, \omega_2, \cdots, \omega_n\right)^{\mathrm{T}}$.

The largest characteristic roots $\lambda_{max} = \frac{1}{n} \sum_{i=1}^{n} \frac{(AW)_i}{W_i}$ can be calculated.

When solving practical problems, due to the complexity of objective phenomena and the mind's cognitive limitations, our understanding of the problem is subjective and involves one-sidedness and fuzzy judgment; the structure of the judgment matrix is, therefore, may not fully meet the requirements of consistency, and often includes some deviation. The constructed judgement matrix does not always meet consistency condition. If the judgement matrix passes the consistency test, the calculated index weight can be adopted; if the consistency test is not passed, the judgement matrix needs to be adjusted. The consistency index is divided into complete consistency (CI) and satisfactory consistency (CR), $CI = \frac{\lambda_{max} - n}{n-1}$.

When $CI = 0$, the judgment matrix is considered to be completely consistent. $CI \neq 0$, it is considered that the judgment matrix is not completely consistent. $CR = \frac{CI}{RI}$, RI is the average random consistency index, and the value of RI are shown in Table 4 for the judgment matrix of n = 1–14.

Table 4. The mean random consistency index.

Order	1	2	3	4	5	6	7	8	9	10	11	12	13	14
RI	0	0	0.52	0.86	1.10	1.26	1.34	1.40	1.43	1.49	1.51	1.54	1.56	1.58

6.4. Identification

When the evaluation level is divided in order, the maximum membership degree is no longer applicable, so the credible degree criterion is usually adopted. For positive sequence partitioning, the confidence level is usually assumed to be λ ($\lambda > 0.5$), and the value of λ is 0.6 or 0.7. Set:

$$k_0 = \min_{k}\left\{ k : \sum_{l=1}^{k} \mu_{il} \geq \lambda, k = 1, 2, \cdots, p \right\} \tag{11}$$

7. Case Study

7.1. Overview

This section applies the established index system to evaluate the sustainable development of the Harbin-Dalian PDL, a typical HSR project in China. Harbin-Dalian PDL is put into operation by the end of 2012, marking the basic formation of "Four Vertical" in the "Four Vertical and Four Horizontal" of Chinese railway mainline. Harbin-Dalian PDL had finished the connection with existing HSR line of the Beijing-Shanghai and Beijing-Guangzhou. Since the operation of Harbin-Dalian PDL, it has not only shortened the distance of Heilongjiang, Jilin, and Liaoning provinces, but it has also strengthened regional economic integration.

Harbin-Dalian PDL runs through three provinces in Northeast China, 4 sub-provincial cities, and 6 prefecture-level cities. The total length stands at 921 km, including 553 km in Liaoning Province, 270 km in Jilin Province, and 81 km in Heilongjiang Province. Harbin-Dalian PDL starts from Dalian, through Yingkou, Anshan, Liaoyang, Shenyang, Tieling, Siping, Changchun, and Songyuan, and finally to Harbin (see Figure 4). According to the statistics of 2015, the 10 cities account for 40.90% of tourism resources, 46.53% of the population, and 59.87% of the total GDP in Northeast China [55]. Since its opening, the railway has effectively stimulated tourism in the cities along its route and across Northeast China.

Figure 4. Images of Harbin-Dalian PDL in China.

Harbin-Dalian PDL is the most advanced technology integration of China's HSR, and its operation has become the most powerful explanation for China's export of "China Railway High-speed" to countries along "the Belt and Road".

Through the investigation and analysis of the HSR construction project, the comprehensive evaluation index can be divided into 5 grades: V = {very poor, poor, qualified, good, excellent}, correspond to V = $\{v_1, v_2, v_3, v_4, v_5\}$. The direct choice of experts is given according to the multi-layered set of factors, and the number of experts supported as judgement of the index. In light of Saaty's 1–9 ratio scale estimation, V = {1, 3, 5, 7, 9}.

The secondary indices are quantified based on the basic data and expert scoring (by an expert panel consisting of 10 experts in the industry). The single measure vectors of the third indices (See Table 5) are obtained in light of the scores and the membership degree equation.

Thus, according to the vector measures, the measurement matrix of the secondary index is established as follows:

$$I_1 : \overline{\mu}_1 = \begin{bmatrix} 0 & 0 & 0.2 & 0.5 & 0.3 \\ 0 & 0 & 0.2 & 0.5 & 0.3 \\ 0 & 0 & 0 & 0.5 & 0.5 \\ 0 & 0 & 0 & 0.5 & 0.5 \\ 0 & 0 & 0.1 & 0.5 & 0.4 \\ 0 & 0 & 0 & 0.7 & 0.3 \\ 0 & 0.1 & 0.3 & 0.5 & 0.1 \\ 0 & 0 & 0 & 0.6 & 0.4 \\ 0 & 0 & 0.2 & 0.5 & 0.3 \\ 0 & 0 & 0 & 0.7 & 0.3 \end{bmatrix} \quad I_2 : \overline{\mu}_2 = \begin{bmatrix} 0 & 0 & 0.1 & 0.5 & 0.4 \\ 0 & 0.1 & 0.2 & 0.4 & 0.3 \\ 0 & 0 & 0 & 0.6 & 0.4 \\ 0 & 0.4 & 0.5 & 0.1 & 0 \\ 0 & 0 & 0.2 & 0.6 & 0.2 \\ 0 & 0 & 0.2 & 0.6 & 0.2 \end{bmatrix}$$

$$I_3 : \overline{\mu}_3 = \begin{bmatrix} 0 & 0 & 0 & 0.6 & 0.4 \\ 0 & 0 & 0 & 0.8 & 0.2 \\ 0 & 0 & 0.1 & 0.6 & 0.3 \\ 0 & 0 & 0.2 & 0.6 & 0.2 \\ 0 & 0 & 0.1 & 0.6 & 0.3 \\ 0 & 0 & 0 & 0.5 & 0.5 \\ 0 & 0.1 & 0.2 & 0.6 & 0.1 \\ 0 & 0 & 0.1 & 0.7 & 0.2 \\ 0 & 0 & 0.1 & 0.5 & 0.4 \\ 0 & 0.1 & 0.1 & 0.5 & 0.3 \end{bmatrix} \quad I_4 : \overline{\mu}_4 = \begin{bmatrix} 0 & 0 & 0.1 & 0.7 & 0.2 \\ 0 & 0 & 0 & 0.6 & 0.4 \\ 0 & 0 & 0.1 & 0.6 & 0.3 \\ 0 & 0 & 0 & 0.7 & 0.3 \\ 0 & 0 & 0 & 0.6 & 0.4 \\ 0 & 0 & 0.1 & 0.6 & 0.3 \end{bmatrix}$$

Table 5. The weights of hierarchy and expert scoring results.

Total Index	First Index	Secondary Index	Third Index	Evaluation Results of 10 Experts				
				Very Poor	Poor	Qualified	Good	Excellent
X	F_1 (0.300)	S_1 (0.341)	T_1 (0.305)	0	0	2	5	3
			T_2 (0.37)	0	0	2	5	3
			T_3 (0.325)	0	0	0	5	5
		S_2 (0.361)	T_4 (0.087)	0	0	0	5	5
			T_5 (0.285)	0	0	1	5	4
			T_6 (0.306)	0	0	0	7	3
			T_7 (0.225)	0	1	3	5	1
			T_8 (0.097)	0	0	0	6	4
		S_3 (0.298)	T_9 (0.6)	0	0	2	5	3
			T_{10} (0.4)	0	0	0	7	3
	F_2 (0.339)	S_4 (0.25)	T_{11} (0.5)	0	0	1	5	4
			T_{12} (0.5)	0	1	2	4	3
		S_5 (0.75)	T_{13} (0.532)	0	0	0	6	4
			T_{14} (0.127)	0	4	5	1	0
			T_{15} (0.172)	0	0	2	6	2
			T_{16} (0.169)	0	0	2	6	2
	F_3 (0.251)	S_6 (0.75)	T_{17} (0.08)	0	0	0	6	4
			T_{18} (0.095)	0	0	0	8	2
			T_{19} (0.091)	0	0	1	6	3
			T_{20} (0.139)	0	0	2	6	2
			T_{21} (0.157)	0	0	1	6	3
			T_{22} (0.125)	0	0	0	5	5
			T_{23} (0.215)	0	1	2	6	1
			T_{24} (0.098)	0	0	1	7	2
		S_7 (0.25)	T_{25} (0.75)	0	0	1	5	4
			T_{26} (0.25)	0	1	1	5	3
	F_4 (0.110)	S_8 (0.5)	T_{27} (0.199)	0	0	1	7	2
			T_{28} (0.386)	0	0	0	6	4
			T_{29} (0.415)	0	0	1	6	3
		S_9 (0.5)	T_{30} (0.182)	0	0	0	7	3
			T_{31} (0.428)	0	0	0	6	4
			T_{32} (0.39)	0	0	1	6	3

7.2. Weight Calculation of Second Grade Index

The weights of the secondary indices are calculated using information entropy. Below is the calculation of weight of process evaluation (F_1):

$$I_1 : \overline{\mu_1} = \begin{bmatrix} 0 & 0 & 0.2 & 0.5 & 0.3 \\ 0 & 0 & 0.2 & 0.5 & 0.3 \\ 0 & 0 & 0 & 0.5 & 0.5 \\ 0 & 0 & 0 & 0.5 & 0.5 \\ 0 & 0 & 0.1 & 0.5 & 0.4 \\ 0 & 0 & 0 & 0.7 & 0.3 \\ 0 & 0.1 & 0.3 & 0.5 & 0.1 \\ 0 & 0 & 0 & 0.6 & 0.4 \\ 0 & 0 & 0.2 & 0.5 & 0.3 \\ 0 & 0 & 0 & 0.7 & 0.3 \end{bmatrix}$$

Using Equation (5): $v_{11} = 0.5528$, $v_{12} = 0.5528$, $v_{13} = 0.6990$, $v_{14} = 0.6990$, $v_{15} = 0.5903$, $v_{16} = 0.7347$, $v_{17} = 0.4926$, $v_{18} = 0.7077$, $v_{19} = 0.5528$ and $v_{110} = 0.7347$.

Using Equation (6): $\omega_{11} = 0.0875$, $\omega_{12} = 0.0875$, $\omega_{13} = 0.1107$, $\omega_{14} = 0.1107$, $\omega_{15} = 0.0935$, $\omega_{16} = 0.1163$, $\omega_{17} = 0.0780$, $\omega_{18} = 0.1120$, $\omega_{19} = 0.0875$ and $\omega_{110} = 0.1163$.

Thus, level indices can be obtained under F_1 category weights:

$$\overline{\omega_1} = (0.0875\ 0.0875\ 0.1107\ 0.1107\ 0.0935\ 0.1163\ 0.0780\ 0.1120\ 0.0875\ 0.1163).$$

The same way can be obtained under F_2, F_3, F_4 category weights:

$$\overline{\omega_2} = (0.1692\ 0.1021\ 0.2232\ 0.1692\ 0.1678\ 0.1678);$$
$$\overline{\omega_3} = (0.1131\ 0.1251\ 0.0975\ 0.0938\ 0.0975\ 0.1117\ 0.0842\ 0.1041\ 0.0943\ 0.0787);$$
$$\overline{\omega_4} = (0.1598\ 0.1806\ 0.1442\ 0.1692\ 0.1906\ 0.1806\ 0.1442).$$

7.3. Measure Calculation of First Grade Index

Using Equation (7), the measurement vector of the first index under process evaluation (F_1) is:

$$\mu_1 = \overline{\mu_1} \times \overline{\omega_1} = \begin{bmatrix} 0.0875 \\ 0.0875 \\ 0.1107 \\ 0.1107 \\ 0.0935 \\ 0.1163 \\ 0.0780 \\ 0.1120 \\ 0.0875 \\ 0.1163 \end{bmatrix}^T \times \begin{bmatrix} 0 & 0 & 0.2 & 0.5 & 0.3 \\ 0 & 0 & 0.2 & 0.5 & 0.3 \\ 0 & 0 & 0 & 0.5 & 0.5 \\ 0 & 0 & 0 & 0.5 & 0.5 \\ 0 & 0 & 0.1 & 0.5 & 0.4 \\ 0 & 0 & 0 & 0.7 & 0.3 \\ 0 & 0.1 & 0.3 & 0.5 & 0.1 \\ 0 & 0 & 0 & 0.6 & 0.4 \\ 0 & 0 & 0.2 & 0.5 & 0.3 \\ 0 & 0 & 0 & 0.7 & 0.3 \end{bmatrix} = (0\ 0.0078\ 0.0853\ 0.5577\ 0.3490)$$

The measurement vector of the first index under evaluation of economic benefits (F_2) is:

$$\mu_2 = \overline{\mu_2} \times \overline{\omega_2} = \begin{bmatrix} 0.1692 \\ 0.1021 \\ 0.2232 \\ 0.1692 \\ 0.1678 \\ 0.1678 \end{bmatrix}^T \times \begin{bmatrix} 0 & 0 & 0.1 & 0.5 & 0.4 \\ 0 & 0.1 & 0.2 & 0.4 & 0.3 \\ 0 & 0 & 0 & 0.6 & 0.4 \\ 0 & 0.4 & 0.5 & 0.1 & 0 \\ 0 & 0 & 0.2 & 0.6 & 0.2 \\ 0 & 0 & 0.2 & 0.6 & 0.2 \end{bmatrix} = (0\ 0.0779\ 0.1890\ 0.4776\ 0.2547)$$

The measurement vector of the first index under effect evaluation (F_3) is:

$$\mu_3 = \bar{\mu}_3 \times \bar{\omega}_3 = \begin{bmatrix} 0.1131 \\ 0.1251 \\ 0.0975 \\ 0.0938 \\ 0.0975 \\ 0.1117 \\ 0.0842 \\ 0.1041 \\ 0.0943 \\ 0.0787 \end{bmatrix}^T \times \begin{bmatrix} 0 & 0 & 0 & 0.6 & 0.4 \\ 0 & 0 & 0 & 0.8 & 0.2 \\ 0 & 0 & 0.1 & 0.6 & 0.3 \\ 0 & 0 & 0.2 & 0.6 & 0.2 \\ 0 & 0 & 0.1 & 0.6 & 0.3 \\ 0 & 0 & 0 & 0.5 & 0.5 \\ 0 & 0.1 & 0.2 & 0.6 & 0.1 \\ 0 & 0 & 0.1 & 0.7 & 0.2 \\ 0 & 0 & 0.1 & 0.5 & 0.4 \\ 0 & 0.1 & 0.1 & 0.5 & 0.3 \end{bmatrix} = (0\ 0.0152\ 0.0828\ 0.6070\ 0.3011)$$

The measurement vector of the first index under sustainability evaluation (F_4) is:

$$\mu_4 = \bar{\mu}_4 \times \bar{\omega}_4 = \begin{bmatrix} 0.1598 \\ 0.1806 \\ 0.1442 \\ 0.1906 \\ 0.1806 \\ 0.1442 \end{bmatrix}^T \times \begin{bmatrix} 0 & 0 & 0.1 & 0.7 & 0.2 \\ 0 & 0 & 0 & 0.6 & 0.4 \\ 0 & 0 & 0.1 & 0.6 & 0.3 \\ 0 & 0 & 0 & 0.7 & 0.3 \\ 0 & 0 & 0 & 0.6 & 0.4 \\ 0 & 0 & 0.1 & 0.6 & 0.3 \end{bmatrix} = (0\ 0\ 0.0448\ 0.6350\ 0.3201)$$

Thus the measurement matrix of the first index is:

$$\bar{\mu} = \begin{bmatrix} \mu_1 \\ \mu_2 \\ \mu_3 \\ \mu_4 \end{bmatrix} = \begin{bmatrix} 0 & 0.0078 & 0.0853 & 0.5577 & 0.3490 \\ 0 & 0.0779 & 0.1890 & 0.4776 & 0.2547 \\ 0 & 0.0152 & 0.0828 & 0.6070 & 0.3011 \\ 0 & 0 & 0.0448 & 0.6350 & 0.3201 \end{bmatrix}$$

7.4. Determining the Classification Weight of First Grade Index

The first index judgment matrix is established using Saaty's 1–9 scale, and AHP is applied to calculate the weights as the final results (see Table 5).

7.5. Calculation of Comprehensive Measure Vector

Point multiplication of the first index weight and the first measurement matrix results in judgment matrix as follows:

$$B = \omega_i^0 \times \bar{\mu} = \begin{bmatrix} 0.300 \\ 0.339 \\ 0.251 \\ 0.110 \end{bmatrix}^T \times \begin{bmatrix} 0 & 0.0078 & 0.0853 & 0.5577 & 0.3490 \\ 0 & 0.0779 & 0.1890 & 0.4776 & 0.2547 \\ 0 & 0.0152 & 0.0828 & 0.6070 & 0.3011 \\ 0 & 0 & 0.0448 & 0.6350 & 0.3201 \end{bmatrix}$$
$$= (0\ 0.0326\ 0.1154\ 0.5514\ 0.3018)$$

Thus the score is calculated as:

$$S = B \times A = (0\ 0.0326\ 0.1154\ 0.5514\ 0.3018) \times (1\ 3\ 5\ 7\ 9) = 7.2508$$

The calculation results show that the overall score of The Harbin-Dalian PDL is 7.2508, and the sustainable evaluation result is good.

7.6. Confidence Level Recognition

Confidence level recognition is performed using Equation (11) and the calculated comprehensive measurement vector. Here, λ is set as 0.7:

When $\lambda = 0.7$, $k_0 = \min \sum_{l=1}^{k} \mu_{il} \geq 0.7$, $k = 5$; it shows that the confidence level recognition is good.

7.7. Result Discussion

According to the evaluation of the primary evaluation indices, the measurement vectors of process evaluation (F_1), economic benefits (F_2), effect evaluation (F_3) and sustainability evaluation (F_4) were all good. Thus, the results were discussed in the four aspects below.

7.7.1. Process Evaluation

In terms of process, the project performs well in decision-making, construction and operation. Good decision-making is revealed in route selection, as the railway links up four important sub-provincial cities and six prefecture-level cities. Moreover, the project lives up to all the technical requirements, especially the solution to frost heaving and snow accumulation. Since the railway was put into service, almost all performance parameters (e.g., ridership, operating cost, and equipment utilization) have reached the desired level.

Nonetheless, the project lags behind in investment control, mainly because of the construction difficulty and the management structure. (1) Construction difficulty: The Harbin-Dalian PDL, as the world's first HSR in severe cold region, was constructed with the strictest standards. Due to the lack of precedent, the project budget was poorly estimated. For example, the cost of frost heaving control was not even included in the budget. (2) Management structure: The government is the sole financier of the project, and any attempt to reform the financing mode requires government approval. The investment is managed in a backward way with unclear responsibilities. What is worse, the investment and financing risk is poorly controlled. Specifically, the ticket pricing mechanism is unreasonable, the funds utilization is not well supervised, the relevant polices and regulations are incomplete, and the local governments fail to provide a good risk control platform.

7.7.2. Economic Benefits

In terms of economic benefits, the project boasts good financial and economic benefits. The PDL has contributed to the regional GDP, optimized resource allocation and utilization, accelerated industrial restructuring, and motivated tourism and other tertiary industries along its route.

However, due to high investment and heavy loans, the early phase of operation is overshadowed by huge pressure of repayment, resulting in poor financial performance. The financial dilemma is partially attributable to the outdated management mode. In Harbin-Dalian PDL, the railway network infrastructure (network) and the passenger/freight transportation (transportation) are still managed by the same entity. This goes against the more efficient and rational management mode of "separation between network and transportation" [56]. The new mode encourages competition and yields more profits.

7.7.3. Effect Evaluation

In terms of effect, the project has exerted fairly good social and environmental effects. The social effect is manifested by the optimized transport capacity and structure in the region, as well as the shorter travel time and rising income of local residents. As for the environmental effect, the pollution, noise, and electromagnetic radiation should be better despite the overall good environmental performance of the railway. In particular, the noise and radiation generated during the operation of the railway has disturbed the daily life of the residents.

To eliminate the noise effect, noise barriers should be installed along the route. The barriers must be able to suppress the noise on both sides of the railway from 65~75 dB to about 30~40 dB, such that the residents living along the route can sleep well at night. When the railway passes through the urban areas, it should be covered like a tunnel. The top cover can be made of cement of suitable thickness, and the supports can be made of steel mesh. Another solution is to build an underground tunnel for the railway in the urban areas. This approach can basically eliminate the negative effect on the environment along the route, but it is too costly to realize. Therefore, the basic principle of HSR construction is to maximize the use of overhead viaducts and avoidance of densely populated areas.

7.7.4. Sustainability Evaluation

In terms of sustainability, the project enjoys a good prospect of sustainable development. The sustainability is showcased by its advanced technology, immense popularity, high ridership, and sound mechanism. Besides, a virtuous circle is formed as the railway promotes the regional economy in Northeast China, which, in return, increases the demand for the railway. Most importantly, the PDL has revitalized the tourism in the region with its obvious time-space compression effect. The railway pulls the tourist spots close to tourist sources and cuts down the cost of tourists. Since the opening of the PDL, the journey time to tourist scenic spots in Northeast China has plunged across the board.

The average journey time to these tourist scenic spots has decreased from 7.4 h before the start-up of the PDL to 4.9 h at present, a drop of 2.5 h, while that between different cities has shrunk from 7.7 h to 5.4 h, a decrease of 2.3 h (See Table 6) [57]. The closer a tourist scenic spot/city is to the railway, the shorter the journey time. The greatest decrease occurred between Dalian and Harbin, located on each end of the railway, with a decline of over 80%. In general, Dalian, Changchun and Harbin experienced the most obvious changes in journey time (-3.6 h), followed by Anshan, Liaoyang, Shenyang, Tieling, and other cities along the route (-2 h). For the cities far away from the railway, the journey time was cut short by less than 1.5 h.

Table 6. A comparison of accessibility indices before and after the operating of HSR [55].

Index	Before the Operating of HSR	After the Operating of HSR
Number of tourist scenic spots in one day	0	53
Number of tourist scenic spots in two days	32	84
Number of tourist cities in one day	0	8
Number of tourist cities in two days	8	20
Average accessibility of tourist scenic spots/h	7.4	4.9
Average accessibility of tourist cities/h	7.7	5.4

7.7.5. Overall Result

The overall score of Harbin-Dalian PDL construction project is 7.2508, indicating that the world's first HSR in severe cold region has a good sustainability. The project strengthens the link between central and southern Liaoning with the hinterland of Northeast China and draws together the high-quality resources in the three north-eastern provinces. The opening and operation of the PDL will bring long-term benefits to tourism, economy, transport capacity and many other fields.

For example, it only takes 1.5 h for passengers to travel on the PDL between Harbin, Changchun, and Shenyang, the three central cities in the region. Hence, these cities can efficiently share their education, medical, entertainment and other resources, achieve coordinated development of infrastructure, and accelerate regional economic integration. The same will happen to all the other cities along the route. The PDL itself is a moving economic corridor in Northeast China, with rapid flow of passengers, logistics, and information. The successful operation of the railway is critical to the revitalization of the old industrial base in the region.

To sum up, the author established an index system for the sustainability evaluation of HSR construction projects, calculated the index values by the unascertained measure, and applied the established model to evaluate the sustainability of Harbin-Dalian PDL. The evaluation results are consistent with the actual construction and operation status of the PDL, an evidence for the rationality and feasibility of the index system. The index system can improve the design, construction, and operation of HSR, and strengthen its ability to resist risks. The development of HSR plays an important role in increasing economic vitality, it is in line with the needs of the sustainable development of China's economy. HSR will become the backbone transportation mode in the modern and integrated transportation system. We will further the planning and construction of HSR network, enrich the railway transportation infrastructure of "the Belt and Road", accelerate China's HSR going out pace.

8. Conclusions

Considering the various influencing factors on HSR sustainability, this paper created an evaluation index system from four primary aspects: process, economic benefits, effect, and sustainability. Then, the unascertained measure was introduced for comprehensive evaluation. The main conclusions and innovation points of this research are listed as follows.

(1) The AHP was adopted to realize simultaneous qualitative and quantitative evaluations of influencing factors. Then, the weights were assigned in a rational and consistent manner, such that the importance of each index was measured correctly. This reflects the significance different among evaluation factors in the sustainability evaluation system.

(2) The sustainability evaluation shows that the Harbin-Dalian PDL project achieved good results in process, economic benefits, effects, and sustainability, thanks to the excellent performance of decision-making, construction and operation. The opening and operation of the PDL will bring long-term benefits to tourism, the economy, transport capacity, and many other fields.

(3) In spite of its good overall sustainability, the Harbin-Dalian PDL project needs to further increase its economic benefits and reduce its negative environmental effect. For this purpose, it is necessary to adopt the management mode of "separation between network and transportation" and apply noise prevention measures like noise barriers, tunnels, and overhead viaducts.

(4) Facing the increasingly rapid development and construction of HSR, the sustainability evaluation of HSR construction projects needs to learn from the relatively mature research methods at home and abroad, in order to make up for the shortcomings of the existing research methods in terms of breadth and accuracy. The scientific evaluation system should be constructed to make the citation and index selection more scientific in the empirical research, and to strengthen the prediction of the impact capacity of the future construction of HSR, not just to analyze the impact of the existing HSR.

All in all, this research lays a solid basis for the sustainability evaluation of HSR construction projects and simplifies the modelling process for designers of HSR.

Acknowledgments: The authors gratefully acknowledge financial support from National Science and Technology Basic Work of Special Key Projects (2017FY10130), supported by National Natural Science Foundation of China (41511130033), supported by Research project of the Chinese Academy of Sciences (Y66Q034).

Author Contributions: This paper presents a team work research result written by the co-authors, Yongzhi Chang, Yang Yang, and Suocheng Dong. Suocheng Dong conceived and designed the study; Yongzhi Chang and Yang Yang analyzed the HSR construction project the data. With cross discussions of the research results, the co-authors have contributed substantially to the work reported.

Conflicts of Interest: The authors declare no conflict of interest.

References

1. Yao, E.J.; Yang, Q.R.; Zhang, Y.S.; Sun, X.A. Study on high-speed rail pricing strategy in the context of modes competition. *Discret. Dyn. Nat. Soc.* **2013**, *2013*, 4–9. [CrossRef]
2. Wang, Y.H.; Deng, X.M.; Marcucci, D.J.; Le, Y.E. Sustainable development planning of protected areas near cities: Case study in China. *J. Urban Plan. Dev.* **2013**, *139*, 133–143. [CrossRef]

3.	Ericson, J.A. A participatory approach to conservation in the Calakmul Biosphere Reserve, Campeche, Mexico. *Landsc. Urban Plan.* **2006**, *74*, 242–266. [CrossRef]

4.	Jansson, Å. Reaching for a sustainable, resilient urban future using the lens of ecosystem services. *Ecol. Econ.* **2013**, *86*, 285–291. [CrossRef]

5.	The State Council of the People's Republic of China. About Print and Distribute "the Notice of National Resources City Sustainable Development Planning (2013–2020)" [EB/OL]. The Central Government Portal Website. Available online: http://www.gov.cn/zwgk/2013-12/03/content_2540070.htm (accessed on 3 December 2013).

6.	Moretti, L.; Moretti, M.; Ricci, S. Upgrading of Florence public transport to incorporate new tramlines. *Ing. Ferrov.* **2017**, *72*, 569–584.

7.	Arduin, J.P.; Ni, J.C. French TGV network development. *Jpn. Railw. Trans. Rev.* **2005**, *40*, 22–28.

8.	Huang, Y.; Ge, Y.J.; Ma, T.; Liu, X.F. Geopolitical space of China's high-speed railway diplomacy. *Prog. Geogr.* **2017**, *36*, 1489–1499.

9.	Wu, H.; Liu, W.; Choul, K.S. The strategic marketing of China high-speed railway: Government behavior or market behavior. *J. Mark. Stud.* **2017**, *25*, 185–194.

10.	Lin, Y.T. Travel costs and urban specialization patterns: Evidence from China's high speed railway system. *J. Urban Econ.* **2017**, *98*, 98–123. [CrossRef]

11.	Li, H.; Zhang, P.Y.; Cheng, Y.Q. Economic vulnerability of mining city—A case study of Fuxin City, Liaoning Province, China. *Chin. Geogr. Sci.* **2009**, *19*, 211–218. [CrossRef]

12.	Hayter, R.; Barnes, T.J. Labor market segmentation, flexibility and recession: A British Colombian case study. *Environ. Plan.* **1992**, *10*, 333–353. [CrossRef]

13.	Randall, J.E.; Ironside, R.G. Communities on the edge: An economic geography of resource-dependent communities in Canada. *Can. Geogr.* **1996**, *40*, 17–35. [CrossRef]

14.	Hilson, G. Sustainable development policies in Canada mining sector: An overview of government and industry efforts. *Environ. Sci. Policy* **2003**, *3*, 110.

15.	Jin, X.F.; Dong, S.C.; Liu, W.; Li, X. Study on relationship between industrial chain extension and evolution of resource-based city—A case study on Tongling city. *Econ. Geogr.* **2010**, *30*, 403–408.

16.	Swaine, M.D. Chinese Views and Commentary on the "One Belt, One Road" Initiative. *China Leadersh. Monit.* **2015**, *1*, 1–24.

17.	Greene, D.L.; Wegener, M. Sustainable transport. *J. Transp. Geogr.* **1997**, *5*, 177–190. [CrossRef]

18.	Hoogma, R.; Kemp, R.; Schot, J.; Truffer, B. Experimenting for sustainable transport: The Approach of Strategic Niche Management. *Technol. Anal. Strateg.* **2002**, *23*, 517–518.

19.	Litman, T. Developing Indicators for comprehensive and sustainable transport planning. *Transp. Res. Rec. J. Transp. Res. Board* **2007**, *43*, 10–15. [CrossRef]

20.	Elvik, R. The non-linearity of risk and the promotion of environmentally sustainable transport. *Accid. Anal. Prev.* **2009**, *41*, 849–855. [CrossRef] [PubMed]

21.	Hickman, R. Automobility in transition: A socio-technical analysis of sustainable transport. *Transp. Rev.* **2012**, *33*, 128–129. [CrossRef]

22.	O'Brien, O.; Cheshire, J.; Batty, M. Mining bicycle sharing data for generating insights into sustainable transport systems. *J. Transp. Geogr.* **2014**, *34*, 262–273. [CrossRef]

23.	Chang, Y.Z.; Dong, S.C. Study on green ecological assessment of high-speed railway construction project based on unascertained measure and AHP. *Teh. Vjesn.* **2017**, *24*, 1579–1589.

24.	Liu, Y.H. *Study on Sustainable Development of China's High-Speed Railways*; Press of University of Chinese Academy of Sciences: Beijing, China, 2014.

25.	Litman, T.; Burwell, D. Issues in sustainable transportation. *Int. J. Glob. Environ.* **2006**, *6*, 331–347. [CrossRef]

26.	Chen, X.Y.; Chang, Y.L. The Study of evacuation passenger service level of Shanghai-Nanjing high-speed railway stations. *Procedia Soc. Behav. Sci.* **2013**, *96*, 265–269. [CrossRef]

27.	Chang, Y.Z.; Dong, S.C. Study on Post Evaluation of High-Speed Railway Based on FAHP and Matlab Simulation Calculation. *Teh. Vjesn.* **2017**, *24*, 1749–1758.

28.	Moretti, L.; Di Mascio, P.; Bellagamba, S. Environmental, human health and socio-economic effects of cement powders. *Int. J. Environ. Res. Public Health* **2017**, *14*, 645. [CrossRef] [PubMed]

29.	Chen, J.H.; Yang, L.R.; Su, M.C.; Lin, J.Z. A rule extraction based approach in predicting derivative use for financial risk hedging by construction companies. *Expert Syst. Appl.* **2010**, *37*, 6510–6514. [CrossRef]

30. Wang, Y.M.; Elhag, T.M.S. Fuzzy TOPSIS method based on alpha level sets with an application to bridge risk assessment. *Expert Syst. Appl.* **2006**, *31*, 309–319. [CrossRef]
31. Cucala, A.P.; Fernández, A.; Sicre, C.; Domínguez, M. Fuzzy optimal schedule of high-speed train operation to minimize energy consumption with uncertain delays and driver's behavioral response. *Eng. Appl. Intell.* **2012**, *25*, 1548–1557. [CrossRef]
32. Chen, S.K.; Leng, Y.; Mao, B.H.; Liu, S. Integrated weight-based multi-criteria evaluation on transfer in large transport terminals: A case study of the Beijing South Railway Station. *Transp. Res. Part A Policy Pract.* **2014**, *66*, 13–26. [CrossRef]
33. Luca, D.M.; Dell'Acqua, G.; Lamberti, R. High-speed rail track design using GIS and multi-criteria analysis. *Procedia Soc. Behav. Sci.* **2012**, *54*, 608–617. [CrossRef]
34. Liu, P.; Yang, L.X.; Gao, Z.Y.; Li, S.K.; Gao, Y. Fault tree analysis combined with quantitative analysis for high-speed railway accidents. *Saf. Sci.* **2015**, *79*, 344–357. [CrossRef]
35. Di Mascio, P.; Loprencipe, G.; Maggioni, F. Visco-elastic modeling for railway track structure layers. *Ing. Ferrov.* **2014**, *69*, 207–222.
36. D'Andrea, A.; Loprencipe, G.; Xhixha, E. Vibration induced by rail traffic: Evaluation of attenuation properties in a bituminous sub-ballast layer. *Procedia Soc. Behav. Sci.* **2012**, *53*, 245–255. [CrossRef]
37. Zoccali, P.; Cantisani, G.; Loprencipe, G. Ground-vibrations induced by trains: Filled trenches mitigation capacity and length influence. *Constr. Build. Mater.* **2015**, *74*, 1–8. [CrossRef]
38. Ministry of Housing and Urban-Rural Development of the PRC; National Development and Reform Commission of the PRC; Ministry of Railways of the PRC. *Railway Construction Project Economic Evaluation Methods and Parameters*, 3nd ed.; China Planning Press: Beijing, China, 2012.
39. Shuai, B.; Huang, W.C.; Li, M.L.; Wang, S.; Li, K.N. Railway benefit assessment summary of the World Bank and the Asian Development Bank. *J. Transp. Syst. Eng. Inf. Technol.* **2017**, *3*, 222–228.
40. Chang, Y.Z.; Dong, S.C. Evaluation of sustainable development of resources-based cities in Shanxi Province based on unascertained measure. *Sustainability* **2016**, *8*, 585. [CrossRef]
41. Chang, Y.Z.; Dong, S.C. Study on evaluation model of international trade in agricultural products based on unascertained measure. *Chem. Eng. Trans.* **2016**, *51*, 673–678.
42. Mill, J. Whether political economy is useful? *Lond. Rev.* **1836**, *2*, 535–572.
43. Kolmogorov, A.H. *Foundations of the Theory Probability*, 2nd ed.; Chelsea Publishing Company: New York, NY, USA, 1956.
44. Zaden, L.A. Fuzzy sets. *Inf. Control* **1965**, *8*, 338–353.
45. Deng, J.L. *The Grey System (Society Economic)*; National Defence Industry Press: Beijing, China, 1985.
46. Wang, Q.Y. *Grey Mathematical Basis*; HUST Press: Wuhan, China, 1996.
47. Wang, G.Y. Unascertained information and its mathematical treatment. *J. Harbin. Univ. Civ. Eng. Archit.* **1990**, *4*, 1–3.
48. Álvaro, F.; Sánchez, J.A.; Benedí, J.M. An integrated grammar-based approach for mathematical expression recognition. *Pattern Recognit.* **2016**, *51*, 135–147. [CrossRef]
49. Wu, H.B.; Xu, J.H.; Ji, Y.; Wu, M. Uncertain flow calculations of a distribution network containing DG based on blind number theory. *IET Gener. Transm. Distrib.* **2017**, *11*, 1591–1597. [CrossRef]
50. Chen, X.; Liu, Y.; Zhao, X.; Liang, Y.X. Coal industry international competitiveness research. *Adv. Sci. Technol. Lett.* **2016**, *121*, 222–226.
51. Guo, C.Y.; Liu, Z.Q.; Zhang, C.Y. Evaluation on the express enterprise's service quality of customers perception based on SPSS and unascertained measure model. *Int. J. U E Ser. Sci. Technol.* **2016**, *9*, 17–26.
52. Li, S.C.; Wu, J.N.; Xu, Z.H.; Li, L.P. Unascertained measure model of water and mud inrush risk evaluation in karst tunnels and its engineering application. *KSCE J. Civ. Eng.* **2017**, *21*, 1170–1182. [CrossRef]
53. Saaty, T.L. *The Analytic Hierarchy Process*; McGraw-Hill: New York, NY, USA, 1980.
54. Javid, P.J.; Nejat, A.; Hayhoe, K. Selection of CO_2 mitigation strategies for road transportation in the United States using a multi-criteria approach. *Renew. Sustain. Energy Rev.* **2014**, *38*, 960–972. [CrossRef]
55. National Bureau of Statistics. *International City Statistical Yearbook 2016*; China Statistics Press: Beijing, China, 2016.
56. Di Mascio, P.; Loprencipe, G.; Moretti, L. Competition in rail transport: Methodology to evaluate economic impact of new trains on track. Sustainability, Eco-Efficiency and Conservation in Transportation Infrastructure Asset Management. In Proceedings of the 3rd International Conference on Transportation Infrastructure (ICIT), Pisa, Italy, 27 April 2014; pp. 669–675.

57. Guo, J.K.; Wang, S.B.; Wang, H.; Liu, T.B. Impact of Harbin-Dalian high-speed railway on the spatial distribution of tourism supply and demand markets in Northeast China cities: Based on the accessibility of the scenic spots. *Prog. Geogr.* **2016**, *35*, 504–514.

Article

A Study on the Analysis of CO_2 Emissions of Apartment Housing in the Construction Process

Jonggeon Lee [1], Sungho Tae [2,*] and Rakhyun Kim [3,*]

[1] Green Remodeling Department, Korea Land & Housing Corporation, Seongnam-daero 54beon-gil, Seongnam 13637, Korea; leejg125@lh.or.kr

[2] School of Architecture & Architectural Engineering, Hanyang University, 55 Hanyangdaehak-ro, Sangrok-gu, Ansan 15588, Korea

[3] Architectural Engineering, Hanyang University, 55 Hanyangdaehak-ro, Sangrok-gu, Ansan 15588, Korea

* Correspondence: jnb55@hanyang.ac.kr (S.T.); redwow6@hanyang.ac.kr (R.K.); Tel.: +82-31-400-5187 (S.T.); +82-31-436-8076 (R.K.); Fax: +82-31-406-7118 (S.T.); +82-31-406-7118 (R.K.)

Received: 25 November 2017; Accepted: 26 January 2018; Published: 31 January 2018

Abstract: Recent research in the construction industry has focused on the reduction of CO_2 emission using quantitative assessment of building life. However, most of this research has focused on the operational stage of a building's life cycle. Few comprehensive studies of CO_2 emissions during building construction have been performed. The purpose of this study is to analyze the CO_2 emissions of an apartment housing during the construction process. The quantity of CO_2 emissions associated with the utilization of selected building materials and construction equipment were used to estimate the CO_2 emissions related to the apartment housing life cycle. In order to set the system boundary for the construction materials, equipment, and transportation used, 13 types of construction work were identified; then the CO_2 emissions produced by the identified materials were calculated for each type of construction work. The comprehensive results showed that construction work involving reinforced concrete accounted for more than 73% of the total CO_2 emissions. The CO_2 emissions related to reinforced concrete work was mainly due to transportation from the supplier to the construction site. Therefore, at the time that reinforced concrete is being supplied, shipping distance and fuel economy management of concrete transportation vehicles should be considered thoroughly for significant reduction of CO_2 emissions.

Keywords: apartment housing; construction process; CO_2 emissions

---

1. Introduction

Much effort has been made to improve environmental regulations in order to properly respond to climate change associated with global warming, both overseas and domestically throughout all industries. In order to prepare for climate change, the government has announced a voluntary reduction goal, called the Intended Nationally Determined Contributions (INDC) to reduce domestic greenhouse gases by 37%, compared to Business as Usual (BAU) [1]. Amid this, the construction industry, as a core industry, has caught the attention of many for its potential contribution to the achievement of this domestic greenhouse gas reduction goal. The construction industry is a large-scale consumption industry that is responsible for about 30–40% of the CO_2 emissions in all industries [2–4]. We are in a state where technology that can reduce CO_2 emissions is becoming a necessity. Technology that could achieve the CO_2 emissions reduction goal through a realistic CO_2 emissions reduction plan, over the entire life cycle of buildings, is needed. In order to reduce the amount of CO_2 emissions that are inevitable over the entire life cycle of buildings, the construction industry is evaluating this via quantitative investigation, and research on methods that could reduce or improve carbon emissions are actively underway [5]. In order to reduce the CO_2 emissions that are released over the life cycle of a building, the CO_2 emissions data must first be analyzed for each stage, and then a building's greenhouse gas evaluation must be performed considering its entire life cycle. However, more research is currently being performed on CO_2 emissions assessment at the operational stage, where more energy is used. In particular, there is very little research on CO_2 emission assessment of construction processes because data analysis is difficult due to the short construction period, and because data is difficult to secure [6,7].

This study focused on the construction stage for the reduction and management of CO_2 emissions of the apartment housing. The main purpose of this study is to suggest a CO_2 emission assessment method for the apartment housing construction process. Another purpose is to analyze the CO_2 emissions characteristics for each of the 13 main work types used during construction of the apartment buildings, using CO_2 emissions evaluation proposals of apartment housing construction projects, and construction records for actual construction sites.

In this study, apartment buildings construction sites were the subjects and the CO_2 emissions characteristics for construction materials and construction equipment were analyzed by the work types that occur in each stage. The following process was used to quantitatively assess the amount of CO_2 emissions from apartment housing construction, and reflects actual data [8,9].

First, the life cycle assessment was divided into three stages: transportation, construction, and disposal, to assess the CO_2 emissions from the construction site. The assessment subjects, which were construction materials and equipment, were the construction materials and equipment invested into each process.

Second, the amount of CO_2 emissions from the transportation stage was limited to the fuel consumption of the transportation vehicles required to transport necessary materials to the construction site. An estimation method was proposed for CO_2 emissions that considered the transportation distance, transportation vehicle type and average fuel efficiency, number of transportation vehicles, and load for each construction material.

Third, the amount of CO_2 emissions during the construction stage was the total fuel and electricity used by construction machines, transportation equipment, the field office, and other facilities that were used on site during construction. The amount of fuel and electricity used during the construction stage for machine equipment and electricity usage was calculated by analyzing data from standard construction estimates and data from the field office.

Fourth, the amount of CO_2 emissions during the disposal stage was estimated by first calculating the amount of construction waste that occurs by work type. This was done by reviewing the construction materials excess, then quantitatively calculating the amount of CO_2 emissions according to the method of handling construction waste, based on the statistical research report on the recycling of construction waste. With regard to the carbon emission coefficient in relation to the amount of construction materials used, and the amount of energy consumed in each stage, the emission factor of the Korea National life cycle inventory database (LCI DB) and the Korea National DB construction material environment information were applied. The flow of this study is shown in Figure 1.

An assessment method was created regarding the amount of energy used and the CO_2 emissions from construction materials and equipment in relation to 13 main types of construction work. This was done by dividing the construction process into three stages: transportation, construction, and disposal. A database was built for each stage, and the amount of CO_2 emissions was estimated using it. The 13 main work types during the apartment construction process were deduced, and an analysis of the CO_2 emissions characteristics of each stage was performed.

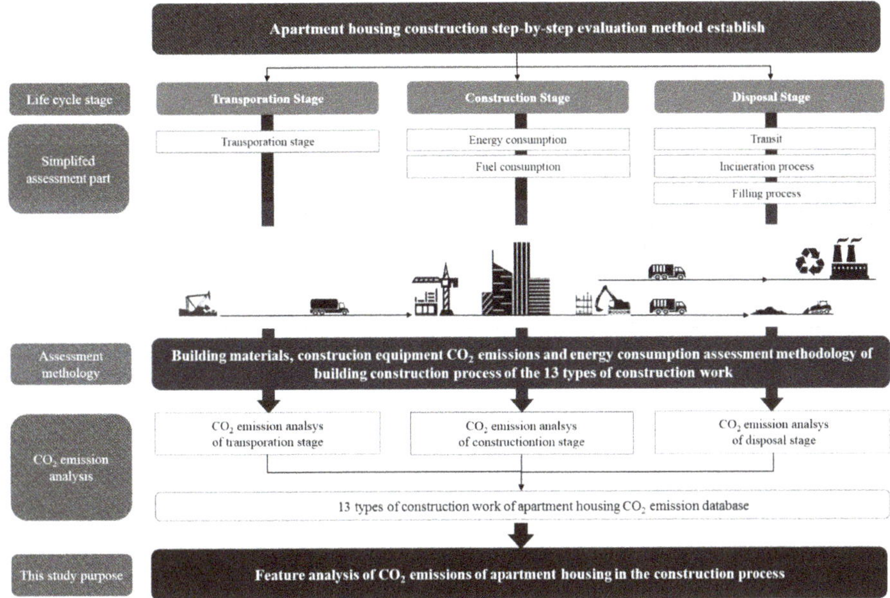

Figure 1. The conceptual flow chart of this study.

2. Literature Review

2.1. Theory of the Building LCA Assessment Method

Life cycle assessment (LCA) is a method that includes record of a list of invested materials and products related to the system life cycle of a product, and which allows examination of the potential environmental impacts related to these. For this study, when the list-analysis results and impact-assessment results were interpreted, research was actively performed according to the main methodology. This methodology has been used for environmental performance assessments of buildings from the 1990s to present day. Overseas LCA studies have proceeded as government-directed projects since 1970, and the majority of the research results on creating a LCI DB basic unit depending on the life cycle stage are already complete [10]. Research is also actively underway in Korea both nationally and publically, on a regular basis. Life cycle assessment is a tool that quantitatively lists the resources and energy that are used and the pollutants that are released during the product system's life cycle. It also systematically assesses their potential adverse effects on the environment. LCA is also used to derive measures to minimize and improve these adverse effects under the supervision of the Korea Ministry of Environment. It is applied to a product's eco-friendly design and other environmental claims, process improvements, and public policy establishment through the ISO 14040 series. LCA is divided into the four stages shown in Figure 2.

The definition of purpose and scope is the basic stage that sets the study's purpose and the system's boundary, which must be defined first in the life cycle assessment. Record analysis is the stage that quantifies the materials that are used and the products related to the product system. During this stage, data that describes all energy and by-product raw materials that are used and calculated, are collected and used in the calculations. Data are collected through a repetitive process, and new data demands and restrictions can be expected in order to achieve the purpose of the study by acquiring more information for each system boundary. Impact assessment uses the list-analysis results, and its purpose is to assess the importance of potential environmental impacts [11,12]. During impact assessment, listed data are generally linked to a specific environmental impact category, and category score, in order to understand their impacts and to provide information in the interpretation stage. An assessment range is set according to the LCA methodology of this type of series, and is generally created according to ISO 21930 as shown in Figure 3.

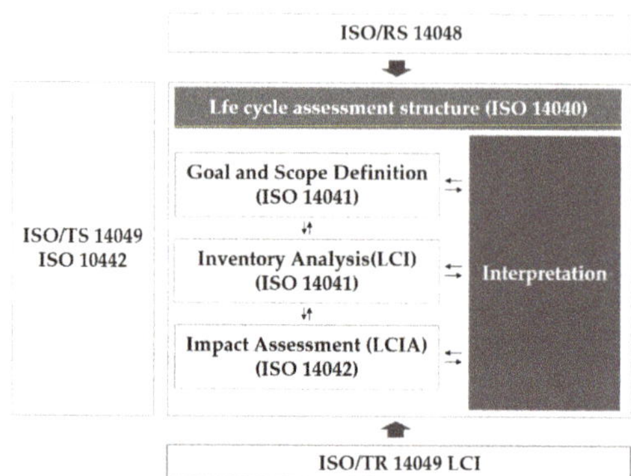

Figure 2. LCA's stages according to ISO 14040.

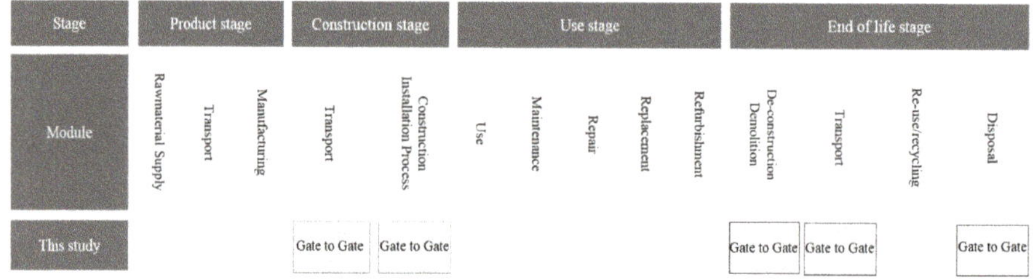

Figure 3. Life cycle stages for the building assessment in ISO 21930.

This study was set on gate-to-gate life cycle assessment, and an analysis was performed on CO_2 emission characteristics by dividing the construction process, from a building's life cycle assessment range, into transportation, construction, and demolition stages [13,14].

2.2. Research on the Construction Process

Preceding literature was reviewed regarding methods to estimate the amount of CO_2 emissions from construction equipment and materials used during the apartment construction process in Table 1. Kim et al. presented an estimation study on the amount of energy consumed and CO_2 released, for each work type during the apartment design stage. Chung et al. calculated the amount of energy consumed and CO_2 released in the building establishment stage and construction stage. After considering the results in preceding literature on the calculation methods for CO_2 emissions, it was found that most existing studies applied the LCA model based on input-output analysis [15,16]. The input-output analysis method enables easy abundant data calculation. It is an analysis method that describes the inter-industry relation table's connections in units of monetary value, and considers energy or resources according to the flow of monetary amounts and goods. However, because the LCA model based on input-output analysis calculates the amount of CO_2 emissions only by using information on the cost of products or services, its disadvantage is that it can only be used as an assessment outline. It is also limited as it does not consider system boundaries, fuel extraction, processing, production process, waste disposal, and other various factors regarding used materials that should be considered in the construction process. The key preceding research and literature on important factors that should be considered in the construction process, including environment, materials, equipment, transportation, and management were examined. However, there is very little research on CO_2 emission assessment of construction processes because data analysis is difficult due to the short construction period, and data is difficult to secure. Hadjimitsis et al. analyzed the amount of energy used with regard to concrete in the transportation stage, Jo et al. analyzed the weight of the construction work and managing work type component category conclusion [17,18], and Murat et al. suggested that recycling of ferrous and non-ferrous metals, cardboard, plastic and glass maximize the environmental and economic savings. Also, Roh et al. analyzed the CO_2 emission characteristics regarding concrete construction. After analyzing the previous research performed in Korea regarding construction processes, it was found that the research was focused only on analyzing some work types and on the amount of energy used in the construction process [19,20]. Moreover, limitations were seen in the amount of CO_2 emissions that occurred in the transportation of construction materials, construction process, and process of disposing of construction waste, all of which are involved in the construction process, and which were generally not being considered. Although the use of construction equipment and materials, transportation, energy use on site, and other matters were defined in detail, there were still parts that were excluded from consideration. Research on the basis of calculation used in actual CO_2 assessment was not being performed. Most research evaluated only a portion of construction work types, or was based on specific materials or equipment used at the construction site. As a result, research that includes construction materials, construction equipment, construction work types, transportation, construction process, construction waste disposal, and other factors should be conducted. This study looked into the construction process that was divided into transportation, construction, and demolition stages for a detailed assessment. Moreover, this work was focused on analyzing construction equipment, transportation equipment, and construction materials. In addition, an individual method using the Korea National LCI DB was used instead of using existing analyses of CO_2 emissions based on monetary values, to perform an analysis on CO_2 emissions characteristics for each work type during apartment construction. In prior studies regarding the assessment of CO_2 emissions during the construction process, CO_2 emission-impact factors were merely listed, or only extremely limited factors were applied to the building environmental impact assessment and analysis process [21,22].

Table 1. Review of existing literature.

Division	Summary	Analysis Target				Analysis Method
		Transit Equipment	Construction Equipment	Construct Work	Input Material	
Hadjimitsis, D.Gand et al.	Environmental Impact and Energy Consumption of Transport Pavements in Cyprus	■				Budget statement by district Industry-related analysis
Murat K et al.	Life Cycle Assessment and Optimization-Based Decision Analysis of Construction Waste Recycling for a LEED-Certified University Building sustainable				■	Quantity calculation sheet Industry-related analysis
Choi MS et al.	Calculate unit cost for building materials input by type of apartment building construction				■	Quantity calculation sheet Industry-related analysis
Kim JY et al.	Analysis of energy consumption for construction materials and calculation of CO_2 emissions				■	Quantity calculation sheet Industry-related analysis
Jeong YC et al.	A Study on the Appropriateness of the Application of the Input-Output Table according to the Calculation of CO_2 Emission Unit Value by Major Materials				■	Quantity calculation sheet Industry-related analysis
Kim DH et al.	A Study on the Estimation of Energy Consumption and Carbon Dioxide Emission of Building Materials Entered by Construction Type			■	■	Quantity calculation sheet Industry-related analysis
Kim KW et al.	Investigation and analysis of the influence factors of oil consumption on building equipment and materials during construction phase of apartment house		■		■	Industry-related analysis
Kim JW et al.	A Study on the Development of CO_2 Evaluation Method for Concrete Liquefied Transportation in the Construction Phase of Apartment Buildings	■				Industry-related analysis

■: Analysis target in the construction process.

3. Methodology

3.1. Overview

In this study, the construction process of the apartment housing was divided into transportation, construction, and disposal stage for the purpose of analyzing CO_2 emission characteristics of the apartment housing construction process in Figure 4.

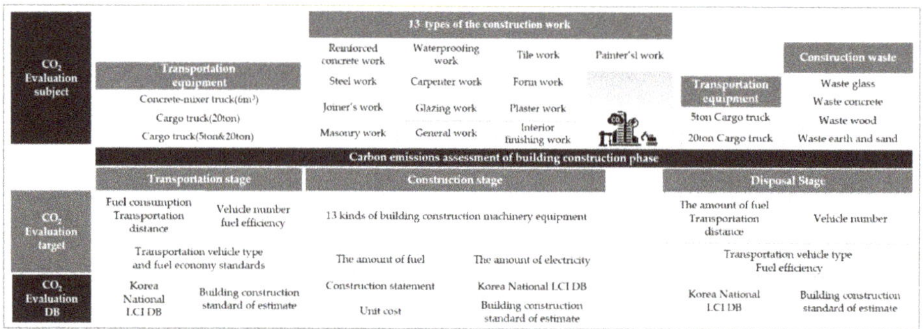

Figure 4. CO_2 emission assessment method proposed in construction process.

In order to estimate the amount of energy used in vehicles transporting materials to the apartment housing construction site, the amount of CO_2 emissions in the transportation stage was limited to the amount of fuel used by transportation vehicles when transporting materials from the supply location to the construction site. The distance for fuel consumption calculation was estimated based on scenario. Based on previous research, a CO_2 emissions-assessment method considering the distance, average fuel efficiency, number of transportation vehicles, and load (depending on the amount of construction materials used) was presented. The amount of CO_2 emissions during the construction stage was the total fuel and electricity used by construction machines and equipment, transportation equipment, the field office, and other facilities used for and during construction on site. The amount of fuel and electricity used during the construction stage for machine equipment and electricity usage was calculated by analyzing data from Korean standard construction estimates and from Korea breakdown cost tables. The amount of CO_2 emissions during the disposal stage was estimated by first calculating the amount of construction waste that occurred (by work type) based on surcharges on the construction materials, then quantitatively calculating the amount of CO_2 emissions according to the method of handling construction waste, based on a statistical research report on the recycling of construction waste. With regard to the CO_2 emission factor regarding the amount of construction materials used and the amount of energy consumed in each stage, the Korean National LCI DB and the Korean National DB of the construction-material environmental information was applied [23,24]. In the transportation stage, distance and fuel for transportation equipment for CO_2 evaluation were analyzed, and Korean standard product and national LCI DB were applied. In the construction phase, the construction specification, the national LCI DB, and the Korean Standard Specification were applied to analyze the amount of fuel and the amount of power applied to evaluate 13 construction types. In the disposal stage, the national LCI DB and the Korean Standard Package were applied to analyze the distance fuel for the waste transportation equipment.

3.2. Estimation Method in the Transportation Stage

The main factors that must be considered in order to estimate the amount of CO_2 emissions in the transportation stage are mainly the amount of construction materials used, and the equipment used for transportation [25]. Also, it considered the transportation distance and the vehicles used to load at the supply and construction sites. This data was used to estimate the amount of energy used, which was then converted to an estimate of the amount of CO_2 emissions. This calculation method includes use of the amount of materials at the construction site to deduce the number of transportation vehicles used, and fuel consumption was calculated by using the transportation distance and vehicle fuel efficiency. To do this, the distance to the construction site (from the supply sites), type of vehicle and standard fuel efficiency, and the number of vehicles used must be considered. Because the length of time that transportation vehicles used at a site is flexible, it is impossible to quantify work capacity that depends on the type of vehicle [26,27].

In this study, the loads of vehicles listed in standard construction estimates were investigated, along with the main transportation vehicles observed at the actual site, in order to calculate the number of transportation vehicles according to the supply output value of each construction material. The amount of fuel that was consumed during the transportation stage was calculated by multiplying the distance to the site by the average fuel efficiency of each vehicle. The assessment range of the transportation stage was limited to the fuel consumption for one-way transportation of construction materials by the types of vehicles used for each work type.

In this study, the transportation vehicles and transportation distance regarding the materials used were set as shown in Table 2, and the number of transportation vehicles was calculated according to the supply output value for each construction material. Equations (1) and (2) show the calculation formulas for fuel consumption and CO_2 emissions, respectively, during the transportation stage [28,29]. The transportation distance was set to 40 km as referenced from the guideline of Korea Ministry of Land, Infrastructure and Transport standard specification for construction [29].

$$DO_t = \sum \left(U_t \times \frac{k}{M_t} \right) \tag{1}$$

DO_t: Amount of Fuel Consumed in the Transportation Stage (ℓ)
U_t: Number of Vehicles Used (n)
k: Distance (km)
M_t: Transportation Vehicle Fuel Efficiency (km/L)

$$C_t = \sum (DO_t \times EF_t) \tag{2}$$

C_t: Amount of CO_2 Emissions in the Transportation Stage (kg-CO_2)
DO_t: Amount of Fuel Consumed in the Transportation Stage (L)
EF_t: Fuel Emission Factor (kg-CO_2/L)

Table 2. Construction equipment calculated in accordance with the input material.

Input Materials	Transportation Vehicle	Transportation Distance
Concrete	Concrete-mixer truck (6 m^3)	40 km
Steel	Cargo truck (20 ton)	40 km
Subsidiary material	Cargo truck (5 ton and 20 ton)	40 km

3.3. Estimation Method in the Construction Stage

In order to estimate the amount of CO_2 emission in the construction stage, this study created categories of fuel consumption and electricity consumption amounts, and calculated the total amount of fuel and electricity used for construction machines and equipment, transportation equipment, field office, and other facilities at the construction site, used for constructing buildings. For the amount of fuel consumed, the number of days and time that equipment was used, according to each work type, was calculated by analyzing the equipment usage conditions from the construction report. The fuel efficiency for each kind of equipment was based on the operating expenses estimated in the standard construction estimate. Electricity consumption was calculated by obtaining the monthly electricity usage statement for the entire construction period, and calculating the electricity consumption according to the duration of each work type based on the progress schedule. If multiple work types proceeded simultaneously, according to the characteristics of the construction process, the relevant electricity consumption was divided up between the number of construction days for the overlapping work types. In this study, for construction processes in which the construction method (including construction conditions by site) and equipment usage differed, data regarding construction machines and equipment, and transportation equipment used at the actual construction site were obtained from the supply output lists from the actual construction site, and from construction reports. Data regarding the amount of fuel used were analyzed. Data were also analyzed regarding electricity consumption from the actual field office. The calculation formula for the amount of CO_2 emissions according to fuel and electricity consumption during the construction stage is shown in Equations (3) and (4), respectively [30].

$$C_{co} = \sum (M_c \times T_c \times EF_c \times U_c) \tag{3}$$

C_{co}: Amount of Fuel CO_2 Emissions in the Construction Stage
M_c: Fuel Efficiency (L/h)
T_c: Time (h)
EF_c: Fuel Emissions Factor (kg-CO_2/L)
U_c: Work Equipment Qty. (n)

$$C_{ce} = \sum (I_c \times EF_c) \tag{4}$$

C_{ce}: Amount of Electricity CO_2 Emissions in the Construction Stage
I_c: Electricity Usage Amount (kwh)
EF_c: Electricity Emissions Factor (kg-CO_2/kwh)

3.4. Estimation Method in the Disposal Stage

In order to estimate the amount of CO_2 emissions in the disposal stage, the amount of construction waste that occurs for each work type was estimated, and the amount of CO_2 emissions, according to the processing method of each construction waste, was calculated. In order to estimate the amount of construction waste that occurs by work type, the waste generation estimation method based on surcharge was used. In this study, the premium rate was applied as a means to predict waste at the construction site, and the surcharge value that was presented in the standard construction estimate issued in 2012 (by the Korea Institution of Construction Technology and the Ministry of Land). Construction waste was divided into construction demolition waste, combustible waste, and nonflammable waste according to the method of handling construction waste, which are handled through intermediate processing agencies, incinerators, and landfills, respectively. When calculating data regarding intermediate processing agencies, because data differs from different companies, the incineration and landfill rate referred to the present processing conditions of waste for each construction material according to the Korea Waste Statistical Yearbook in 2012. The surcharges, weight conversion factor, landfill, and incineration emissions factor were considered for each construction material according to the amount used, in order to estimate the amount of CO_2 emissions that occur in the incineration and landfill process Equation (5). The amount of energy consumed during the construction waste transportation process was calculated using Equations (6) and (7), based on the amount of fuel consumed in the vehicles that transport the waste to landfills and incinerators.

The estimation method for the amount of CO_2 emissions in the construction waste transportation stage was identical to that of the construction process above, and it was set as the amount of energy consumed by transportation equipment according to waste volume. In order to do this, the amount of construction equipment, standard fuel efficiency, and the distance to landfills and incinerators were applied to calculate the amount of fuel used during transportation. The assessment scope of the waste transportation process was limited to the amount of fuel consumed by vehicles providing one-way transportation of construction waste [31,32].

$$C_{dl} = \sum (DM_d \times EF_{dl} \times L_d) \tag{5}$$

C_{dl}: Amount of CO_2 Emissions in Landfill Process of the Demolition Stage (kg-CO_2)
DM_d: Construction Waste Volume (kg)
L_d: Landfill Rate (%)
EF_{dl}: Landfill Emissions Factor (kg-CO_2/kg)

$$C_{di} = \sum (DM_d \times EF_{di} \times I_d) \tag{6}$$

C_{di}: Amount of CO_2 Emissions in Incineration Process of the Demolition Stage (kg-CO_2)
DM_d: Construction Waste Volume (kg)
I_d: Incineration Rate (%)
EF_{di}: Incineration Emissions Factor (kg-CO_2/kg)

$$DO_d = \sum \left(U_d \times \frac{k}{M_d} \right) \tag{7}$$

DO_d: Amount of Fuel Consumed in the Construction Waste Transportation Stage (L)
U_d: Number of Vehicles Used (n)
k: Distance (km)
M_d: Transportation Vehicle Fuel Efficiency (km/L)

$$C_d = \sum (TO_d \times EF_d) \tag{8}$$

C_d: Amount of CO_2 Emissions in the Construction Waste Transportation Stage (kg-CO_2)
TO_d: Amount of Fuel Consumed by Transportation Equipment (L)
EF_d: Fuel Emissions Factor (kg-CO_2/L)

4. Results

4.1. Overview

The architectural overview of the evaluation subject is summarized in Table 3. In order to analyze the CO_2 emission characteristics during the apartment housing construction process, 1004 household, reinforced concrete apartment houses (combined area of 208,393 m^2) were selected in Seoul, Korea.

Table 3. Architectural Overview.

Description	Contents	Description	Contents
Business name	Seoul M Urban Development Project	Structural system	Reinforced concrete, bearing wall structure
Local district	Semi-residential area	Building type	Flat-type and tower-type Apartment
Architectural drawing	Apartment house	Household number	1004 Household
Building scale	16 stories above ground/2 stories below ground	Land area	5,633,600 m^2
Number of buildings	13 buildings	Total floor area	20,839,380 m^2

This study targeted the 13 types of construction work such as, Reinforced concrete works, Steel works, Carpenters works, Waterproofing works, Tile works, Glass works Painter's works, Form works, Plaster works, Interior finishing works, General works, Joiner's works and Masonry works. Then, the CO_2 emissions characteristics for each work type and the main material-types were deduced. In order to compare analysis results, the emission amount was calculated by dividing it by the total area.

4.2. Analysis of CO_2 Emission Characteristics by Work Type

4.2.1. Transportation Stage

In order to deduce the amount of fuel consumed by each work type in the transportation stage, the amount of CO_2 emissions in the transportation stage was estimated by considering fuel efficiency regarding the number of vehicles used, transportation distance, and each kind of equipment. The transportation equipment count, according to the load sizes of equipment that transported materials to the construction site, was analyzed as shown in Table 4. The results showed that most of the material was reinforced concrete in the case of 20-ton trucks and 6 m^3 concrete-mixer trucks. For 5-ton trucks, most of the material was for tile work. The total amount of CO_2 emissions from the transportation stage analysis results was 5.46 kg-CO_2/m^2 in as shown Figure 5. The amount of CO_2 emissions from reinforced concreted work accounted for 89% of all work types, which was analyzed to display most of the CO_2 emission characteristics. The analysis results for the CO_2 emission characteristics of other kinds of construction, were analyzed in the order waterproof construction, tile work, and plastering; and the amount of CO_2 emissions for each construction type was 0.08 kg-CO_2/m^2, 0.07 kg-CO_2/m^2, and 0.06 kg-CO_2/m^2, respectively. After analyzing 34,500 transportation vehicles used at the construction site for each work type, the amount of CO_2 emissions from the 30,422 concrete transportation vehicles used for the reinforced concrete work accounted for around 75% of the amount of CO_2 emissions for the entire transportation stage. Therefore, it was found that the load distance between the construction site and the concrete manufacturer (when concrete is collected for reinforced concrete work), and the fuel efficiency of concrete transportation vehicles must be managed in order to reduce the amount of CO_2 emissions in the construction process.

4.2.2. Construction Stage

In the construction stage, the amount of CO_2 emissions can be analyzed using the amount of fuel and electricity consumed for each work type. In order to do this, equipment that was used at the actual construction site was categorized as shown in the construction stage data in Table 4, and the number of construction equipment work types that were used was analyzed in detail. Results showed that most of the equipment was used on reinforced concrete work, and that the 50-ton mobile crane was used in more work types than was other equipment. Electricity consumption was estimated by multiplying the emissions factor by the amount of electricity used in the field office, and the resulting value was estimated by considering the total area and the total value of the calculated fuel consumption and electricity consumption. After analyzing the construction process, the amount of fuel consumed was found to be 0.32 kg-CO_2/m^2, the amount of electricity consumed was 2.97 kg-CO_2/m^2, and the total amount of carbon emissions in the construction stage was 3.29 kg-CO_2/m^2 as shown in Figure 6. After analyzing the reinforced concrete work from the construction stage; fuel and electricity consumption accounted for around 31% of all work types. Among other work types, waterproof construction took up 9%, and plastering took up 8%. With regard to the amount of CO_2 emissions in the construction stage, high CO_2 emission characteristics were expected according to the construction method and equipment that was used at the actual site, but the amount of emissions was relatively low compared to the preceding transportation stage.

Table 4. Input breakdown by work type in the apartment housing construction.

Stage	Emission Factor	Machine Equipment	Equipment Specification	Unit	Form Work	Rein-Forced Concrete Work	Steel Work	Masonry Work	Water Proofing Work	Painter's Work	Tile Work	Masonry Mason's Work	Joiner's Work	Glazing Work	Carpenter Work	Interior Finishing Work	Finishing's Work
Transportation stage	Oil	Cargo truck	5 ton	N	-	8	4	2	-	-	296	-	14	21	-	126	59
			20 ton	N	41	1618	-	578	474	368	237	96	-	-	219	-	-
		Mixer truck	6 m^3	N	-	30,422	-	-	-	14	-	-	-	-	-	-	-
Construction stage	Oil	Backhoe	0.6 m	N	15	2	-	-	-	-	-	-	-	-	-	-	-
			0.8 m	N	12	-	-	-	-	-	-	-	-	-	-	-	-
			1.8 m	N	3	-	-	-	-	-	-	-	-	-	-	-	-
		Pump car	36 m	N	-	221	-	-	-	-	-	-	-	-	-	-	-
			52 m	N	-	9	-	-	-	-	-	-	-	-	-	-	-
		Mobile crane	100 ton	N	-	18	-	-	-	-	-	-	-	-	-	-	-
			50 ton	N	-	62	-	2	-	1	1	3	2	1	-	1	-
		Dozer	6 P	N	2	-	-	-	-	-	-	-	-	-	-	-	-
	Electricity	Site electric power consumption		kwh	46,725	190,965	11,642	26,410	40,630	54,851	52,820	32,504	32,504	24,378	28,441	46,725	42,662
Disposal stage	Oil	Cargo truck	20 ton	N	-	286	1	29	24	20	13	5	-	-	11	-	-
	Landfill waste	Construction wastes		kg	-	1228	-	2370	-	180	-	-	-	-	-	-	-
		Combustible wastes		kg	-	4123	-	876	-	7071	-	-	-	-	-	-	5982
		Incombustible wastes		kg	-	-	-	-	-	-	13,334	-	-	208	-	14,439	-
	Incineration waste	Construction wastes		kg	-	-	-	-	-	-	-	-	-	-	-	-	-
		Combustible wastes		kg	-	-	1176	-	-	-	-	-	-	-	-	2815	-
		Incombustible wastes		kg	-	-	-	-	-	-	-	-	-	-	-	199	-

4.2.3. Disposal Stage

For transporting construction waste, the vehicle loads for the amount of construction waste involved in each work type was considered in Table 4 and the vehicle loads were analyzed. After analyzing the input breakdown by work type, it was found that reinforced concrete work accounted for a large proportion. For the landfill and incineration processes, the detailed amounts involved in construction waste for each work type were analyzed in Table 4. The landfill-incineration emissions factor regarding construction waste, and data regarding the landfill-incineration rate, were analyzed by referencing a statistical survey report on construction waste recycling, and the Korea statistical yearbook for waste from the Korea Ministry of Environment. After analyzing the demolition stage of this study, the total amount of CO_2 emissions was found to be 0.076 kg-CO_2/m^2 as shown in Figure 7, which was the lowest figure among the three stages. Moreover, because waste from rebar, concrete, and related products are abundant during work with reinforced concrete, it was found that higher amounts of CO_2 emissions resulted from work with reinforced concrete than from other work types. Because the amount of CO_2 emissions from the landfill process was 98%, it accounted for the majority of the demolition stage emissions, and analysis showed that the amount of CO_2 emissions from waste produced from plaster boards took up a high proportion among the other construction wastes in the landfill process.

4.3. Comprehensive Analysis

An analysis of the CO_2 emission characteristics for each stage of the apartment construction process by work type is shown in Table 5. The total CO_2 emission characteristics by each stage in the construction process are shown in Figures 5–8.

Table 5. CO_2 emission by work type analysis in the apartment housing construction.

(Unit: kg-CO_2/m^2)			
Stage Classification	Transportation Stage	Construction Stage	Disposal Stage	Total
Form work	0.001	0.258	0.000	0.259
Reinforced concrete work	5.053	1.198	0.043	6.294
Steel work	0.003	0.001	0.001	0.005
Masonry work	0.125	0.127	0.006	0.258
Water proofing work	0.077	0.195	0.005	0.277
Plaster work	0.061	0.264	0.004	0.329
Tile work	0.070	0.253	0.006	0.329
Masonry mason's work	0.016	0.157	0.001	0.174
Joiner's work	0.001	0.156	0.000	0.157
Glazing work	0.002	0.119	0.001	0.122
Carpenter work	0.036	0.138	0.007	0.181
Interior finishing work	0.013	0.225	0.001	0.239
Painter's work	0.006	0.205	0.003	0.214
Total	5.463	3.295	0.078	8.838

An assessment of each stage of the construction process showed that reinforced concrete work was the work type with the highest proportion of CO_2 emissions, and after analyzing the total CO_2 emissions characteristics of the entire construction process, most of the CO_2 emissions of reinforced concrete work stemmed from the transportation stage. An assessment of the amount of CO_2 emissions in the construction process by each stage showed that the total emissions from reinforced concrete work was 8.83 kg-CO_2/m^2, which was around 73% of the entire amount of CO_2 emissions. Of this, 89% was from the transportation stage, and the total proportion of CO_2 emissions during the construction stage was about 36%. Also, the construction waste transportation proportion was around 57% in reinforced concrete work at the disposal stage, and although the amount of CO_2 emissions from all work types was small, an analysis showed that CO_2 emissions were abundant during the process of incinerating wallpaper and other materials used in wood work and other interior finishing work. The amount of CO_2 emissions released from plaster boards and insulation materials in the landfill process was abundant in the order reinforced concrete work, wood work, masonry work, and tile work. By analyzing CO_2 characteristics from the apartment housing construction process, we believe that activities to reduce CO_2 emissions must begin by managing the load distance between the construction site and concrete manufacturer when collecting concrete for reinforced concrete work, and by managing the fuel efficiency of concrete transportation vehicles. It is expected that the assessment process of CO_2 emission during the construction phase based on the apartment housing evaluation done in this research would be applicable in other studies in South Korea as well as other countries. However, the regional applicability range could be comparatively limited as the established data base of the research is based on the actual data of multi-unit dwellings that are built in South Korea.

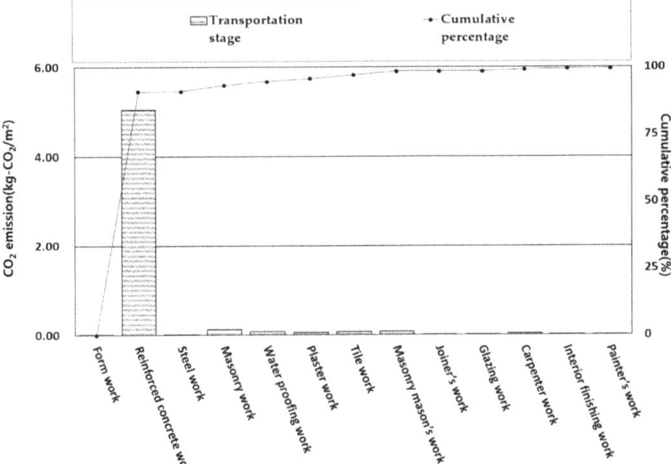

Figure 5. Work type CO_2 emission in transportation stage.

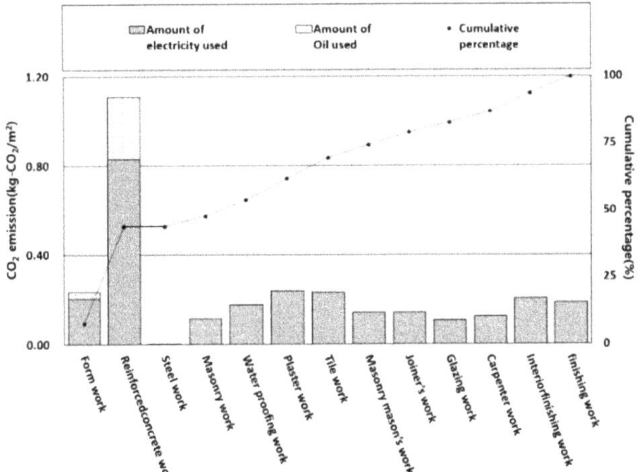

Figure 6. Work type CO_2 emission in construction stage.

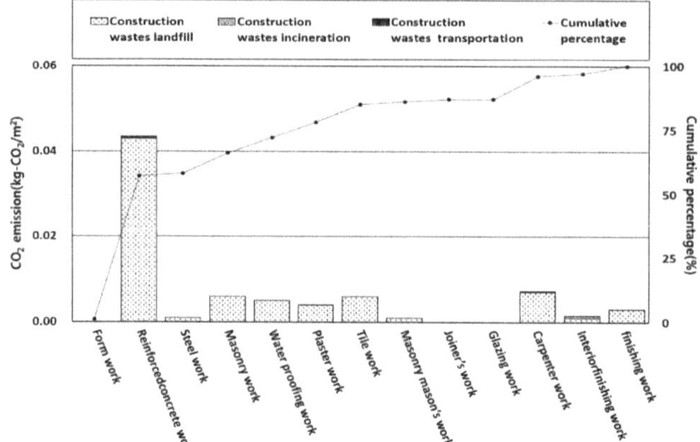

Figure 7. Work type CO_2 emission in disposal stage.

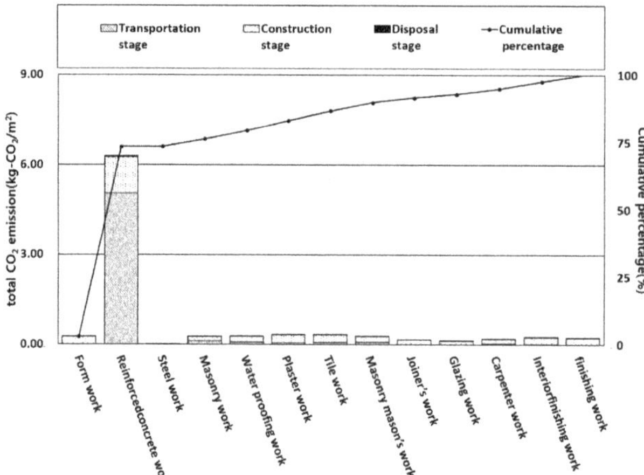

Figure 8. Work type CO_2 emission in construction process.

5. Conclusions

This is part of a study for the reduction and management of CO_2 emissions from construction processes during the entire life cycle of buildings. The purposes of this study was to suggest a CO_2 emissions assessment method for each stage of the apartment housing construction process, and to analyze CO_2 emissions characteristics by stage for each of the 13 types of the construction work using construction records from actual work sites. From these, the following conclusions were obtained.

1. In order to analyze the total amount of CO_2 emissions for each work type of the apartment housing construction process, this study obtained the establishment breakdown and construction records of actual apartment houses, construction records, and other construction books from which to analyze CO_2 emission characteristics.
2. System boundaries were established in order to analyze the CO_2 emission characteristics of the apartment construction process. In order to do this, the construction process was divided into transportation, construction, and disposal stages, and the amount of CO_2 emissions was assessed using the assessment method presented in this study.

3. After analyzing construction CO_2 emissions by stage for apartment construction, the total amount of CO_2 emissions was estimated to be 8.83 kg-CO_2/m². The amount of CO_2 emissions from the transportation stage, construction stage, and disposal stage, was 5.46 kg-CO_2/m², 3.29 kg-CO_2/m², and 0.07 kg-CO_2/m² respectively.

4. The proportion of total CO_2 emissions of reinforced concrete work was about 73% of the total amount of CO_2 emissions from the apartment construction process, which is an overwhelming proportion. The amount of CO_2 emissions from reinforced concrete work was found to be mainly from the transportation stage.

5. From this, we deduced that the load distance between the construction site and the concrete manufacturer (when collecting concrete for reinforced concrete work), and the fuel efficiency of concrete transportation vehicles, must be managed to significantly reduce the amount of CO_2 emissions resulting from construction.

6. We believe that in order to reduce the amount of carbon emissions in the construction stage in the future, priority should be given to reducing the concrete load distance in reinforced concrete work, which produces the most CO_2 emissions, and to considering the use of high-efficiency heavy transportation equipment.

As a result of analyzing CO_2 emission characterized by construction type, it was found that in the case of reinforced concrete construction, which has the highest CO_2 ratio, the amount of emission was higher in the transportation phase than in the construction phase. This proved that the assumption that transportation distance greatly contributes in CO_2 emission during the construction work. Therefore, when the work of reinforced concrete construction is supplied and received, it is analyzed that management of fuel mileage is necessary for CO_2 emission reduction due to construction work.

In order to reduce carbon emissions during the construction process, it is considered that priority should be given to the reduction of the shipment distance of concrete and the use of high-efficiency transportation equipment in the reinforced concrete construction with the greatest CO_2 emission.

Acknowledgments: This research was supported by a grant (Code 18CTAP-C129766-02) from Construction Technology Research Program (CTIP) funded by Ministry of Land, Infrastructure and Transport.

Author Contributions: All authors contributed substantially to all aspects of this article.

Conflicts of Interest: The authors declare no conflict of interest

References

1. Yi, I.S.; Seo, K.S. *Social Indicators in Korea*; National Statistical Office: Seoul, Korea, 2010.

2. Cabeza, L.; Rincon, L.; Vilarino, V.; Perez, G.; Castell, A. Life cycle assessment (LCA) and life cycle energy analysis (LCEA) of buildings and the building sector: A review. *Renew. Sustain. Energy Rev.* **2014**, *29*, 394–416. [CrossRef]

3. Kim, R.H.; Tae, S.H.; Roh, S.H. Development of low carbon durability design for green apartment buildings in South Korea. *Renew. Sustain. Energy Rev.* **2017**, *77*, 263–272. [CrossRef]

4. Baek, C.H.; Tae, S.H.; Shin, S.W. Life Cycle CO_2 Assessment by Block Type Changes of Apartment Housing. *Sustainability* **2016**, *8*, 752. [CrossRef]

5. Pachauri, R.K.; Reisinger, A. *IPCC, 2007: Climate Change 2007: Synthesis Report*; Contribution of working Groups I, II and III to the Fourth Assessment Report of the Intergovermental Panel on Climate Change; IPCC: Geneva, Switzerland, 2008; p. 104.

6. Khasreen, M.M.; Banfill, P.F.; Menzies, G.F. Life-Cycle Assessment and the Environmental Impact of Buildings. *Sustainability* **2009**, *1*, 674–701. [CrossRef]

7. Greenhouse Gas Inventory & Research Center of Korea. *2013 National Greenhouse Gas Inventory*; Report of Korea; Ministry of Environment: Sejong-si, Korea, 2013.

8. Korea Ministry of Environment. *National Greenhouse Gas Emissions Reduction Roadmap 2020*; Ministry of Environment: Sejong-si, Korea, 2014; p. 6.

9. Eggleston, H.S.; Buendia, L.; Miwa, K.; Ngara, T.; Dan, T.K. *2006 IPCC Guidelines for National Greenhouse Gas Inventories*; Prepared by the National Greenhouse Gas Inventories Program; IGES: Kanagawa, Japan, 2006.

10. Korea Environmental Industry & Technology Institute. *Life Cycle Assessment Theory and Practice*; Korea Environmental Industry & Technology Institute: Seoul, Korea, 2012.

11. Pennington, D.W. Life cycle assessment Part 2: Current impact assessment practice. *Environ. Int.* **2004**, *30*, 721–739. [CrossRef] [PubMed]

12. International Organization for Standardization. *ISO 14040, Environmental Management-Life Cycle Assessement-Principles and Framework*; International Organization for Standardization: Geneva, Switzerland, 2006.

13. Yoe, H. Development of a life cycle assessment tool for construction and maintenance of asphalt pavement. *J. Clean. Prod. (UK)* **2009**, *17*, 283–296.

14. KICT (Korea Institute of Civil Engineering and Building Technology). Life Cycle Analysis Annual Report. Available online: http://www.kict.re.kr/ (accessed on 27 September 2013).

15. Kim, D.H.; Kwon, B.M.; Choi, Y.O.; Lee, G.H. The Estimation of the Energy Consumption and CO_2 Emission at the Construction Stage in the Apartment Housing, Korea living institute. *Archit. Inst. Korea* **2006**, *2*, 328–334.

16. Chung, Y.C.; Kim, S.Y.; Jang, Y.J.; Kim, T.H.; Kim, G.H. Application of Input-Output Table to Estimate of Amount of Energy Consumption and CO_2 Emission Intensity in the Construction Materials Focusing on Input-Output Tables Published in 2005, 2007. *J. Korea Inst. Build. Constr.* **2011**, *11*, 247–255.

17. Hadjimitsis, D.G.; Themistocleous, K.; Achilleos, C.; Kallis, S.; Neofytou, P.; Neocleous, K.; Pilakoutas, K. Environmental Impact and Energy Consumption of Transport Pavements in Cyprus. In Proceedings of the 11th International Conference on Environmental Science, Crete, Greece, 3–5 September 2009; pp. A419–A425.

18. Jo, A.R.; Kim, C.W.; Jo, H.H.; Kang, G.I. Deduction of the Primary Management Works for Reduction of the Environmental loads at the Construction Phases. *Korea Inst. Build. Constr.* **2013**, *13*, 1–5.

19. Kucukvar, M.; Egilmez, G.; Tatari, O. Life Cycle Assessment and Optimization-Based Decision Analysis of Construction Waste Recycling for a LEED-Certified University Building. *Sustainability* **2016**, *8*, 89. [CrossRef]

20. Roh, S.J.; Tae, S.H.; Lee, J.H.; Lee, J.S.; Ahan, J.H.; Shin, S.W. A Study on the Development of CO_2 Assessment Management System for Construction Site Korea Concrete Institute. *Archit. Inst. Korea* **2011**, *23*, 1.

21. Nansai, K.; Moriguchi, Y.; Tohno, S. Embodied energy and emission intensity data for japan Input-Output tables (3EID). *Natl. Inst. Environ. Stud. (JP)* **2003**, *37*, 2005–2015.

22. Korea LCI Data Base Inpormation. Available online: www.edp.or.kr/lcidb (accessed on 29 January 2018).

23. KICT (Korea Institute of Civil Engineering and Building Technology). *Standard Estimating Data for Building Construction*; Korea Ministry of Land, Infrastructure and Transport: Sejong, Korea, 2013.

24. Pierucci, A. LCA evaluation methodology for multiple life cycles impact assessment of building materials and components. *Tema Tempo Mater. Archit.* **2015**, *1*, 1–6.

25. Korea Ministry of Land, Infrastructure and Transport. *A Study on the Calculation of Carbon Emissions in Each Facilities*; Korea Ministry of Land, Infrastructure and Transport: Sejong, Korea, 2011; pp. 288–319.

26. Korea Ministry of Land, Infrastructure and Transport. *Energy Saving Designing Standard*; Korea Ministry of Land, Infrastructure and Transport: Sejong, Korea, 2013.

27. U.S. Department of Transportation. *Transportation's Role in Reducing U.S. Greenhouse Gas Emissions, Report to Congress*; U.S. Department of Transportation: Washington, DC, USA, 2012; p. 2.

28. Ortiz, O.; Castells, F.; Sonnemann, G. Sustainability in the construction industry: A review of recent developments based on LCA. *Constr. Build. Mater.* **2009**, *23*, 28–39.

29. Korea Ministry of Land, Infrastructure and Transport. *Standard Specification for Construction*; Korea Ministry of Land, Infrastructure and Transport: Sejong, Korea, 2013.

30. Flower, D.J.M. Green House Gas Emissions due to Concrete Manufacture. *Int. J. Life Cycle Assess.* **2007**, *12*, 282–288. [CrossRef]

31. Korea Ministry of Environment. *Second Basic Plan Building Construction Waste Recycling for Ecological Conservation and Resources Recycling Strengthening Infrastructure*; Korea Ministry of Environment: Gwacheon-si, Korea, 2011.

32. Denison, R.A. Environmental life-cycle comparisons of recycling, landfilling, and incineration: A review of recent studies. *Annu. Rev. Energy Environ.* **1996**, *21*, 191–237. [CrossRef]

Article

Bio-Inspired Sustainability Assessment for Building Product Development—Concept and Case Study

Rafael Horn [1,*], Hanaa Dahy [2,3], Johannes Gantner [1], Olga Speck [4] and Philip Leistner [5]

1. Department of Life Cycle Engineering, Fraunhofer Institute for Building Physics IBP, 70563 Stuttgart, Germany; johannes.gantner@ibp.fraunhofer.de
2. BioMat Department: Bio-Based Materials and Materials Cycles in Architecture, Institute of Building Structures and Structural Design, University of Stuttgart, 70174 Stuttgart, Germany; h.dahy@itke.uni-stuttgart.de
3. Faculty of Engineering, Department of Architecture (FEDA), Ain Shams University, 11517 Cairo, Egypt
4. Plant Biomechanics Group, Botanic Garden, University of Freiburg, 79104 Freiburg, Germany; olga.speck@biologie.uni-freiburg.de
5. Institute for Acoustics and Building Physics, University of Stuttgart, 79569 Stuttgart, Germany; philip.leistner@iabp.uni-stuttgart.de
* Correspondence: rafael.horn@ibp.fraunhofer.de; Tel.: +49-711-970-3188

Received: 30 November 2017; Accepted: 29 December 2017; Published: 8 January 2018

Abstract: Technological advancement culminating in a globalized economy has brought tremendous improvements for mankind in manifold respects but comes at the cost of alienation from nature. Human activities nowadays are unsustainable and cause severe damage especially in terms of global depletion and destabilization of natural systems but also harm its own social resources. In this paper, a sustainability assessment method is developed based on a bio-inspired sustainability framework that has been developed in the project TRR 141-C01 "The biomimetic promise". It is aims at regaining the advantages of societal embeddedness in its environment through biological inspiration. The method is developed using a structured approach including requirement specification, description of the inventory models on bio-inspiration and sustainability assessment, creation of a bio-inspired sustainability assessment model and its validation. It is defined as an accompanying assessment for decision support, using a six-fold two-dimensional structure of social, economic and environmental functions and burdens. The method is applied and validated in 6 projects of TRR 141 and its applicability is exemplarily shown by the assessment of "Bio-flexi", a biobased and biodegradable natural fiber reinforced plastic composite for indoor cladding applications. Based on the findings of the application the assessment method itself is proposed to be advanced towards an adaptive structure and a consequent outlook is provided.

Keywords: bio-inspiration; sustainability assessment; function; resource; burden; Design for sustainability; life cycle thinking; bio-flexi

---

1. Introduction

The technological and cultural development of mankind includes rapid-growing interconnectivity of markets as a result, culminating in a globalized economy is the main driver for the tremendous improvements that mankind is profiting from in manifold respects. It enabled certain autonomy from the dependence on natural cycles and dealt as one of the main drivers towards what is known as civilization [1]. The Neolithic revolution can be identified as the first step of this emancipation process, offering a way to decouple availability and demand for food in time. Coming along with innovative transportation methods and economic development trade was established which reduced the spatial dependency. Although the resulting human activities already had strong impacts on their environments, these remained mainly local and did not affect the global ecosystem as such [2]. With the discovery of fossil energy, the availability of energy underwent a similar process, enabling shifting energy in time and space and thus overcoming spatial dependency. This final step of emancipation from the restrictions of natural cycles facilitated the globalized economy of the modern society. Although providing indisputable advantages in almost every sphere of life for humans, this development comes at a cost of both the alienation of humans from nature itself and of the current global depletion and destabilization of natural systems [1,3]. Regarding the results of men's worldwide activities; a point is reached where potentially irreversible impacts on systems of global relevance are likely to be threatening human life [4,5]. Thus, solutions have to be created to address these challenges in a way that allows ongoing societal prosperity.

Over the history of sustainability living nature has played a major role in its understanding and application of frameworks to strive for and oftentimes was referred to as source of inspiration [6–9]. To facilitate the transfer of advantageous aspects from biology to technology, the functions of biological systems deal as fundamental basis for the assessment system structure. Therefore, a methodological approach to assess innovations has been developed through conflating bio-inspiration and sustainability based on the reintegration of basic bio-inspired principles into material systems of humankind. Its goal is to effectively develop sustainable construction products, which requires an adaptive assessment system to accompany product development. This is achieved through the abstraction of basic principles of biological systems on artificial systems and the deriving of a set of indicators and according weighting based on these transposed rules. While the conceptual framework has already been described, its concretization and exemplary application is subject to this work [10,11]. The framework has been put into practice by applying a structured approach to design and specify the elements of the sustainability assessment including their interrelation on a quantitative basis. This assessment system and its development are the core of the publication.

2. Development of the Assessment System

The proposed assessment system has been developed following a structured approach. It grounds on the bio-inspired sustainability framework developed and is applied in the Collaborative Research Centre "TRR141: Biological Design and Integrative Structures—Analysis, Simulation and Implementation in Architecture", funded by the German Research Foundation DFG [10]. The framework is inspired by one of the basic principles of living systems, namely the autopoietic model describing self-maintaining through the fulfilling of elementary functions using available resources [12,13]. As the area of application is restricted to the built environment and mainly deals with the development of innovative, bio-derived products, the assessment has been designed for this field and its applicability is restricted to it.

The development of an assessment system usually consists of a consolidation of existing and specifically developed fragments through a scientific approach including a validation of the assessment framework and its underlying calculation scheme. The central underlying existing frameworks applied in this context are bio-inspiration and life cycle based sustainability assessment, coming together through the scientific process of biomimetics. Within the Collaborative Research Center (TRR 141) using an interdisciplinary team of experts is jointly working on the development of bio-inspired innovations for the construction sector aiming among others at sustainability of the solutions. As this requirement is an integral part of the TRR 141, it proves an ideal development and application environment for a Bio-inspired Sustainability Assessment (BiSA) model. The method development is conducted based on the following steps:

I.   Requirement specification
II.   Initial situation

    a.   Sustainability
    b.   Bio-inspiration

III. Synthesis creation
IV. Application
V. Validation
VI. Adaption

The requirements (I) to the accompanying assessment of bio-inspired product development are defined in Section 2.1. This includes both general requirements for assessment systems and specific requirements concerning sustainability, bio-inspiration and decision support in product development. In Section 2.2 (IIa, sustainability) and Section 2.3 (IIb, biology) the fundamental frameworks are presented focusing on the adaptions to the state of the art that are required and the specifically developed schemes. The BiSA system derived as synthesis from the underlying basic concepts is described in Section 2.4 (III) and applied to a case study in Section 3. Based on this application (IV) the compliance to the requirements are discussed in Section 4 (V) and recommendations (VI) are presented in Section 5, giving an insight in the planned adaption and improvement of the assessment.

2.1. Requirements to a Bio-Inspired Sustainability Assessment

The development of a comprehensive sustainability assessment model for a targeted development of sustainable products underlies certain requirements in terms of methodology and applicability. The general requirements refer to basic principles for scientific methods such as consistency, comparability, reproducibility and falsifiability [14]. Besides the general requirements that apply for all assessment systems, specific context-related requirements have to be considered for evaluating the assessment system. These are related to decision support systems, sustainability assessment systems and bio-inspired systems [11].

Although mainly focused on the management of decision making processes, there have been several approaches to specify requirements for decision support systems (DSS) in decision theory that provide universal requirements [15–17]. Their common denominator is the emphasis of an adaptive and flexible applicability of DSS. A DSS should therefore also be capable of supporting semi-structured and unstructured decisions, for all levels of decision makers, regardless of their proficiency and throughout all phases of the decision making process [17]. For sustainability assessment systems, the systemic framework shown in Figure 1 is applied [18,19]. It provides a semi-quantitative scale of seven criteria covering the most relevant aspects that are prevailing in scientific literature [20–23]. All criteria are staggered in three levels and providing a scorecard of the assessment system.

The classification of an assessment system as bio-inspired can be described based on the intrinsic system properties of effectivity, adaptability and resilience [10]. Effectivity is defined with regard to the required effort by the practitioner required to generate the desired information yield. As this is an aspect that can only be investigated through practitioner monitoring through application, it requires a minimum number of applied studies with integrated BiSA. The system is classified as adaptive when it offers flexibility and expandability in an indicator and weighting scheme and realizes this through ongoing self-evaluation and adaption. If the assessment model is able to absorb changes in input in terms of reasonable system deflections, it is regarded as being resilient [24].

2.2. Properties of Sustainability Assessment Methods

Sustainability can foremost be understood as societal paradigm and its perception as such has potentially a high influence on almost every decision that is taken, starting from everyday decisions up to global politics. Its constructive ambiguity together with its level of abstraction as well as the complexity of cause-action-relation, interconnectivity and multidimensionality are leading to the point that the meaning of the term is commonly changed and shaped due to subjective perceptions [25]. Although sustainability nowadays shows ubiquitous appearance there is still a lack of consensus when it comes to defining detailed concepts going beyond the overall agreement shaped in "Our common future" [26]. The sustainability development goals (SDGs) can be seen as a milestone in the effort of consensus finding but still does not offer a comprehensive and quantifiable catalogue of indicators capable of assessing the sustainability especially when it comes to dedicatedly developing sustainable products. Overall the multitude of concepts, interpretations and respective methods and models to assess sustainability gives a hint that the paradigm of sustainability is still evolving and its shape is still to be found [27].

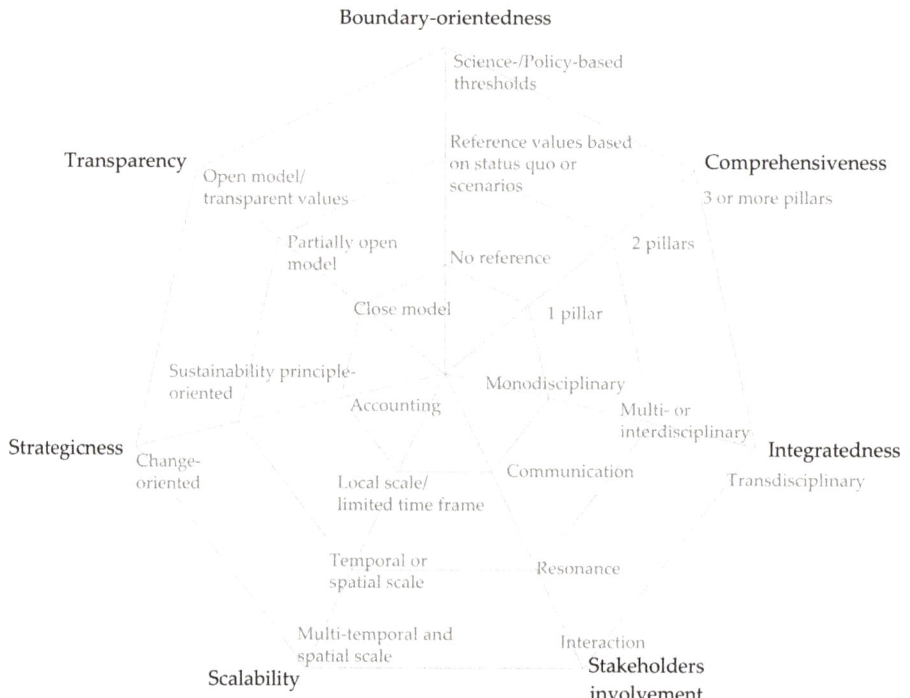

Figure 1. Requirements depicted as spider chart to assess the capability of sustainability assessment methods to address sustainability (adapted from Sala et al. (2015), figure licensed under CC BY NC ND) [19].

In the following, concepts and assessment methods are presented focusing on their consistency and suitability to a quantified assessment of products over their life cycle. While there are numerous concepts available, most are originally restricted to a schematic level and have to be transferred and differentiated to fully apply quantified life cycle thinking and thus provide comparable and specific results on a level that facilitates detailed decision support [28]. The concept of cradle to cradle, for example, is presented as a design framework for sustainable products but offers several inconsistencies when combined with quantified life cycle thinking [29,30]. The same does apply to the concept of natural capitalism, which only monetarizes all environmental resources and is therefore a method for single point creation in Life Cycle Assessment (LCA) than a sustainability concept. If enhanced by human and man-made capital as applied in the triple-bottom-line LCA, all the pillars can be addressed, while still their interpretation is restricted to monetary quantities [31,32].

There are numerous approaches available which are neither consistent nor comparable among each other on a quantitative basis. As the investigation of existing sustainability assessments has been extensively conducted by several recent publications, these are chosen as a basis for the assessment of the research situation [33–35]. Guinée has provided an extensive meta-assessment of Life Cycle

Sustainability Assessment (LCSA) studies in scientific literature including a comprehensive list of general recommendations and potential improvements that are lacking for existing studies and should be tackled in the LCSA assessment development [35]. Among others, the following key points have been stated: general need for data and methods, especially for social indicators; communication of results; integration of beneficial aspects; avoiding of double counting and inconsistent application. As these points are still unsolved, there is clear evidence for a demand of further methodological development for improved sustainability assessments, dealing with these issues and thus improving both broadness and depth as well as communication [35].

2.3. Biological Idea Generators for Sustainability Assessment

Particularly because the described sustainability assessment is inspired by biology and tailored to the construction sector, different concepts of learning from nature are presented and illustrated by means of selected examples from the building sector. Learning from nature is linked with the hope of learning from biological solutions that seem to be optimized in the evolutionary process over the last 3.8 billion years. In principle, three levels of learning from nature can be distinguished: (i) learning from the results, (ii) the processes and (iii) the principles of biological evolution [36]. These three levels have a common systematic approach of knowledge transfer but differ in the type of the transferred knowledge.

The first level of learning from living nature is the study of the form-function relationships of biological role models. Taking into account that even the transfer of an inspiring idea is a conscious process, the transferred inspiration leads to a bio-inspired product. A famous example is the plant-inspired reinforced concrete developed by the French gardener Joseph Monier in 1867 [37]. Based on an inspiration, additional knowledge transfer is possible, such as the transfer of morphology leading to a biomorphic product such as the Crystal Palace, a cast-iron construction built by the gardener Sir Joseph Paxton being inspired by the ribbed leaves of water lilies [38] and the transfer of a functional principle resulting in a biomimetic product [11]. Special attention should be paid to the transfer of a function or in other words the statement that the biological role model and the technical product possess the same function as for example the self-cleaning surfaces of lotus leaves and the façade paint Lotusan® or different functions such as the façade shading system Flectofin® inspired by the pollination mechanisms of the bird-of-paradise flower [39,40]. The meaning of function is thereby different whether used in the field of biology or technology. Biological functions are understood in the sense of traits evolved to increase the organism's fitness and contribute to the evolutionary success [41]. In contrast, technical functions are defined in the sense of a specific process, action or task [42]. Examples for the second level of learning from evolutionary processes are the optimization algorithms based on growth rules of trees (Computer Aided Optimization) and bones (Soft Kill Option) and the evolutionary algorithms, which lead to biomimetically optimized products [11]. The third level of learning from nature is based on the principles of biological evolution such as multifunctionality, hierarchy, robustness (fault tolerance), resilience (failure tolerance), redundancy, self-X-functions, adaptation, consistency, modularity, sudden transitions (i.e., leaf drop), gradual transitions, growth, opportunism, metabolism under mild environmental conditions (enzymes) [43] and resource efficiency [8]. In ecology, the term "resources" refers to essential environmental factors that can be subdivided into biotic (e.g., food, host, reproductive partners) and abiotic factors (e.g., space, light, water) [44].

In summary, it can be said that despite the inspiratory flow and knowledge transfer from biology to technology, bio-inspired products are not necessarily sustainable as a side effect. The challenge is that there is no biological model and no method for a straightforward transfer into any model of the paradigm of sustainability. This is due to the fact that living nature itself as a result of biological evolution cannot be comprehensively described through the concept of sustainability. It is a man-made teleological and anthropocentric paradigm with the goal of preserving the status quo for the next generations [45,46]. The paradigm of sustainability is of teleological nature and therefore to be distinguished from biological systems, where teleology is seen as an insufficient concept to describe reproduction and evolution [47]. In contrast, biological evolution is seen a blind process characterized by the dynamics of evolutionary adaptations on basis of mutation, recombination and selection in an ever-changing environment with the result of multifunctional and optimized structures or processes after several generations [41]. On the one hand, the concept of teleology is a useful element of explaining adaption, when using the goal-directedness to explain the composition and processes of systems [47]. On the other hand, a teleological approach can put us on the wrong foot as explained in the review "If bone is the answer, then what is the question?" describing the increasing understanding of adaptive bone architecture over time [48]. Thus, the principles that facilitate adaption, especially the principles of biological evolution may serve as idea generators and may have great potential to contribute to sustainable solutions, precisely because the challenging situation that the stable preservation of certain ecological systems requires constant changes. However, this proposed transferability cannot be seen as an automatic transfer and has to be investigated thoroughly. Furthermore, what is called the social pillar of sustainability has no counterpart in biological systems and no direct conclusions concerning social aspects may be drawn from nature.

Although nature does not bear a fully comprehensible set of role models for a bio-derived understanding of sustainability, it undoubtedly is a great source of inspiration in terms of multiple aspects. The dynamic adaptation and the efficient utilization of locally and currently available resources but especially the fact that biological systems have been optimized in the course of evolution are fundamentals that qualify biological systems as role models for innovation. This mainly bears the potential for environmentally optimized solutions and offers economic potentials as well as these are related when it comes to efficiency. If one looks at the interaction between sustainability and biology from the perspective of the assessment of sustainable development, the question arises as to what commonalities this can be built on. Although the differences are also reflected in the different definitions of function and resource in biology and technology, the ratio between function and resource seems very promising.

2.4. Bio-Inspired Sustainability Assessment

One fundamental question when it comes to deriving solutions from biological systems is if living nature actually does provide a fully comprehensive counterpart to what is described as sustainability. The underlying proposition is that if nature is chosen as direct role model, its solutions should have been created considering the same framework conditions that are applied for sustainability assessments. If not, any direct transfer from nature cannot be stated as to create sustainable solutions by itself and the overall concept of sustainability has to be accepted as to be at least partially independent from our understanding of nature and thus artificial. To derive a robust answer to this question, both the concept of sustainability and the fundamental principles of biological systems have to be investigated. However, there are two main inconsistencies when trying to directly derive biological systems to sustainability metrics. First, the prevailing sustainability concept is explicitly anthropocentric and thus does not relate to the nature of biological systems. This becomes apparent by several intrinsic properties of sustainability paradigms such as the explicit focus on mankind in the UN but also when applied as assessment framework [21,26,49]. The second inconsistency arises from the concept of social sustainability that is still under discussion [50,51].

With regard to these considerations sustainability as bio-inspired concept is defined through the interdependence of system functions and the therefore required depletion of resources. A system is defined as sustainable, when a specific set of functions is fulfilled while simultaneously ensuring the ongoing availability of resources in time. The concept depicted in Figure 2 shows the predominant transformation direction of the prevailing economic metabolism, which is to create social functionality by depleting environmental resources driven by economic facilities. The graph is meant to show the dynamics of this metabolism indicating a sustainable system when its shape is kept stable, ensuring an ongoing provision of resources for an ongoing creation of function. Economy is interpreted as means to an end transforming resources into functions, enabling business models and thus facilitating the application of new products. Besides this main flow direction many processes are motivated otherwise and a general rule cannot be derived. The concept deals as a template to depict mechanisms of actions in terms of the fulfilment of functions including its intended and unintended effects. Nevertheless, it is quite uncommon for processes to dedicatedly create an environmental function or not utilize the environmental resources in a depleting manner. Furthermore, the predominant role of Economy is depicted as connecting element between Society and Environment. To keep this metabolism sustaining it is crucial not to exploit the resources to an extent that prohibits the ongoing fulfilment of societal functions. A sustainable system according to this scheme is achieved if the dynamic societal metabolism is maintained and is kept stable under dynamic conditions. It enhances the existing models through the integration of positive aspects and offers a shell like structure that is able to integrate different assessment methods.

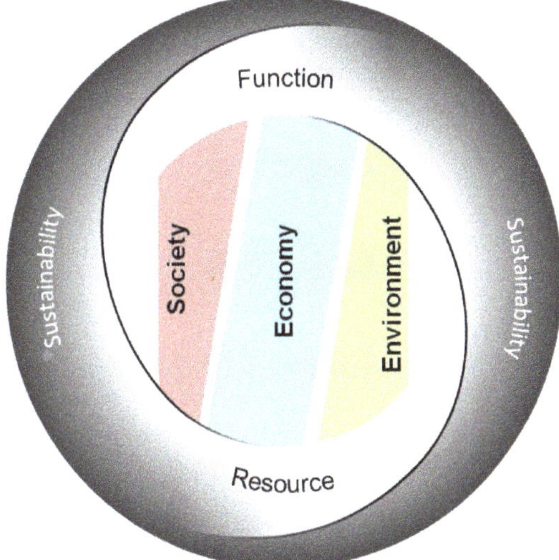

Figure 2. Bio-inspired Sustainability Assessment depicted as conceptual structure, showing the three dimensions called society, economy and environment and the two aspects, namely function and resource as integral parts of the assessment. Societal functions are fulfilled through the transformation of environmental, economic and societal resources (own figure).

In the following, a quantifiable BiSA model is presented based on a six-fold structure including the three dimensions of sustainability for both intended and unintended aspects called functions and burdens:

- Environmental burden: the unintended impacts of the assessed system to the environment based on an environmental life cycle assessment
- Environmental function: the dedicated design functions aiming at positively impacting the environment, calculated based on environmental life cycle assessment
- Economic burden: the life cycle related costs calculated based on the life cycle assessment model completed through process immanent costs and nonmaterial costs
- Economic function: the economic function from the point of view of the shareholders, calculated as economic profitability
- Social burden: the unintended effects of the assessed system on human society, calculated as impact on human capabilities and health, calculated based on environmental life cycle assessment

- Social function: the primary design function restricted to the intended building physical function of the assessed system

The aspects are focused on the development of bio-inspired and bio-based products in the construction sector but are not restricted to these. The assessment aspects are chosen in terms of consistency and applicability with the main intension to provide feedback on decisions during product development. Its bio-inspiration lies primarily in its intrinsic setup inspired by biological systems and the overall structure of resource-function-relationships. The underlying physical model is created as a life cycle inventory system based on the GaBi database and supplemented by economic and country related information on process level. This model provides a consistent quantitative basis for 5 of 6 aspects. Moreover, it is applicable for both early and advanced development phases as it provides generic data but can include specific primary data as well. This facilitates a flexible structure and level of detail allowing the specification of the system as precisely as possible while still being able to estimate the coarsely defined aspects. The model has been implemented as semi-automatic tool to provide feedback for specific questions within the embedding project but has not been created as software for automatic application yet. However, an increased level of automation is envisaged for the next project phase.

2.4.1. Environmental Burden

The depletion of natural resources is the main source of human prosperity and as such of central relevance for the assessment of bio-inspired sustainability. Building upon the treatment of resources in biological systems, similarities can be identified mainly in terms of the dependency from physical sources. As mentioned before, all ecosystems are dependent on biotic (living) and abiotic (nonliving) environmental factors. In the course of the earth's history it has been shown, that especially after mass extinction source–sink dynamics influence the variation in habitat quality affecting biodiversity, population growth and number of organisms [52]. Even though it is repeatedly claimed that nature does not produce any waste, this is not the case on closer inspection. For example, most of the crude oil produced today originates from dead marine organisms, buried underneath sedimentary rocks. The deposits are therefore nothing more than landfills for fossilized organic materials or in other words natural waste. Thus, the use of fossil fuels such as coal, natural gas and crude oil hydrate is associated with respective CO_2 emissions.

The environmental burden is calculated by LCA, using a single point value based on a selection of characterization methods. The steps that are to be performed when conducting an LCA according to the pertinent standards cannot be fully applied due to the interactive nature of the BiSA assessment [53–55]. Nevertheless, the functional unit, the system boundaries and consistent specifications of the applied calculation principles such as allocation and cut-off criteria have to be stated. Furthermore, the life cycle inventory models for each of the assessed systems and variants have to be created. This model was created using the Software GaBi 8.2 (thinkstep, Leinfelden-Echterdingen, Germany), which is one of the world's leading LCA software providers and the GaBi SP 34 (thinkstep, Leinfelden-Echterdingen, Germany) database, providing more than 10,000 environmental profiles as a modeling basis [56]. For the assessment of the environmental burden it is possible to directly derive indicators and weighting schemes from the investigation of natural systems. This is, on a quite abstract level, the transfer of biophysical system stability as role model on global scale. A quantifiable concept to address the issue of global biophysical system stability was introduced by Rockström et al. in 2009 and has been since then further refined and continuously updated [5,57]. It identifies the main biophysical systems that are threatened by human activities and provides a framework to quantify planetary boundaries that should not be exceeded by mankind on global scale if the global ecosphere is to be kept intact. The concept provides an approach to address the manifold depletion of nature by man and is subject to the ongoing development to include new insights of scientific discourse. The planetary boundaries have been transferred to deal as life cycle assessment weighting scheme by several authors [58–62]. The approach proposed by Sala et al. is chosen and adapted using the presented values for distance-to-target normalization due to the fact that a single value for the environmental burden is required [59,63]. In addition to this approach, the areas of protection are differentiated in sink and source related categories. Sink related categories are summarized as global biophysical system stability and correlate directly to the biological system they depict. As the ongoing availability of both biotic and abiotic resources is crucial for the metabolism stated above, the depletion of resources is classified in the area of global resource stock. Strictly seen the availability of resources for mankind is mainly underground and does not directly contribute to the stability of the biophysical system but is crucial for the concept of scarce resource utilization prevailing in biological systems. Due to significant methodological improvements since the publication of the distance to target values, the abiotic depletion potential is covered by the anthropogenic stock extended abiotic depletion potential (AADP) model and for land use the biotic production indicator as published in LANCA 2.0 (Fraunhofer IBP, Stuttgart, Germany) is applied, which is freely available in the updated version [64–67]. Nevertheless, the assessment of impacts on the global resource stock still bears strong potentials for improvement, especially in terms of temporal and spatial differentiation. Table 1 shows the considered categories and the according normalization factors as well as the chosen methods to quantify the impacts of each category. The results of each impact category over the whole life cycle are multiplied with the normalization factor based on the planetary boundary concept. The normalized values are then added, creating a single point that can be directly compared to the one of the reference system, which is created similarly.

Table 1. Categories and weighting structure for the assessment of the environmental burden derived from [59].

Area of Protection	Impact Category	Abbreviation	Impact Assessment Model	Normalization
	Climate change	GWP	IPCC	4.81×10^{13}
	Ozone depletion potential	ODP	CML	1.34×10^{8}
	Photochemical ozone formation	POF	ReCiPe	2.80×10^{11}
Global biophysical system stability	Freshwater eutrophication	EUTF	ReCiPe	1.76×10^{10}
	Marine eutrophication	EUTM	ReCiPe	1.95×10^{11}
	Freshwater ecotoxicity	FRTOX	UseTox	4.46×10^{12}
	Acidification	AC	TRACI	3.83×10^{11}
	Terrestrial eutrophication	EUTT	TRACI	1.22×10^{12}
	Land use	LU	LANCA	1.00×10^{15}
	Water depletion	WD	WSI	4.81×10^{13}
Global resource stock	Resource depletion	AADP	AADP	3.70×10^{9}
	Biodiversity depletion	-		-

2.4.2. Environmental Function

As defined in the framework, functions are considered to be intended properties of the assessed system. For environmental functions this requires an explicitly specified positive impact on the environment. A very well-known biological example is mutualism, a relationship between different species with positive impact insofar as that both individuals benefit. The biological barter can be a resource-resource-relationship (e.g., mycorrhizal associations between plant roots and fungi) or a service-resource relationship (e.g., birds disperse plant seeds of fleshy fruits that they have eaten before) or a service-service-relationship (e.g., sea anemones and anemone fishes protect each other from their respective predators). In addition to these mutual relationships, a large number of closed material cycles (e.g., carbon cycle, nitrogen cycle, sulphur cycle and phosphorus cycle) are known, in which the starting material is finally available again through periodic transformation of chemical compounds.

Most technical systems do not have a specific environmental function and are consequently not assessed for this aspect. However, for systems which specifically intend to improve the accessibility of their wastes, such as waste treatment systems, or for systems that intend to transform their wastes to be suitable for others in a mutual form, the environmental function can be assessed and quantified. As the same system is investigated, the same assessment structure as for environmental burden is applied and defined as positive intended impacts. Life cycle related indirect effects such as recyclability in general are considered for burden assessment, as they are not regarded as intended design function [42].

2.4.3. Economic Burden

In contrast to environmental burdens, the economic ones are directly related to the success of a product meeting a demand (or providing a function) under ideal market conditions. In real markets, this is interleaved with numerous adaptations such as subsidies, taxes, cross funding, market distortions and many more. However, the economic burdens and its underlying structure are a key aspect of any artificial system. They depict the resource demand of the product scaled by the scarcity of its constituting elements in the actual economic conditions. Scarcity is created as an artificial value based on availability and demand. It is modified by the adaptations stated above and is represented by the price of each element, which deals as intermediary and as such facilitates flexible handling of resources. Even though most nonhuman biological systems do not have an intermediating currency its resource demand is also strongly affected by the scarcity of the constituting elements such as water, solar energy, space or trace minerals.

The economically expressed scarcity in terms of a monetary cost structure determines the aspect of economic burden. It depicts the restrictions and framework conditions that are imposed through the embeddedness of the product in the economic system. While the relevance of each constituting element is different to the environmental burden contribution, the system that is taken into regard is ought to be consistent. The method of Life Cycle Costing (LCC) provides such a consistent framework and can be regarded as consolidated in sustainability science [68]. However, it does imply several fundamental differences due to time relatedness and the consideration of nonmaterial elements and is therefore applied in a multistage adaptation, starting with process immanent costs (PIC) and including further nonmaterial information (FNI) if available. Table 2 depicts the categories that are differentiated in the economic burden assessment. As they all are assessed in monetary values, they can be added without weighting. The assessment of the different cost categories is an integral part of the enhanced Life Cycle Inventory (LCI) model, that has been created through coupling the GaBi database with statistical cost data through the mapping of flows to sectors based on European statistics [56,69]. Thus, the cost structure can automatically be derived for the complete LCI model, providing the same system boundaries and level of detail than for the environmental assessment.

Table 2. Categories and weighting structure for the assessment of the economic burden.

Area of Protection	Impact Category	Impact Assessment Model
Variable Costs	Resource costs	PIC
	Electricity costs	PIC
	Other energy costs	PIC
	Labor costs	PIC
	Machine costs	FNI
	Disposal costs	FNI
Fixed costs	Process related investments	FNI
	Infrastructure costs	FNI

In contrast to most Life Cycle Costing models, the model does not include time-related price change or discount rates, as this is not yet possible in the enhanced LCI model. This however ensures consistency with the environmental burden aspect, as there are no discounting issues considered as well. However, this simplification comes along with several drawbacks in comparability to other economic analyses including potential communication issues.

2.4.4. Economic Function

Economic viability is a crucial prerequisite for any technical system and is usually defined as feasibility in terms of a business model or economic product life cycle. While it is oftentimes not the ultimate purpose, economic viability is an essential means to function fulfilment, which is why the economic function is regarded as separate aspect.

Even though no direct analogy can be drawn between nonhuman biological systems and economic viability, its basic function offers some similarities between evolutionary fitness and economic success [7]. The behavior of companies is to some extent comparable to evolutionary mechanisms of selection and niche occupation, as indicated through analogies in terminology [70]. The two main areas of protection are derived from this analogy and are defined as profitability (efficiency) and competitiveness (fitness). While these aspects are addressed in detail in business practices, they are oftentimes not regarded within product development, especially in early development phases. Therefore, a basic system shown in Table 3 is proposed based on the framework presented above, combining PIC and FNI. On the basic level, the material and energy cost optimum is related to a potential market price based on existing competing products. These values are then complemented through further nonmaterial information on labor costs, investment goods and detailed information on the market situation including potential market prices as well as a potential product price and market volume estimation.

Table 3. Categories and weighting structure for the assessment of the economic function.

Area of Protection	Impact Category	Impact Assessment Model
Profitability	Production costs	PIC
	Current market price	FNI
	Cost reduction potential	PIC + FNI
Competitiveness	Current market price	FNI
	Potential market price	FNI

2.4.5. Social Burden

In contrast to economic and environmental aspects, the definition of social burdens cannot be derived from a state of the art assessment methodology. The main difference can be identified in the ambiguity of social impacts in terms of goal and scope and the fact that the physical quantification for most categories is not applicable due to its immaterial nature [71]. While for social life cycle assessment no method has prevailed yet, a guideline providing general recommendations was published by the United Nations Environment Programme [50,51,72]. Furthermore, Sureau et al. have investigated 14 different frameworks for social LCA and identified high diversity between the approaches as well as substantial demand for further development [49]. From an epistemological point of view, the underlying scientific paradigm is differing between post-positivism- and interpretivism-oriented approaches [73]. While interpretivism-based approaches are aggregating impacts that are mainly chosen case specific with regard to stakeholder groups, interpretivism-oriented approaches are developed in analogy to environmental LCA, trying to provide quantifiable, generally valid impact pathways to be applied to life cycle system models [74]. While the thereby developed methods differ in their approaches to choose and define indicators and assess its inherent impacts, their common denominator is the identification of effects of a system's life cycle with regard to social aspects based on explicitly or implicitly chosen frameworks.

Looking at nonhuman biological systems, ethics, morality and altruism are concepts that do not seem to be relevant in most biological (non human) systems, even though exceptions are known [75,76].

Though still subject to scientific controversy, morality and altruism and the consequent concept of ethics may be an integral part of the evolutionary success of humanity and are therefore key to its understanding and its ongoing success [77,78]. The assessment of social burdens is therefore interpreted as the depletion of societal resources, which are defined as human health and human capabilities. These are the basic prerequisites for humans to live a self-determined life and for society to prosper [79]. Human health is assessed through the LCA model, applying the same distance to target normalization approach as for environmental burdens for the human health related impact categories (Table 4) [59]. Again, the proposed impact categories are replaced by updated version if available, which in the case of human health applies to the USETOX model [80]. The concept of human capabilities to address social impacts has been developed based on the capability approach as framed by Sen [79,81]. It is applied in a hybrid approach using the Social Hotspots Database (SHDB) for quantification of the impact categories based on risk levels, which are available on country and sector level [82]. The risks are linked with the product system using a bottom-up approach to assign working time to each unit process of the model based on statistical data according to an updated and extended version of the Life Cycle Working Environment (LCWE) method. The chosen method can be characterized as environmental LCI database method according to the differentiation provided by Chhipi-Shrestha et al. (2014) [73]. The model is integrated in the GaBi software and thus applicable using the enhanced LCI model, providing working time in seconds for both aggregated and unit processes. While the method is still under development, it has already successfully been applied in several research projects [83–85]. The indicators are each calculated as both total working seconds under high and very high risk of each category and its share related to the overall working time. The indicators have been chosen based on the framework introduced by Reitinger et al., 2011, which adds fairness to the categories introduced by Sen [79,86]. To prevent double counting, the aspect of health and safety is covered by the LCA impact assessment and not included in the capability assessment.

Table 4. Categories and weighting structure for the assessment of the social burden.

Area of Protection	Impact Category	Impact Assessment Model	Considered Categories/Normalization Values
Human capabilities	Political freedoms	SHDB	Freedom of Association, Collective Bargaining, and Right to Strike
	Economic facilities	SHDB	Wage Assessment; Poverty; Labor Laws
	Social opportunities	SHDB	Children Out of School; Child Labor; Working Time; Forced Labor
	Transparency guarantees	SHDB	High Conflict Zones; Legal System; Corruption
	Protective Security	SHDB	Access to Improved Sanitation; Access to Hospital Beds; Access to Improved Drinking Water
	Fairness	SHDB	Gender Equity; Migrant Workers; Indigenous Rights
Human health	Human toxicity, cancer effects	USETOX (V 2.01)	9.16×10^4
	Human toxicity, non-cancer effects	USETOX (V 2.01)	1.13×10^6
	Particulate matter/Respiratory inorganics	USETOX (V 2.01)	6.86×10^{10}
	Ionizing radiation	Human Health effect model (V1.09)	2.04×10^{12}

2.4.6. Social Function

The function of a technical system can ultimately be defined as to serve a specific, desired purpose. While these functions can be distinguished between design functions, use functions and service functions only design function can be considered for product development as this is the function that was designed as a means of achieving its end [42]. It is therefore a crucial aspect to be considered when designing or developing products to address the specific function or functions.

In biological systems, the concept of function is defined differently and mainly applied with regard to evolution and fitness [41,87]. Nevertheless, it is possible to investigate and quantify physical properties of biological systems with regard to their technical design function. Especially in biomimetic science, the identification of these functional principles is key to successfully transfer its core working principles to technical solutions [88]. To define these technical design functions with focus on the area of application of this model, the building physical functions specified by Moro are applied as design functions (see Table 5) [89]. As most products are only focusing on one or a few of these functions, the use of a generalizable quantification system for all categories was not applied. The presented impact categories and the according building physical functions shall be used as catalog to choose the design function including multifunctional properties, each of which then should be assessed individually based on the according building physical properties. When a product is developed as load bearing element, for instance, its design has to be chosen to bear not less load than the reference system and its dimensioning has to be chosen accordingly.

Table 5. Categories and weighting structure for the assessment of the social function, restricted to building physical functions [89].

Area of Function	Impact Category
Load bearing	Primary support structure Secondary support structure Tertiary support structure
Enveloping	Vapor balance control Indoor acoustics conditioning Protection against infiltration Privacy, glare and sun protection Fire protection Noise protection Illumination Natural ventilation Thermal conditioning Electricity control and supply
Supply and disposal	Water supply Lighting Cooling supply Heating supply

3. Application

The proof of general compliance to the requirements that have been identified as relevant to qualify the BiSA requires ongoing application. As the assessment system is developed as part of an ongoing project, its ongoing application is ensured through the ongoing project integration. However, to depict the basic functionality of the assessment the biobased and biodegradable composite Bio-flexi will be presented in the following. This specific exemplary application is chosen and presented including technology introduction, technical description, assessment results and discussion.

3.1. Bio-Flexi—A Biobased and Biodegradable Composite

The Bio-flexi innovation is a biocomposite fiberboard manufactured from annually generated agricultural residues' fibers in the form of a flexible high-density fiberboard. The raw agro-fibers were bonded till 90% of mass load, without pre-chemical modification, by a thermoplastic elastic binder (TPE) using classic plastic-industry machinery [90]. The fibers applied were chosen from the agricultural residues stream, namely straw, which is the cheapest non-wood lignocellulosic natural fiber abundantly available worldwide from the cereal crops by-products' streams, mainly out of wheat, rice and maize. As the fibers are not edible, the discussion on competition with food is not relevant here. Other than availability, these lignocellulosic fiber types were applied to actively replace mineral-based flame retardants in plastics, depending on the high natural silica contents naturally present in their chemical composition reaching up to 20% of mass load in certain types as in the case with rice straw. The selected TPE binder was purposely chosen to be biodegradable under industrial compost conditions to give an opportunity to have multiple end-of life options after the end of the useful life time of the developed biocomposite material. These combined parameters were applied to increase the positive environmental feedback of this development during and after the useful life time of application in the building industry. It has at least two end-of-life options as it can be recycled to a number of recycling cycles then industrially composted if further recycling cycles would not be feasible, which is a solution that helps in minimizing wastes' accumulation. Waste accumulation minimization is accordingly hereby achieved twice: once during the production phase, as it is mainly based on agricultural residues fibers and secondly after the end of its useful life time. These end-of-life options are rarely available in the contemporary fiberboards market worldwide, the thing that promotes a wider application of this development in the contemporary building industry replacing a wide range of non-recyclable petro-based building products.

Usage of agro-fibers with these high mass-loads (up to 90% mass load) will help replacing the slow-renewable wood and improve forestry practices. In addition to these ecologic values, the elastic nature of the developed fiberboard enables the possibility of achieving attractive free-form architectural interior designs using cheap alternatives and available production techniques. Flat horizontal applications like flooring systems as in sport halls, yoga mats and others as well as in vertical applications as partitions and interior fittings are possible. Usage of thin veneer covering layers is necessary in different applications to give a covering aesthetic feature, to offer a final reinforcing layer and to close open fiber pores to guarantee durability. The high density developed boards can be provided with minimal thicknesses starting from 1–2 mm and can be transported in the form of rolls to minimize transportation and storage costs. In Figure 3, flexibility and possible applications in both flat- and curved-morphologies are emphasized.

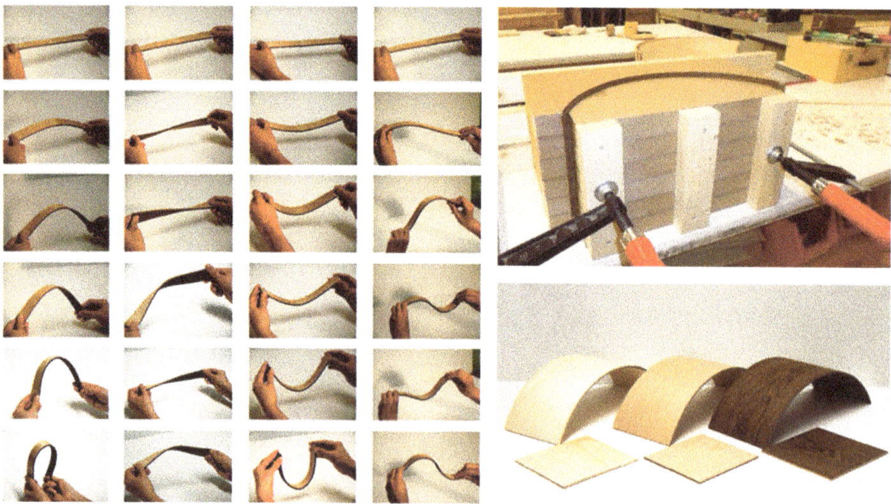

Figure 3. Illustration of the flexibility of the Bio-flexi panel when veneered from one side and how it can be veneered from both sides to fix the geometry intentionally (republished: [91]).

3.2. Technical Characterization

The Bio-flexi panel was extruded using a double-screw extrusion machine, in which four heating canals existed to control heating temperatures throughout the longitudinal mixing path of the compounded mixture. At the heating canals as well as the feeding canal, gas absorbers were integrated to absorb water vapor arising from the natural fibers, which previously captured natural atmospheric humidity to optimize the mixture and eliminate any inhomogeneity or plasticity (Figure 4).

Figure 4. Illustration of the production of Bio-flexi in mass-production scale indicating the control and feeding units in Naftex GmbH company, Wiesmoor, Germany (Photo: Dahy, H.).

The developed Bio-flexi HDF product was mechanically tested to evaluate the transportation safety without distortion and the usage possibility in flooring systems in respect to residual indentation after DIN EN 433 and indentation resistance after DIN EN 1516, to evaluate if the product can be applied in the scope of flat flooring in sport halls [92]. To validate the possibility of applying this material in flooring systems, the fiberboard was tested under static loads to measure thickness losses, as a step to measure its residual indentation and indentation resistance. Measurement of residual indentation after DIN EN 433 simulates the static furniture loads. The result indicated that Bio-flexi at 80% fiber-load by mass has a residual indentation of 0.14 mm, which is comparable with other elastic flooring materials as Linoleum that lies between 0.07–0.4 mm. To validate the resistance to indentation of elastic surfaces for sport areas, DIN EN 1516 test standards were applied to determine that the permanent change in the flooring plate thickness was only 0.02 mm that fits in the range set in this standard not exceeding 0.5 mm permanent thickness loss. This indicates that the developed fiberboard can be applied in flooring systems in sport halls, sport activity areas and in cushioning services [91].

Under the raising awareness of the environmental possible drawbacks of all newly developed building materials, thermoset matrix application was eliminated here and a thermoplastic elastomeric binder (TPE) was applied instead. Recyclability is here guaranteed without further experimental proof dependent on the thermoplasticity of the binder and the high heat-resistance of the natural fibers reaching to 220 °C depending on the flame-resistant silica loaded contents, which should enable multiple recycling cycles before the fiber deteriorates. However, in the area of Natural Fiber Reinforced

Polymer Composites (NFRP) recycling virgin thermoplastic binders of the same or another compatible base as well as virgin natural fibers are needed to be added in small ratios in each recycling cycle to guarantee preserving the same original quality of the first produced series. The composting option was otherwise experimentally proved through soil burial tests that were conducted for 15 months, where the samples were buried in a chosen field plot in the middle of the Stuttgart city in South Germany. Compostability conditions were set so that aerobic bacteria at a level of maximum −8 cm under the soil's surface were activated, to measure if or not the test samples will start decaying. Weight reduction was monitored and visual qualitative inspection took place each 3 months. By the end of the test, plant roots were observed growing within the samples' bodies and a final weight reduction of around 41% after 15 months of soil burial were measured. Through these results, it was concluded that the Bio-flexi fiberboard has also the tendency to be industrially composted as a second end-of life option in addition to its recyclability. In Figure 5, the closed cycle graph of the Bio-flexi according to the cradle to cradle® design conception is shown including the two main product circles that are impacted by the Bio-flexi life cycle.

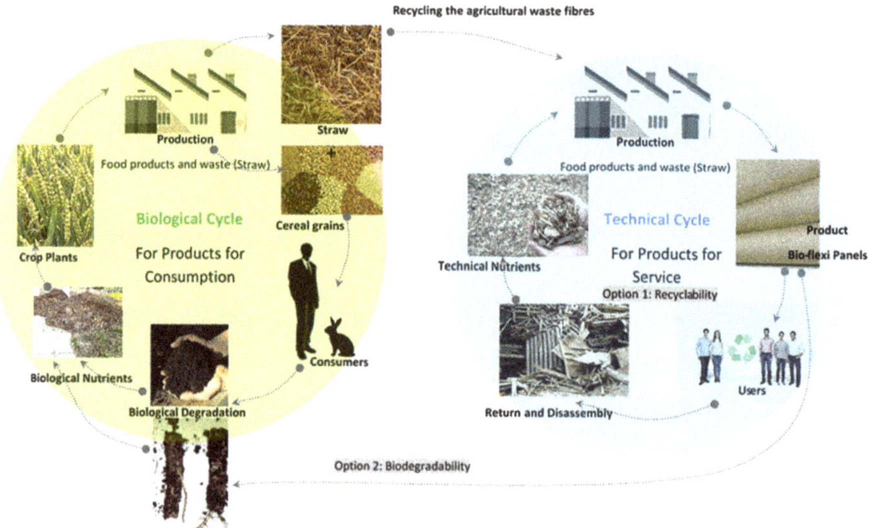

Figure 5. Graph indicating the closed proposed cycle of the Bio-flexi HDF fiberboard after the cradle to cradle® concept (Photo: Dahy, H.).

3.3. Bio-Inspired Sustainability of Bio-Flexi

The assessment of Bio-flexi is performed based on a system model that has been created in analogy to the Life Cycle Inventory model in LCA. For the modelling, the LCA software GaBi was used [56]. In analogy to LCA, the goal and scope definitions are presented in the following. Goal of the assessment is to investigate the sustainability of Bio-flexi in a sports facility flooring application compared to a conventional reference system, for which a polyurethane based flooring material is chosen. One square meter of flooring material is chosen as functional unit. As the complementing build up is assumed to be similar for both systems, the comparison is restricted to these surface layer materials, also providing the basic functions of shock absorption and cushioning in a comparable way. The service life is defined as 20 years and no difference in maintenance is considered.

The technical characterization has been transferred to a Life Cycle Inventory model using primary data provided by the manufacturers including the life cycle phases A1–A3, C3 and D [93]. As the technology provides two EoL-options, a sensitivity analysis has been conducted, resulting in two different scenarios that depict the most realistic options. Scenario one is using a maximum amount of recycling and scenario two is treated through composting. In addition to the information given in the technical characterization, several additional processes were added to complement the life cycle with regard to recycling and composting. For the recycling option a grinding process has been applied to facilitate the recirculation of recyclate to the virgin raw material stream before the extrusion process takes place. A percentage of 20% was specified as maximum recyclate rate. For the composting option, an industrial composting plant model was chosen to estimate the according material decomposition rate in real application. As coarse material is removed showing a relatively low decomposition rate due to its small surface area, a grinding process is assumed here as well. For the remaining materials thermal utilization is assumed, using a dynamic energy mix based on German lead scenarios to calculate the benefits beyond system boundary [56]. The system boundary of the two scenarios is shown in Figure 6.

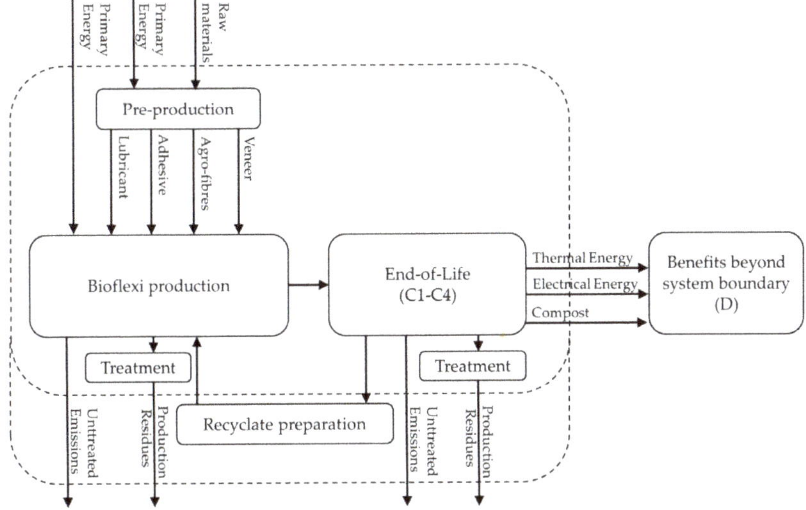

Figure 6. System boundary of the investigated life cycle inventory model. The main material and energy flows for production, End-of-Life and Benefits are shown. The recyclate preparation is only considered for the recycled scenario and therefore depicted separately.

The model was used as basis to perform a BiSA according to the assessment structure described above. As there is no decidedly specified environmental function, the environmental function is not investigated. For the economic assessment, a simplified approach is chosen due to the fact that Bio-flexi is still a product under development and a number of non-material information is not yet available. The social function in terms of building physical design functions is assumed to be comparable to the conventional reference product as indicated by divers tests [91]. In the following, the aspect environmental burden is presented in detail due to its relevance related to the motivation of the developers. Figure 7 depicts the overall environmental burden in normalized numbers, showing the overall environmental impact with regard to the planetary boundaries.

While the impact of the most relevant category for PUR is strongly reduced in both Bio-flexi scenarios, the overall impacts of all categories are almost compensating these savings for the recycled scenario and even overcompensating the savings for the composted scenario, having an increased in normalized impact by 17%. The increased impact mainly origins in the agricultural system, which especially impacts on eutrophication and acidification. Overall, the recycling option already is comparable to the reference system and offers further saving potential especially when the recycling rate can be further improved.

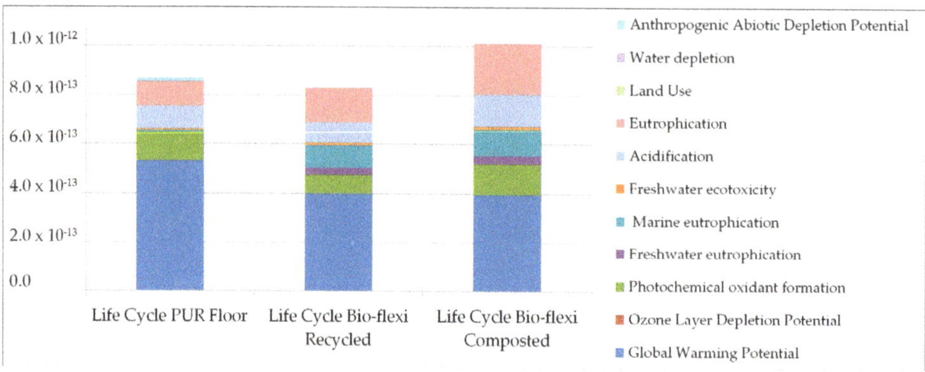

Figure 7. Environmental burden of the reference system and the two scenarios of Bio-flexi (Recycled and composted) as normalized results for the considered life cycle phases.

The overall sustainability assessment result is depicted in Figure 8. The segments are scaled in relation to the reference system using the radius as scaling element. Diagram (1) shows the recycled scenario, diagram (2) on the right hand side depicts the composted scenario. While for both economic and environmental burden only small changes can be identified, the social burden offers significant savings for both scenarios.

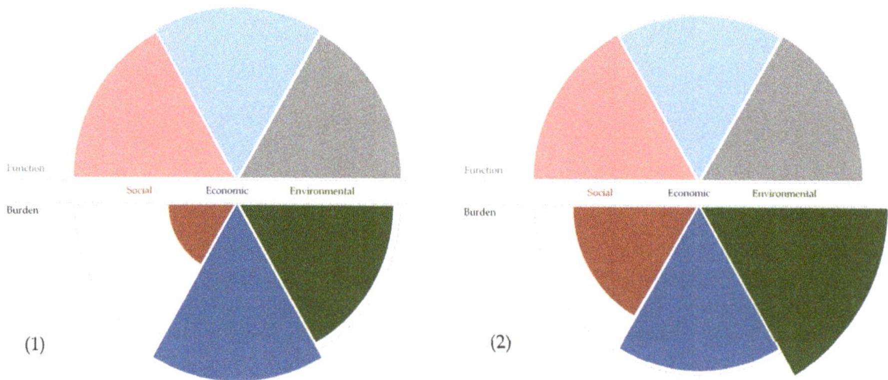

Figure 8. Bio-inspired sustainability of Bio-flexi as pie charts, including the six aspects of sustainability, each depicted by a specific color. The red elements depict the social aspects, the blue elements depict the economic aspects and the green element depicts the environmental aspects. As there is no environmental function of both reference and assessed system, the environmental function element is greyed out. Each aspect is shown in relation to the reference product, depicted by a grey circle line which is identical for both graphs. The relative value is depicted as change in pie element radius and therefore linear. (1) The recycled scenario shows Bio-flexi with a maximum of recycling compared to a conventional reference product. (2) The composted scenario shows Bio-flexi with a maximum of composting compared to a conventional reference product.

For both economic burden and function, the assessment was restricted to the process immanent cost model, where both scenarios provide a similar cost structure. The overall production costs are furthermore comparable to the conventional flooring material. For social burdens, a significant reduction of the impact on human capability has been identified. This mainly origins in the fact that the Bio-flexi production including its upstream material chain takes place mainly in Germany, while the fossil based reference product includes significant share of work with higher risk of human capability reduction mainly in the raw material extracting countries. The impact on human health does not provide significant saving potentials for the composted scenario and is increased by 51% due to the impacts occurring in the composting process. For the recycled scenario, a reduction of 25% of normalized impact in comparison to the reference system is determined. Overall, the recycled scenario offers a higher potential with regard to bio-inspired sustainability, although this does not apply to each aspect concurrently. The main improvement could be identified in the reduction of social burdens and global warming potential, while no significant change could be identified for economic aspects for both function and burden.

The biobased and biodegradable composite Bio-flexi appears to be able to compete in terms of bio-inspired sustainability with its conventional, fossil-based reference in the application as flooring system in sports facilities. In contrast to the reference, however, Bio-flexi bears several additional optimization potentials and is expected to be generally beneficial when further developed under consideration of the decision support provided by the BiSA system. Especially with regard to recyclability, improvement potentials have been identified, as the recycled Bio-flexi scenario provides significant improvements in environmental and social burden compared to both the reference system and the composted Bio-flexi scenario. While product development continues, economic function and burden assessment can be specified further, focusing on nonmaterial information. Nevertheless, the already competitive process immanent costs indicate potential profitability as cost reduction potentials are oftentimes identified in ongoing product development.

4. Discussion

In this paper, the discussion is focused on a critical interpretation of the presented method itself. The key requirements specified in Section 2.1 are therefore chosen as a basis and deal as structure for the following investigation of the BiSA method. The main question to be answered can therefore be raised as follows: To which extent are the self-imposed goals of the developers achieved?

The general requirements have been set as consistency, comparability, reproducibility and falsifiability. Consistency of assessment systems is especially relevant with regard to system models. While the aim of the developers was to achieve full consistency of all aspects through the utilization of one system model, this could not be applied to the social function, as the building physical properties that provide the core of this aspect are not included in the life cycle inventory model that deals as a basis for the other aspects. Nevertheless, 5 out of 6 aspects could be modelled in a consistent way. Both comparability and reproducibility are met with the restriction of a not completely open model, which is caused by the fact that the underlying background databases are not publicly available. While consistency, comparability and reproducibility can be investigated on case study level, the falsifiability of sustainability in general is a critical requirement, as it simultaneously deals as a paradigm and as a scientific concept [94,95]. While the falsifiability of sustainability science and its scientific nature was questioned by Neumayer and Ziegler as well as Ott complemented the discourse and identified sustainability as a hybrid science that is falsifiable in the wider sense of conjecture and refutation [96]. In this sense, the presented framework offers a conjecture of BiSA open to refutation.

The basic requirements to decision support systems are met, even though their application has to prove true in future application. The proposed assessment system is able to support semi-structured and unstructured decisions and is applicable by decision makers throughout all phases of the decision making process. The requirements to sustainability assessment systems are classified based on the framework depicted in Figure 9 based on the experiences of previous applications. Overall, the assessment was developed by means of a comprehensive sustainability assessment with regard to the meta-assessment scheme, resulting in a classification of full strategicness, comprehensiveness and integratedness. A medium classification in terms of transparency, boundary-orientedness, scalability and stakeholders' involvement can be specified.

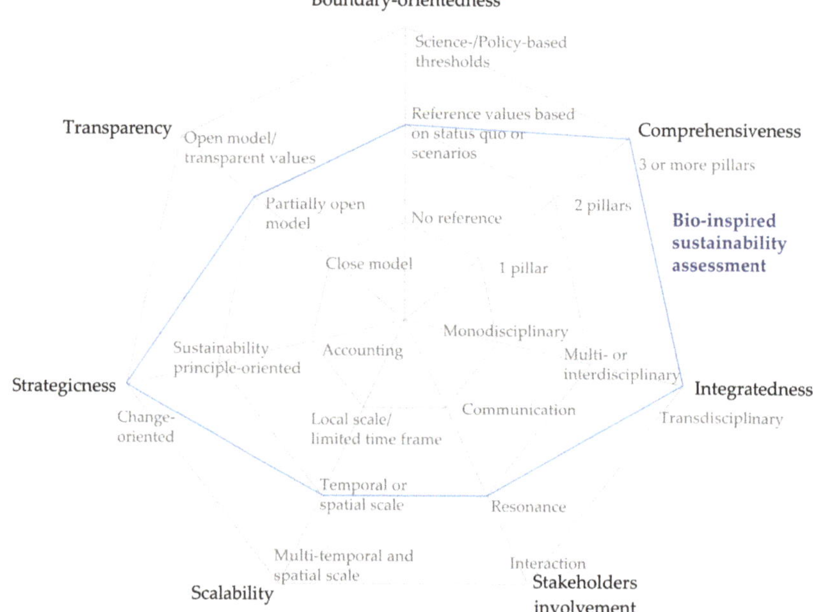

Figure 9. Classification of the BiSA based on the findings of the first applications (adapted from Sala et al. (2015), figure licensed under CC BY NC ND) [19].

The requirements of bio-inspired systems are mainly focusing on the practical application properties. Therefore, the meeting of the bio-inspired requirements can only be validated through ongoing application, including feedback of practitioners. Nevertheless, the general structure of the assessment system is designed to be applicable by developers effectively and it is planned to adapt the assessment structure based on the feedback of the practitioners. However, these specifications cannot be validated with the amount of studies available. The resilience has been investigated through performing sensitivity analyses and already gave a hint that the social burden assessment shows unintended changes and potentially low resilience with regard to modelling artifacts in the background database. As far as this can be identified by the assessing experts, the resulting distortions can be limited. In any case, this indicates further research demand in the field of social burden assessment based on the LCWE method.

5. Conclusions

The method of BiSA was presented in detail and applied to a first case study. The integration of bio-inspiration into sustainability and sustainability assessment proved to lead to new insights and a better connection to actual development processes. Furthermore, the connection of system properties derived from biology showed to provide improved access to the overall sustainability performance for practitioners. The results are depicted in a reduced approach, providing a single value for each aspect, including burdens and functions. However, several main points stay unresolved due to the limited amount of conducted studies so far. Since the project in which the evaluation was developed will be continued, an ongoing application of the assessment model is expected. Thereby, several key points will be investigated:

- Are the bio-inspired requirements useful as criteria for meta-assessment and does the BiSA meet them?
- How could BiSA support the degree of target attainment in terms of developing sustainable solutions in the building sector?
- Is it possible to directly integrate further success principles of evolution (multifunctionality and change of function) in the assessment structure?
- Is it useful (or eligible) to determine the detailed structure or should the model itself be adaptive?
- Is it possible to fully address the meta-assessment classification requirements to sustainability assessments through further model development?
- Is it possible to gain a deeper understanding of the biological model systems in the framework of a further development of the BiSA?

Furthermore, a comprehensive investigation on the interaction of multifunctionality (both simultaneously and temporally separated) and resource demand will be performed. This will include the assessment of functionality as multidimensional set of functions and underlying properties enabling self-maintenance of living organisms making use of resources under environmental stress. The concept of bio-inspired sustainability provides a consistent framework including a sustainability model and a thereupon developed assessment system including all three dimensions of sustainability for both intended functions and unintended burdens. Its assessment is performed based on a consistent model on the basis of life cycle thinking and aims to reintegrate the principles of nonhuman biological systems into product development to purposefully develop sustainable solutions. While providing a first implementation, the model will be further developed towards a tool with a higher degree of automation and adapted in conventional, bio-inspired and biobased practical application.

Acknowledgments: The authors especially thank the German Research Foundation DFG for funding the CRC-Transregio 141 "Biological Design and Integrative Structures—Analysis, Simulation and Implementation in Architecture", in which the bio-inspired sustainability assessment has been developed. The co-author, H. Dahy developer of Bio-flexi, would like to thank the University of Stuttgart for the funded technology transfer project (HDF flexible BIO-Faserplatte), for the further development of the product and the Agency of Renewable Resources (FNR) under the German Ministry for Consumer Protection, Food and Agriculture (BMEL) for funding the project PLUS-funding number FKZ: 22008413 in which among others, upgrading systems for mass-production of NFRP have been developed. In addition, the authors thank the project partner companies Naftex GmbH—Germany where the extrusion took place, think-blue company—Germany for providing the extrusion tool and K. Westermann GmbH + Co. KG Company—Germany for assisting in the veneering finishing of the Bio-flexi.

Author Contributions: Rafael Horn has developed the bio-inspired sustainability assessment and performed the assessment of the case study; Rafael Horn, Johannes Gantner and Philip Leistner have written and harmonized the assessment related chapters, Hanaa Dahy provided the information on Bio-flexi and wrote the technology-related application chapters, Olga Speck wrote the biology-related chapters and supported the assessment development with biological expertise.

Conflicts of Interest: The authors declare no conflict of interest.

References

1. Dickens, P. *Reconstructing Nature: Alienation, Emancipation and the Division of Labour*; Routledge: London, UK; New York, NY, USA, 1996.
2. Von Storch, H.; Stehr, N. Climate change in perspective. *Nature* **2000**, *405*, 615. [CrossRef] [PubMed]
3. Heidegger, M. *The Question Concerning Technology*; Garland Publishing: New York, NY, USA, 1977.
4. Cruzen, P.J. Geology of mankind: The Anthropocene. *Nature* **2002**, *415*, 23. [CrossRef] [PubMed]
5. Rockström, J.; Steffen, W.; Noone, K.; Persson, Å.; Chapin, F.S.I.; Lambin, E.; Lenton, T.M.; Scheffer, M.; Folke, C.; Schellnhuber, H.J.; et al. Planetary Boundaries: Exploring the Safe Operating Space for Humanity. *Ecol. Soc.* **2009**, *14*, 32. [CrossRef]
6. Von Carlowitz, H.C. *Sylvicultura Oeconomica Oder Haußwirthliche Nachricht und Naturgemäßige Anweisung zur Wilden Baum-Zucht*; Johan Friedrich Braun: Leipzig, Germany, 1713.

7. Hawken, P.; Lovins, A.B.; Lovins, L.H. *Natural Capitalism: Creating the Next Industrial Revolution*, 1st ed.; Little, Brown and Co.: Boston, MA, USA, 1999.

8. Von Gleich, A. Das bionische Versprechen: Ist die Bionik so gut wie ihr Ruf? *Ökologisches Wirtsch.* **2007**, *22*, 21–23.

9. Reap, J.; Baumeister, D.; Bras, B. Holism, Biomimicry and Sustainable Engineering. In *Energy Conversion and Resources*; ASME: New York, NY, USA, 2005; pp. 423–431.

10. Horn, R.; Gantner, J.; Widmer, L.; Sedlbauer, K.P.; Speck, O. Bio-inspired Sustainability Assessment: A Conceptual Framework. In *Biomimetic Research for Architecture and Building Construction: Biological Design and Integrative Structures*; Knippers, J., Nickel, K.G., Speck, T., Eds.; Springer: Cham, Switzerland, 2016; Volume 8, pp. 361–377.

11. Speck, O.; Speck, D.; Horn, R.; Gantner, J.; Sedlbauer, K.P. Biomimetic bio-inspired biomorph sustainable? An attempt to classify and clarify biology-derived technical developments. *Bioinspir. Biomim.* **2017**, *12*, 11004.

12. Margulis, L.; Sagan, D.; Sagan, C. *Life, Biology*; Autopoietic Definition of Life; Encyclopædia Britannica, Inc.: Chicago, IL, USA, 2017; Available online: https://www.britannica.com/topic/life#ref1014068 (accessed on 3 November 2017).

13. Varela, F.G.; Maturana, H.R.; Uribe, R. Autopoiesis: The organization of living systems, its characterization and a model. *Biosystems* **1974**, *5*, 187–196. [CrossRef]

14. Popper, K.R. *Logik der Forschung*, 11th ed.; Mohr Siebeck: Tübingen, Germany, 2005.

15. Parker, C. An approach to requirements analysis for decision support systems. *Int. J. Hum. Comput. Stud.* **2001**, *55*, 423–433. [CrossRef]

16. Steele, K.; Stefánsson, H.O. Decision Theory. In *The Stanford Encyclopedia of Philosophy*, 2016th ed.; Zalta, E.N., Ed.; Metaphysics Research Lab., Stanford University: Stanford, CA, USA, 2016.

17. Sprague, R.H. A Framework for the Development of Decision Support Systems. *MIS Q.* **1980**, *4*, 1–26. [CrossRef]

18. Sala, S.; Farioli, F.; Zamagni, A. Life cycle sustainability assessment in the context of sustainability science progress (part 2). *Int. J. Life Cycle Assess.* **2013**, *18*, 1686–1697. [CrossRef]

19. Sala, S.; Ciuffo, B.; Nijkamp, P. A systemic framework for sustainability assessment. *Ecol. Econ.* **2015**, *119*, 314–325. [CrossRef]

20. Finkbeiner, M.; Schau, E.M.; Lehmann, A.; Traverso, M. Towards Life Cycle Sustainability Assessment. *Sustainability* **2010**, *2*, 3309–3322. [CrossRef]

21. UNEP/SETAC Life Cycle Initiative. *Towards a Life Cycle Sustainability Assessment: Making Informed Choices on Products*; United Nations Environment Programme: Nairobi, Kenya, 2011.

22. Kloepffer, W. Life cycle sustainability assessment of products. *Int. J. Life Cycle Assess.* **2008**, *13*, 89–95. [CrossRef]

23. Sala, S.; Farioli, F.; Zamagni, A. Progress in sustainability science: Lessons learnt from current methodologies for sustainability assessment: Part 1. *Int. J. Life Cycle Assess.* **2013**, *18*, 1653–1672. [CrossRef]

24. Folke, C.; Carpenter, S.; Elmqvist, T.; Gunderson, L.; Holling, C.S.; Walker, B. Resilience and Sustainable Development: Building Adaptive Capacity in a World of Transformations. *AMBIO J. Hum. Environ.* **2002**, *31*, 437–440. [CrossRef]

25. Moore, F.C. Toppling the Tripod: Sustainable Development, Constructive Ambiguity and the Environmental Challenge. *Cons. J. Sustain. Dev.* **2011**, *1*, 141–150.

26. World Commission on Environment and Development. *Report of the World Commission on Environment and Development: Our Common Future*; Oxford University Press: Oxford, UK, 1987; Available online: http://www.un-documents.net/our-common-future.pdf (accessed on 3 January 2018).

27. Zamagni, A.; Pesonen, H.-L.; Swarr, T. From LCA to Life Cycle Sustainability Assessment: Concept, practice and future directions. *Int. J. Life Cycle Assess.* **2013**, *18*, 1637–1641. [CrossRef]

28. Parent, J.; Cucuzzella, C.; Revéret, J.-P. Revisiting the role of LCA and SLCA in the transition towards sustainable production and consumption. *Int. J. Life Cycle Assess.* **2013**, *18*, 1642–1652. [CrossRef]

29. Braungart, M.; McDonough, W. *Cradle to Cradle: Einfach Intelligent Produzieren*; Piper: München, Germany; Zürich, Switzerland, 2014.

30. Bjørn, A.; Hauschild, M.Z. Cradle to Cradle and LCA—Is there a Conflict? In *Glocalized Solutions for Sustainability in Manufacturing, Proceedings of the 18th CIRP International Conference on Life Cycle Engineering, Technische Universität Braunschweig, Braunschweig, Germany, 2–4 May 2011*; Hesselbach, J., Herrmann, C., Eds.; Springer: Berlin/Heidelberg, Germany, 2011; pp. 599–604.

31. Onat, N.C.; Kucukvar, M.; Tatari, O. Integrating triple bottom line input–output analysis into life cycle sustainability assessment framework: The case for US buildings. *Int. J. Life Cycle Assess.* **2014**, *19*, 1488–1505. [CrossRef]

32. Hacking, T.; Guthrie, P. A framework for clarifying the meaning of Triple Bottom-Line, Integrated, and Sustainability Assessment. *Environ. Impact Assess. Rev.* **2008**, *28*, 73–89. [CrossRef]

33. Onat, N.; Kucukvar, M.; Halog, A.; Cloutier, S. Systems Thinking for Life Cycle Sustainability Assessment: A Review of Recent Developments, Applications, and Future Perspectives. *Sustainability* **2017**, *9*, 706. [CrossRef]

34. Petti, L.; Serreli, M.; Di Cesare, S. Systematic literature review in social life cycle assessment. *Int. J. Life Cycle Assess.* **2016**, *18*, 1549. [CrossRef]

35. Guinée, J. Life Cycle Sustainability Assessment: What Is It and What Are Its Challenges? In *Taking Stock of Industrial Ecology*; Clift, R., Druckman, A., Eds.; Springer International Publishing: Cham, Switzerland, 2016; pp. 45–68.

36. Von Gleich, A.; Pade, C.; Petschow, U.; Pissarskoi, E. *Bionik: Aktuelle Trends und Zukünftige Potenziale*; Institut für ökologische Wirtschaftsforschung: Berlin, Germany, 2007.

37. Frey, E.; Masselter, T.; Speck, T. Was ist bionisch? *Eine Anal. Ideenflusses Biol. Tech. Naturwissensch. Rundsch.* **2011**, *64*, 117–126.

38. Coineau, Y.; Kresling, B. *Erfindungen der Natur*; Tessloff: Nürnberg, Germany, 1989.

39. Barthlott, W.; Neinhuis, C. Purity of the sacred lotus, or escape from contamination in biological surfaces. *Planta* **1997**, *202*, 1–8. [CrossRef]

40. Lienhard, J.; Schleicher, S.; Poppinga, S.; Masselter, T.; Milwich, M.; Speck, T.; Knippers, J. Flectofin: A hingeless flapping mechanism inspired by nature. *Bioinspir. Biomim.* **2011**, *6*, 45001. [CrossRef] [PubMed]

41. Walsh, D.M. Fitness and Function. *Br. J. Philos. Sci.* **1996**, *47*, 553–574. [CrossRef]

42. Achinstein, P. Function Statements. *Philos. Sci.* **1977**, *44*, 341–367. [CrossRef]

43. Verein Deutscher Ingenieure VDI 6223/1. *Bionik: Bionische Materialien, Strukturen und Bauteile; Biomimetics: Biomimetic Materials, Structures and Components VDI 6223 (Berlin: Beuth)*; Beuth: Berlin, Germany, 2013.

44. Lexikon der Biologie. *"Ressourcen"*; Spektrum Akademischer Verlag: Heidelberg, Germany, 1999; Available online: http://www.spektrum.de/lexikon/biologie/ressourcen/56391 (accessed on 29 November 2017).

45. Antony, F.; Grießhammer, R.; Speck, T.; Speck, O. Natur–(k)ein Vorbild für nachhaltige Entwicklung. In *Bionik: Patente aus der Natur*; Kesel, A., Zehren, D., Eds.; Meiners-Druck: Bremen, Germany, 2013; pp. 164–170.

46. Speck, O. Das bionische Versprechen: Ein Beitrag zur Bildung für nachhaltige Entwicklung. In *Bionik: Patente aus der Natur*; Kesel, A., Zehren, D., Eds.; Meiners-Druck: Bremen, Germany, 2015; pp. 135–146.

47. Walsh, D.M. Teleology. In *The Oxford handbook of Philosophy of Biology*, 1st ed.; Ruse, M., Ed.; Oxford University Press: Oxford, UK, 2010; pp. 113–137.

48. Huiskes, R. If bone is the answer, then what is the question? *J. Anat.* **2000**, *197*, 145–156. [CrossRef] [PubMed]

49. Sureau, S.; Mazijn, B.; Garrido, S.R.; Achten, W.M.J. Social life-cycle assessment frameworks: A review of criteria and indicators proposed to assess social and socioeconomic impacts. *Int. J. Life Cycle Assess.* **2017**, *30*, 181. [CrossRef]

50. Benoît, C.; Mazijn, B. *Guidelines for Social Life Cycle Assessment of Products*; United Nations Environment Programme: Nairobi, Kenya, 2009.

51. Jørgensen, A. Social LCA—A way ahead? *Int. J. Life Cycle Assess.* **2013**, *18*, 296–299. [CrossRef]

52. Moosbrugger, V. Das große Sterben vor 65 Millionen Jahren. In *Katastrophen in der Erdgeschichte—Wendezeiten des Lebens*; Hansch, W., Ed.; Städtische Museen: Heilbronn, Germany, 2003; pp. 144–153.

53. International Organization for Standardization. *Umweltmanagement—Ökobilanz—Grundsätze und Rahmenbedingungen (ISO 14040:2006)*; Beuth: Berlin, Germany, 2009.

54. International Organization for Standardization. *Umweltmanagement—Ökobilanz—Anforderungen und Anleitungen (ISO 14044:2006): Environmental Management—Life Cycle Assessment—Requirements and Guidelines*, 1st ed.; Beuth: Berlin, Germany, 2009.

55. International Reference Life Cycle Data System (ILCD). *ILCD Handbook—General Guide on LCA—Provisons and Action Steps/General Guide for Life Cycle Assessment: Provisions and Action Steps*; Publications Office: Luxembourg, 2010.

56. Thinkstep, A.G. *GaBi TS: Sofware-System and Databases for Life Cycle Engineering*; Thinkstep AG: Leinfelden-Echterdingen, Germany, 2017.

57. Steffen, W.; Richardson, K.; Rockstrom, J.; Cornell, S.E.; Fetzer, I.; Bennett, E.M.; Biggs, R.; Carpenter, S.R.; de Vries, W.; Wit, C.A.D.; et al. Planetary boundaries: Guiding human development on a changing planet. *Science* **2015**, *347*, 1259855. [CrossRef] [PubMed]

58. Ryberg, M.W.; Owsianiak, M.; Richardson, K.; Hauschild, M.Z. Challenges in implementing a Planetary Boundaries based Life-Cycle Impact Assessment methodology. *J. Clean. Prod.* **2016**, *139*, 450–459. [CrossRef]

59. Sala, S.; Benini, L.; Crenna, E.; Secchi, M. *Global Environmental Impacts and Planetary Boundaries in LCA: Data Sources and Methodological Choices for the Calculation of Global and Consumption-Based Normalisation Factors*; European Commission: Brussels, Belgium, 2016.

60. Sandin, G.; Peters, G.M.; Svanström, M. Using the planetary boundaries framework for setting impact-reduction targets in LCA contexts. *Int. J. Life Cycle Assess.* **2015**, *20*, 1684–1700. [CrossRef]

61. Bjørn, A.; Hauschild, M. Integrating Planetary Boundaries into the Life Cycle Assessment Framework for Assessing Absolute Environmental Sustainability of Products and Systems. In Proceedings of the Third International Resilience Science and Policy Conference, Montpellier, France, 5–9 May 2014.

62. Rockström, J.; Steffen, W.; Noone, K.; Persson, A.; Chapin, F.S.; Lambin, E.F.; Lenton, T.M.; Scheffer, M.; Folke, C.; Schellnhuber, H.J.; et al. A safe operating space for humanity. *Nature* **2009**, *461*, 472–475. [CrossRef] [PubMed]

63. Pizzol, M.; Laurent, A.; Sala, S.; Weidema, B.; Verones, F.; Koffler, C. Normalisation and weighting in life cycle assessment: Quo vadis? *Int. J. Life Cycle Assess.* **2017**, *22*, 853–866. [CrossRef]

64. Vidal Legaz, B.; Maia De Souza, D.; Teixeira, R.; Antón, A.; Putman, B.; Sala, S. Soil quality, properties, and functions in life cycle assessment: An evaluation of models. *J. Clean. Prod.* **2017**, *140*, 502–515. [CrossRef]

65. Schneider, L.; Berger, M.; Finkbeiner, M. Abiotic resource depletion in LCA—Background and update of the anthropogenic stock extended abiotic depletion potential (AADP) model. *Int. J. Life Cycle Assess.* **2015**, *20*, 709–721. [CrossRef]

66. Klinglmair, M.; Sala, S.; Brandão, M. Assessing resource depletion in LCA: A review of methods and methodological issues. *Int. J. Life Cycle Assess.* **2014**, *19*, 580–592. [CrossRef]

67. Bos, U.; Horn, R.; Beck, T.; Lindner, J.P.; Fischer, M. *LANCA—Characterization Factors for Life Cycle Impact Assessment: Version 2.0*; Fraunhofer Verlag: Stuttgart, Germany, 2016.

68. Swarr, T.E.; Hunkeler, D.; Klöpffer, W.; Pesonen, H.-L.; Ciroth, A.; Brent, A.C.; Pagan, R. Environmental life-cycle costing: a code of practice. *Int. J. Life Cycle Assess.* **2011**, *16*, 389–391. [CrossRef]

69. Eurostat. *Statistics on Industrial Production and International Trade (Prom), Annual Detailed Data by Prodcom List (According to NACE)*; Eurostat: Luxembourg, 2016; Available online: http://epp.eurostat.ec.europa.eu/newxtweb/ (accessed on 24 October 2017).

70. Callejas, D.G. Biology and Economics: Metaphors that Economists usually take from Biology. *Ecos Econ.* **2007**, *11*, 153–164.

71. Jørgensen, A.; Le Bocq, A.; Nazarkina, L.; Hauschild, M. Methodologies for social life cycle assessment. *Int. J. Life Cycle Assess.* **2008**, *13*, 96–103. [CrossRef]

72. Sala, S.; Vasta, A.; Mancini, L.; Dewulf, J.; Rosenbaum, E. *Social Life Cycle Assessment: State of the Art and Challenges for Product Policy Support*; Publications Office: Luxembourg, 2015.

73. Chhipi-Shrestha, G.K.; Hewage, K.; Sadiq, R. 'Socializing' sustainability: A critical review on current development status of social life cycle impact assessment method. *Clean Technol. Environ Policy* **2015**, *17*, 579–596. [CrossRef]

74. Iofrida, N.; de Luca, A.I.; Strano, A.; Gulisano, G. Can social research paradigms justify the diversity of approaches to social life cycle assessment? *Int. J. Life Cycle Assess.* **2016**, *20*, 498. [CrossRef]

75. Okasha, S. Biological Altruism. In *The Stanford Encyclopedia of Philosophy*, 2013rd ed.; Zalta, E.N., Ed.; Metaphysics Research Lab., Stanford University: Stanford, CA, USA, 2013.

76. Auletta, G.; Leclerc, M.; Martínez, R.A. (Eds.) Biological Evolution: Facts and Theories: A Critical Appraisal 150 Years after "The Origin of Species". In Proceedings of the 3rd ICSD International Conference, Rome, Italy, 3–7 March 2009; Gregorian & Biblical Press: Roma, Italy, 2011.

77. Alexander, R. Biology and the moral paradoxes. *J. Soc. Biol. Syst.* **1982**, *5*, 389–395. [CrossRef]

78. FitzPatrick, W. Morality and Evolutionary Biology. In *The Stanford Encyclopedia of Philosophy*; Stanford University: Stanford, CA, USA, 2016.

79. Sen, A. *Development as Freedom*, 1st ed.; Knopf: New York, NY, USA, 2001.

80. Fantke, P.; Bijster, M.; Hauschild, M.Z.; Huijbregts, M.; Jolliet, O.; Kounina, A.; Magaud, V.; Margni, M.; McKone, T.E.; Rosenbaum, R.K.; et al. *USEtox® 2.0 Documentation (Version 1.00)*; USEtox®Team: Bordeaux, France, 2017.

81. Alkire, S. Dimensions of Human Development. *World Dev.* **2002**, *30*, 181–205. [CrossRef]

82. Aulisio, D. *All SHDB Tables, Indicators, Characterizations-Update; Social Hotspots Database*; York, ME, USA, 2013. Available online: http://socialhotspot.org/wp-content/uploads/2013/03/All-SHDB-Tables-Indicators-Characterizations-UPDATE.pdf (accessed on 8 January 2018).

83. Barthel, L.-P. *Methode zur Abschätzung Sozialer Aspekte in Lebenszyklusuntersuchungen auf Basis Statistischer Daten*; Fraunhofer Verlag: Stuttgart, Germany, 2015.

84. Albrecht, S.; Endres, H.-J.; Knüpffer, E.; Spierling, S. Biokunststoffe—Quo vadis? *UmweltWirtschaftsForum* **2016**, *24*, 55–62. [CrossRef]

85. Gantner, J.; Beck, T.; Horn, R. *CommONEnergy Deliverable 5.7: Social Impact Assessment of Shopping Mall Retrofitting*; European Commission: Brussels, Belgium, 2017.

86. Reitinger, C.; Dumke, M.; Barosevcic, M.; Hillerbrand, R. A conceptual framework for impact assessment within SLCA. *Int. J. Life Cycle Assess.* **2011**, *16*, 380–388. [CrossRef]

87. Buller, D.J. (Ed.) *Function, Selection and Design*; State University of New York Press: Albany, NY, USA, 1999.

88. Verein Deutscher Ingenieure VDI 6220/1. *Bionik—Konzeption und Strategie—Abgrenzung Zwischen Bionischen und Konventionellen Verfahren/Produkten*; Biomimetics—Conception and Strategy—Differences between Biomimetic and Conventional Methods/Products; ICS 07.080; Beuth: Berlin, Germany, 2012.

89. Moro, J.L. *Baukonstruktion: Vom Prinzip zum Detail*; Springer: Berlin, Germany, 2009.

90. Dahy, H.; Knippers, J. Flexible High-Density Fiberboard and Method for Manufacturing the Same. EP3166765A1; Filed 8. July 2014, Published 13 January 2016. Based on Same Patent, Also Applied as WO2016005026A1; CN106604806A; US20170144327, EP2965882 A1. Available online: https://patentscope.wipo.int/search/en/detail.jsf?docId=WO2016005026 (accessed on 8 January 2018).

91. Dahy, H. Biocomposite materials based on annual natural fibres and biopolymers—Design, fabrication and customized applications in architecture. *J. Constr. Build. Mater.* **2017**, *147*, 212–220. [CrossRef]

92. Dahy, H. *Agro-Fibres Biocomposites Applications and Design Potentials in Contemporary Architecture: Case Study: Rice Straw Biocomposites*; Forschungsbericht Nr. 38; Universität Stuttgart: Stuttgart, Germany, 2015. Available online: http://dx.doi.org/10.18419/opus-113 (accesses on 8 January 2018).

93. Deutsches Institut für Normung e. V. DIN EN 15978. *Nachhaltigkeit von Bauwerken—Bewertung der Umweltbezogenen Qualität von Gebäuden—Berechnungsmethode*; Deutsche Fassung EN 15978-2011; Beuth: Berlin, Germany, 2012.

94. Kuhn, T.S. *The Structure of Scientific Revolutions*, 2nd ed.; University of Chicago Press: Chicago, IL, USA, 1970.

95. Ziegler, R.; Ott, K. The quality of sustainability science: A philosophical perspective. *Sustain. Sci. Pract. Policy* **2017**, *7*, 31–44.

96. Dietz, S.; Neumayer, E. Weak and strong sustainability in the SEEA: Concepts and measurement. *Ecol. Econ.* **2007**, *61*, 617–626. [CrossRef]

Article

Diagnosis of Sustainable Business Strategies Implemented by Chilean Construction Companies

Carlos Giannoni [1], Luis Fernando Alarcón [1] and Sergio Vera [1,2,*]

[1] Department of Construction Engineering and Management, School of Engineering, Pontificia Universidad Católica de Chile, Santiago 7820436, Chile; cmgiannoni@uc.cl (C.G.); lalarcon@ing.puc.cl (L.F.A.)

[2] Center for Sustainable Urban Development (CEDEUS), Pontificia Universidad Católica de Chile, Santiago 7530092, Chile

* Correspondence: svera@ing.puc.cl; Tel.: +562-2354-4245

Received: 30 November 2017; Accepted: 27 December 2017; Published: 30 December 2017

Abstract: Construction companies need to formulate sustainable construction business strategies to create a competitive advantage and remain in the market. This requires that construction firms incorporate sustainability into their business model. However, the current situation of the firm must be known before following the path to be a sustainable construction firm. Therefore, the aim of this research is to identify sustainable business strategies and their level of implementation in Chilean construction companies. A survey was designed and applied to 245 construction firms to provide statistically valid and reliable information, thus supporting both the senior managers' decision-making process and the companies' strategic planning. The main results show that the companies do not pursue business strategies that promote profound organizational changes; instead, they focus their short-term efforts on urgent market demands. This is evidenced by the lack of the function of sustainability management as a permanent role in the organization. Also, this study found that only 32% of Chilean construction companies implement business strategies towards sustainability. Construction firms with higher turnover and subjected to stricter regulations, such as construction companies working in the mining sector, incorporate more sustainable business strategies across their organizations. The lack of a sustainability-oriented vision can affect the transformation of strategy into a competitive advantage, a step that is necessary to support both the company's permanence in the market and its long-term sustainability.

Keywords: sustainable business strategies; sustainable construction; competitive advantage; corporate sustainability

---

1. Introduction

Sixty percent of the most important goods and services that sustain life in the world's ecosystems have been degraded or consumed in an unsustainable manner [1]. As developing countries progress and the population increases, the demand for goods and services will continue to grow [2]. Additionally, humans need a large number of buildings and infrastructure to sustain life and develop civilization. The construction, operation, maintenance and demolition of this infrastructure causes many environmental problems [3]. The construction industry generates negative impacts related to the processes of raw material extraction, material manufacturing, and infrastructure construction, operation and demolition. These impacts can be summarized as the consumption of non-renewable resources, the decline of biological diversity, the destruction of forest zones, the loss of agricultural areas, the destruction of natural spaces, global warming, and water, air and soil contamination [4]. To mitigate this situation, sustainable construction can promote sustainable development that eliminates or mitigates these negative impacts. Essentially, sustainable development is a transformation process in which the exploitation of resources, the course of investments, the direction of technological development and institutional change are harmonized to improve current and future potential so that human needs and aspirations can be fulfilled [5].

The introduction of sustainable development in construction is a new challenge that seeks to meet people's needs while considering the limited resources of the planet. Through the implementation of sustainable construction strategies, it is possible to contribute to the sustainable development of society and companies alike. To anticipate and adapt to change, the construction industry needs to devote more attention to developing new management strategies, techniques and management methods by incorporating new practices required by the market, organization, projects and processes. In recent years, authors such as Robichaud and Anantatmula [6] have noted the crucial role of sustainable construction and the need for modern methods to support decision-making processes. Construction firms' late adaptation to new environmental requirements [7] and stakeholders' demands [8] could entail reduced responsiveness and increased decision-making time, potentially resulting in increased costs.

According to the traditional view [9], delays in decision-making entail high costs. In the short run, costs become exponential and prices rise, thus encouraging competition. In the same way, decisions involve more variables that must be considered; associations and interactions among them are necessary to understand the relationship of a sustainable business model, in which the economic value is subject to social and environmental dimensions [10]. This vision of the construction business facilitates a long-term business strategy definition that favors the company's position and business sustainability. For example, Sfakianaki [11] pointed out that long-term recurrent cost reduction and potential increase in asset valuations will be driven by incorporating environmental and social aspects into the business model. Simultaneously, it reverses the cost curve by increasing brand value and conferring competitive advantages while generating business differentiation and mitigating risks.

Sustainable construction is an emerging field that aims both to apply general concepts of sustainable development to conventional construction practices and to propose models and strategies that include new social, environmental and cultural considerations [12]. Although knowledge in this field is continuously expanding, the sustainable model of the construction industry is not yet a standard industry practice.

1.1. Knowledge Gap

According to Epstein and Roy [13], managers have widely acknowledged the importance of formulating corporate sustainability strategies. For companies that have decided to assume economic, social and environmental responsibility, the question is not whether corporate sustainability strategies should be applied, but how they should be applied [14]. Nevertheless, transforming the concept into action and genuine initiatives remains a challenge. In this context, it is challenging to choose the right sustainability strategies [13,14], which are a consequence of distinguishing these strategies and knowing which are important for business sustainability in the long term. Companies lack a strategic approach to the integration of sustainability because of poor information and substantial uncertainty, meaning that failing to consider sustainability in strategic decision processes can increase the complexity of decisions and uncertainties [15]. From the firm's perspective, it is important to decrease uncertainty and achieve consistency with the purpose of promoting stability conditions. The integration of corporate sustainability into strategic management offers a potential approximation to address these challenges [15]. This complexity is especially related to long-term vision in the context of sustainable development and stakeholder commitment [16]. Thus, different types of strategies are related to different sustainability levels [17], which can be identified both in the literature and in practice [18]. However, this identification is less clear in the construction industry because this industry has failed to consistently prioritize its strategies. Decision-makers need clear sustainable criteria to evaluate their strategies.

The knowledge of which sustainable strategies will be applied in the construction sector will enable design strategies, clarify sustainability concepts and facilitate the fulfillment of the proposed objectives through systematic application. The diagnosis of sustainable construction strategies is useful for identifying the sustainable strategies that are applied in the construction sector and the needs to be met to achieve a sustainable construction industry in Chile.

The purpose of this research is to diagnose sustainable construction strategies implemented in Chile by construction companies, uncover which strategies are applied and analyze the main associations that favor the development of sustainable construction firms. Therefore, this study contributes to mapping the strategic behavior of construction firms in terms of corporate sustainability and identifies the gaps that need to be solved to advance toward a sustainable construction industry.

This paper aims to answer the following questions:

1. What competitive strategies are most frequently implemented to promote sustainability of construction companies?
2. How the dimensions of the business; organization, market, projects and processes, explain the current state of sustainability of construction in Chile?

1.2. Background

A strategy is a set of consistently made business decisions [19]. Business strategies aligned with sustainability reflect the nature and extension of sustainable development opportunities regarding value creation for companies [20]. Sustainable business development considers sustainable practices as the cornerstone of the company's survival; thus, the implementation of sustainable actions within the firm's business strategy can become a source of competitive advantage [21]. A company's competitiveness reflects its long-term relationships both within the industry and with its competitors. A competitive business is constantly informed about the conditions required to add value through a strategy that enables the fulfillment of organizational goals through social responsibility [20]. Thus, corporate social responsibility (CSR) defines companies that integrate social and environmental concerns in both their business operations and their interactions with stakeholders [22]. The construction industry has also incorporated sustainability in its corporate strategies through sustainable construction that targets the responsible and healthy creation and management of the built environment [23]. Sustainable construction is an integral process that seeks to maintain harmony between nature and the built environment by creating human settlements [4]. Its purpose is to achieve a balance among the economic, social and environmental dimensions of sustainability [24].

To standardize the concepts discussed in this paper, the authors propose the following definitions:

1. Sustainable strategies are the set of decisions that create a balanced relationship between the sustainable model and expected economic benefit.
2. Sustainable construction strategies reflect a construction company's awareness of the social, economic and environmental effects of its activities. These strategies are implemented as actions in the short, medium and long term by developing capabilities and skills that ensure sustainable competitive advantages.

1.3. Sustainable Construction Strategies

As corporate sustainability-based decisions are made at the strategic level, the scientific interest has also increased regarding the strategic management issue in relation to the integration of corporate sustainability into a company's strategy, vision and culture [25,26]. The same tendency is observed for sustainable construction strategies. There is increasing scientific and academic interest in studying the strategies implemented by construction firms. Although there have been few studies about this issue, Table 1 summarizes the most important ones, grouping them by topic and dimensions. Table 1 shows that sustainable strategies have been mostly studied on the project dimension, and there are no studies on the marketing dimension, although a wide range of sustainable strategies can be implemented in all dimensions of the sustainable business model.

Moreover, Table 1 shows that the construction industry employs partially sustainable strategies; thus, strategies in other industries that can also be applied to construction companies are not considered. For instance, Table 1 shows that several authors have studied the incorporation of sustainable materials in construction projects but, to the best of the authors' knowledge, there are no studies regarding a sustainable supply chain and sustainable business models in the construction sector, an issue that has been studied in other industries by many several authors such as Lloret [20], Reefke and Sundaram [27], Sarkis et al. [28], Seuring and Müller [29], Walton et al. [30], Al-Saleh and Taleb [31] and Richardson [32].

Table 1. Progression of knowledge in the sustainable construction strategies.

Subdimensions (Business Strategies)	Authors	Market	Organization	Project	Processes
			Dimension		
Sustainable materials in the construction sector	[33,34]				●
Green marketing (includes construction)	[35–39]	●			
Waste and recycling in construction	[40–44]				●
Sustainable leadership in the construction sector	[45,46]		●		
Sustainable knowledge management in construction projects	[47–52]			●	
Sustainable clients in the construction sector	[24,53–55]			●	

● shows that research has been carried out.

A set of strategies—namely, market, organization, project, and process strategies—can be grouped into dimensions that represent the large corporate bodies, thus facilitating their study and understanding. It is key to establish which of these dimensions can add value to the sustainable business model by creating capacities (processes), developing skills (project), providing timely response (market) and establishing the business model (organization). Numerous authors agree on the importance of studying these four dimensions of sustainable business. Baumgartner and Ebner [17] attribute importance to market research. Porter and Kramer [56] note that markets define organizations. Robichaud and Anantatmula [6] consider the study of sustainable projects, and [57] analyze the knowledge of processes aimed at companies' sustainable and competitive development. The four dimensions are briefly described below.

1. Market: Economic development involves the management of a company as a long-term participant in the market and has a positive impact on the economic circumstances of its stakeholders at the local, national and global levels. According to [17,58], sustainability is important because it is a prerequisite for a company's survival. They state that economic development addresses general aspects of the company, which must be respected in the environmental and social contexts to stay in the market in the long term.

2. Organization: The management of organizational change implies a continuous process of renewal, direction, structure and ability to satisfy internal and external changing needs [59]. This is especially important because sustainability requires deeper and permanent organizational change [60]. A downward change is necessary to create the required structure, provide a sustainable vision and encourage participation by all employees. Scheneider et al. [61] state that structural changes in the organization are effective only as long as they are associated with individual changes.

3. Project: Robichaud and Anantatmula [6] indicate that traditional methodologies of construction management, which are usually described as linear and fragmented processes, can cause further setbacks to a project, specifically in the construction of sustainable projects. It is less expensive to address environmental issues early in the project's life cycle than during its implementation. Indeed, the time at which these decisions are made can significantly affect performance rates, short-term construction costs, and long-term operational costs.

4. Processes: Construction companies should continuously improve their operational processes to differentiate their products or services and achieve sustainable development and competitiveness in the long term [57]. Some examples of the construction process are services, acquisitions, supply chains, and financial aspects (see details in Table 1).

1.4. Adaptation of the Model of Increasing Competitive Advantage

The dimensions of market, organization, project and processes should contain a set of strategies that contribute to generate the construction company's competitive advantages and better environmental performance. The phases of strategic business development that increase competitive advantage through the dimensions of market, organization, project and processes can be seen in Figure 1. This figure, based on the model proposed by Fergusson and Langford [62], illustrates the relationship among environmental management, business strategy and competitive advantage. The left side of Figure 1 shows four dimensions to achieve leadership in the construction industry through business differentiation, trade opportunities, client requirements and compliance with legislation. In the lower x-axis, the model shows that the development of environmental strategies will allow growth and performance improvement, whereas the curve represents the opportunity to increase competitive advantage. The right side of Figure 1 displays the homologation of the dimensions studied in this paper: market, organization, project and processes. The dimensions of this model are consistent with those proposed in this research, emphasizing the environmental, social and economic sustainability for a better understanding of the relationship between sustainable strategy development and competitive advantages. Therefore, strategies have been replaced with the aim of evolving toward corporate sustainability to achieve not only the construction industry's leadership but also advancement toward medium- and long-term corporate sustainability models. For this reason, it is important to identify and understand strategies that contribute to create competitive advantages and developing a sustainable business model.

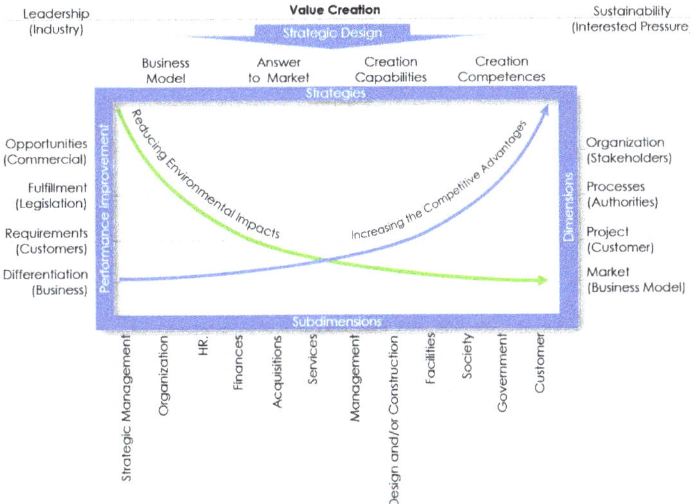

Figure 1. Increasing competitive advantage. Adaptation of the relationship among environmental strategy development, performance improvement and competitive advantage (Adapted from [62]).

2. Methodology and Data

Construction companies needs to develop sustainable competitive strategies to respond to new regulations and increasing social and environmental demands, thus they remain in the market.

This research is based on the data from Chilean construction companies, which is motivated by the need of exploring the main business strategies that affect their corporate sustainability. This research was exploratory, descriptive and correlational. This research is exploratory because of no evidence about the issues investigated in the local construction industry has been found; descriptive because the behavior of variables is analyzed; and correlational due to the validity and confidence of the measurement instrument. The design of this investigation is based on a quantitative, transversal and non-experimental approach. Its quantitative character corresponds to the empirical measurement of the variables under consideration, thus allowing the use of statistics as a tool for analysis. Furthermore, this quantitative study aims to standardize the results obtained in the sample for the population of construction industries. The transversal aspect is justified because it is a one-time measurement, and the non-experimental condition is attributable to the lack of deliberate manipulation of independent variables.

This research considers the following specific objectives: (1) establishing the state-of-the-art about the implementation of sustainable business strategies in the construction industry; (2) collecting data from Chilean construction companies; (3) estimating the validity and reliability of the collection data instrument and its content validity; (4) determining the relative importance among dimensions; (5) evaluating the impact of companies' socio-demographic data on sustainable business strategies (sub-dimensions); (6) categorizing which sustainable business strategies (sub-dimensions) are more implemented by Chilean construction industry; (7) identifying the main associations among sustainable business strategies (sub-dimensions); and (8) determining the attitudes of Chilean construction firms toward sustainability.

Figure 2 shows the research methodology to accomplish the specific objectives. The methodology consists in three main stages (state-of-the-art review, data collection and data analysis) divided in three main activities: literature review, survey design and application, and statistical tests and analysis, which are explained below.

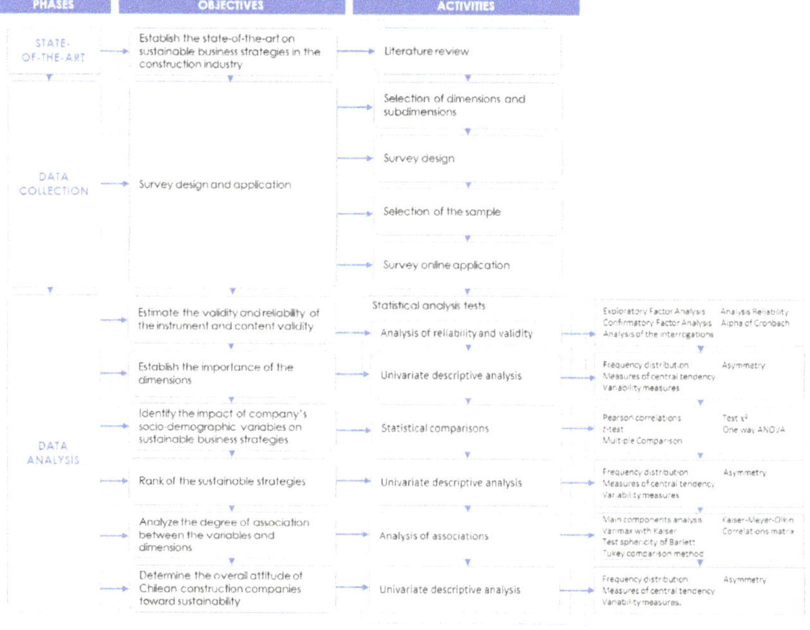

Figure 2. Schematic diagram of the research methodology.

2.1. Literature Review

The literature review was focused on establishing the state-of-the-art about the sustainable business strategies that are implemented by construction and non-construction companies worldwide.

2.2. Survey Design and Application

The instrument used in this research was a survey that collects socio-demographic data (5 questions) and the implementation of sustainable strategies in construction firms (40 questions). The survey was designed not only to provide a diagnosis of the strategies used in Chile's construction industry, through comparison, association and significance but also to determine companies' priorities in terms of the application of sustainable strategies. The socio-demographic variables that are collected are main business activity, position of the interviewee, experience in years, number of employees, and turnover of the construction firms. The 40 questions are measured with a Likert scale (Never, Hardly Ever, Sometimes, Almost Always and Always). The responses in the Likert scale are scored between 1 (Never) and 5 (Always).

The sample design considers a sub-group of the Chilean construction companies that are registered with the Chilean Chamber of Construction (CChC). The survey was administered online for 5 months through the website www.e-encuesta.com. During this period, 75 CEOs of construction companies answered the survey, which corresponds to 30.6% of the construction companies included in the CChC database. The answered surveys were statistically processed using Statistical Package for the Social Sciences software (IBM, New York, NY, USA) to obtain the following statistical results: descriptive analysis, estimate of the instrument's psychometric features, association tests, variance analysis and multiple comparisons.

The measurement instrument was designed to collect information in the following five dimensions, which are divided into 12 sub-dimensions or sustainable construction strategies (Figure 3) and 3 opinion questions.

The measured dimensions and sub-dimensions are set forth below.

1. Market. This dimension focuses on exploring the importance of concerns such as the following: (a) reducing the impact of construction works in the communities; (b) considering policies for managing hazardous waste; (c) protecting air quality both inside and outside of the works; (d) protecting natural resources; (e) creating strategies to encourage sustainable projects; (f) encouraging marketing strategies integrated into the business model; and (g) supporting the development of interest groups in the scope of sustainability. Therefore, the sub-dimensions or sustainable construction strategies studied focus on the main stakeholders of the construction firms (society, government and client).

2. Organization. When companies give importance to the organization dimension, they hold a vision of sustainable organization by engaging in the following activities: (a) clearly defining the role of the sustainable function; (b) developing monitoring processes and measuring practices; (c) communicating the importance of sustainability in the work processes both for the business and for the stakeholders; (d) promoting incentives and rewards; (e) guaranteeing proper environmental work conditions; and (f) developing recycling and other policies, enabling the development and implementation of strategies to address new social, environmental and economic demands. Thus, the sustainable construction strategies studied focus on the main organization roles such as senior management, organization and employees.

3. Project. This dimension is important because the design, construction and sustainable facilities are addressed from a sustainable management perspective, which favors the following considerations: (a) energy efficiency; (b) materials; and (c) construction processes. It also addresses the implementation of sustainable practices, including; (d) recycling; (e) the use of non-toxic chemical products; (f) the reuse of materials and water; and (g) waste management. As consequence, the strategies evaluated can be grouped into management, construction and project facilities.

4. Processes. This dimension is studied to determine the relationship among (a) the financial and social impact of the acquisition of non-sustainable products; (b) the incorporation of sustainable factors into investment decisions; and (c) the implementation of financial-analysis tools to compare costs and assess sustainable solutions or practices. The survey focuses on strategies related to finance, acquisitions and services.

5. Opinion. This dimension has been included in the survey to measure (a) the degree of implementation of sustainable practices; (b) the viability of adopting those practices under the under the current scenario; and (c) the importance that the company assigns to sustainability.

2.3. Statistical Test and Analysis

The answered surveys were statistically processed using Statistical Package for the Social Sciences software (SPSS, release 20). The main statistical analysis and tests were:

-   Univariate descriptive analysis [63]: Analysis that is based on the study of a single variable individually. The techniques used in the univariate analysis were the frequency distribution, measures of central tendency, measures of variability and asymmetry.
-   Analysis of reliability and validity: The reliability of the survey means how effective is this instrument to measure what it is intended to be measured. This is evaluated based on reliability analysis, Cronbach's alpha coefficient [64], and exploratory and confirmatory factorial analysis [63]. The validity corresponds to the consistency of the questions and the suitability of their formulation, which are evaluated by interrogation analysis.
-   Analysis of associations between two variables: this allows determining the existence of some type of association involving two variables and some type of trend or pattern of matching between the different values of the associated variables. The association analysis comprises the following tests: principal components analysis [65], Varimax normalization with Kaiser [66], correlation matrix [67], KMO (Kaiser-Meyer-Olkin) [68], Barlett's sphericity test [69], and Tukey's comparison method [70].
-   Statistical comparisons: this allows identifying differences, correlations and means among the variables studied. The following tests are used to perform multiple comparisons: Pearson correlations [71], χ^2 test [72], *t*-test [73], one-way Analysis of Variance (ANOVA) and multiple comparisons.

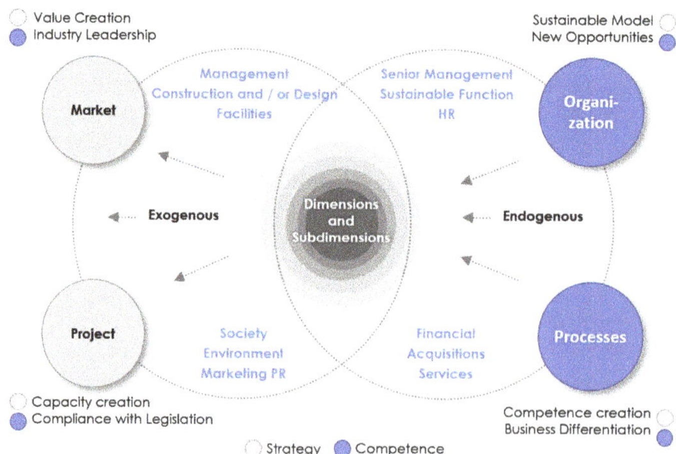

Figure 3. Sustainable strategies in the context of the organization. Overview of strategies leading to the creation of value in the sustainable business model.

3. Results

The presentation of results is primarily based on responses from top-level directors, managers, partners, and superintendents of construction companies. The survey covered the following sectors of the construction industry: industrial, residential buildings, non-residential buildings, mining, road and highways, others. More than 80% of the sample is composed of companies with less than 20 years of experience, with an average of 12.5 years and standard deviation of 9.84. Thirty-two companies were predominant, with a turnover level above US$30 million. In terms of number of employees, there were 30 small companies with 1–100 employees (40%), 21 medium-size companies with 101–500 employees (28%) and 24 large companies with more than 500 employees (32%). Detailed data can be seen in Table 2.

Table 2. Cross tabulation No. of Company Employees vs. Turnover (Million USD).

		Turnover (US$ Million)					
		Under US$1	**1–5**	**5–15**	**15–30**	**Greater than 30**	**Total**
	Between 1 and 10	0	5	0	0	0	5
	Between 11 and 25	2	2	0	1	1	6
	Between 26 and 50	1	3	1	1	0	6
	Between 51 and 100	1	6	0	2	4	13
Number of	Between 101 and 150	0	0	0	0	0	0
company	Between 151 and 200	0	1	2	0	1	4
employees	Between 201 and 250	0	1	2	2	1	6
	Between 251 and 500	0	1	2	3	5	11
	Between 501 and 1000	0	0	1	1	6	8
	More than 1000	0	2	0	0	14	16
	Total	4	21	8	10	32	75
		5.3%	28.0%	10.7%	13.3%	42.7%	100.0%

Several statistical tests were performed to validate the survey and identify the significant sustainable construction strategies implemented by Chilean construction firms. Section 3.1 evaluates reliability, and the validity of the survey which was assessed based on Cronbach's alpha, Kaiser-Meyer-Olkin and Bartlett's sphericity tests. Section 3.2 shows the importance of dimensions evaluated based on the median of the strategies for each dimension. Section 3.3 shows the results of several comparative tests to identify the socio-demographic variables that have the strongest impact on the sub-dimensions. Section 3.4 shows the ranking of sustainable construction strategies implemented by Chilean construction firms. Section 3.5 shows significant associations among sub-dimensions through an analysis of association. Finally, Section 3.6 shows the results of the Opinion dimension about the construction companies' overall level of sustainability.

3.1. Survey Reliability and Validity

The scale's internal consistency was measured by the Cronbach's alpha indicator (0.962), estimating the reliability of 40 items with 71 valid cases. An exploratory factor analysis was performed to estimate the scale validity of the construct. In the correlation matrix, high and significant associations were observed among statements. In addition, a very small determinant was found, indicating that a factor analysis should be performed. The same applies to the Kaiser-Meyer-Olkin measure of sampling adequacy (0.82) and Bartlett's test of sphericity (x2 approx.: 2361.56, gl: 780 and significance: 0.00), whose result is significant. Considering these two indicators, a factor analysis is appropriate. In the exploratory factor analysis, eight dimensions that account for 71.99% of the total variance of sustainability were configured. Although there are eight dimensions in the exploratory analysis, only five are distinguished in the sedimentation graphic, which is consistent with the theoretical dimensions of the scale. Based on the results of the reliability and factor analyses, it is concluded that the survey is both valid and reliable.

3.2. Importance of Dimensions

Table 3 shows the importance given to the dimensions by the respondent. The dimension with the highest median is Project (32.9%), followed by Market (30.8%), Organization (27.6%) and Process (26.4%). The Project dimension is the highest, but with increased dispersion (less homogeneous). Otherwise, the Market dimension has lesser dispersion (more homogeneous data). The Opinion dimension is indicative of the overall sustainability assessment. From the average behavior of the sub-dimensions of the construct, the most highly valued are (1) Sustainability in construction and/or design, but with increased dispersion (less homogeneous), associated with the Market dimension; (2) Sustainability in the environment, also associated with the Market dimension; and (3) Assessment of Sustainability associated with the Opinion dimension, with a slightly increased dispersion (less homogeneous), regarding the Sustainability in the environment. Likewise, Sustainability in Facilities is one of the most highly valued dimensions, which is a consequence of the close relationship with the Sustainability sub-dimension in construction and/or design, where both sub-dimensions are grouped in the Project dimension.

Table 3. Ranking of importance of the dimensions of sustainability.

Dimension	Min.	Max.	Mean [*]	Typical Deviation
Market	12	44	30.89	6.257
Organization	10	45	27.69	8.738
Project	11	50	32.99	8.432
Processes	9	40	26.49	7.880
Opinion	1	4	2.99	0.814

Note: [*] Distribution of mean scores of each dimension.

3.3. Impact of Companies' Socio-Demographic Variables on Sustainable Business Strategies

First, a one-way ANOVA test was carried out to identify which socio-demographic data have a significant effect on which sub-dimensions. Next, multiple comparison tests were carried out to identify significant relationships between socio-demographic variables and sustainable strategies.

(a) One-way ANOVA test: Only one significant relation was found between socio-demographic variables and sub-dimensions given by the impact of company size (turnover and number of employees) on a society's sustainability strategies (F = 2.631 and p = 0.031), including the environmental and social impacts and management of hazardous waste.

(b) Multiple comparison tests: As the one-way ANOVA test showed that turnover is the most important socio-demographic variables, multiple comparison tests were carried out to evaluate the impact of turnover on other sustainable strategies. Table 4 shows the results of these tests. It can be observed that larger companies with turnover above US$30 million show a significant impact on several strategies such as marketing and public relations, role and structure for sustainability in the firm's organization, high-level direction, construction-project management and financial sustainability. In addition, there was a significant impact on small construction companies with turnover between US$1 and US$5 million on human resource and construction management.

3.4. Ranking of Sustainable Strategies

Table 5 ranks the sustainable construction strategies implemented by Chilean construction companies based on the obtained score of each sub-dimension.

Table 4. Multiple Comparison Tests on the Turnover Sub-Dimension.

Sub-Dimensions	US$ Million				
Sustainability	Under 1	1–5	10–15	15–30	Greater than 30
Marketing/PR		(−)			(+) $p = 0.009$
Function		(−)			(+) $p = 0.048$
Senior Management		(−)			(+) $p = 0.012$
HR		(+) $p = 0.041$	(−)		
Management	(−)	(+) $p = 0.031$			
Management			(−)		(+) $p = 0.031$
Construction and/or Design	(+) $p = 0.056$	(−)			
Financial		(−)			(+) $p = 0.022$
Financial				(−)	(+) $p = 0.027$
Acquisition and Supply Chain			(+) $p = 0.052$	(−)	

Note: (+): In that category, it means that scores are higher, (−): In that category, it means that scores are lower.

Table 5. Ranking of importance in the implementation of sustainable strategies.

Sub-Dimensions	Min.	Max.	Mean [*]	Typical Deviation
1. Construction and/or Design	4	20	13.25	3.499
2. Environment	6	15	11.55	1.912
3. Facilities	3	15	10.27	2.522
4. Senior Management	3	15	9.93	3.465
5. Society	3	15	9.93	2.533
6. HR	3	15	9.91	2.548
7. Management	3	15	9.47	3.198
8. Marketing and PR	3	15	9.41	3.146
9. Services	3	14	9.25	2.800
10. Acquisition and Supply Chain	3	15	8.83	3.198
11. Finance	3	13	8.41	2.881
12. Function of Sustainability	3	15	7.85	3.910

Note: [*] Distribution of the mean scores for each sub-dimension.

3.5. Association Among Subdimensions

Table 6 shows the results of the analysis of associations among the sub-dimensions. This table reveals that the most significant associations are: (1) Sustainability in the Facilities with Sustainability in the Construction and/or design (R-value 0.816), in which 66.6% of each sub-dimension's total variability is explained by the other; and (2) Sustainability of Acquisition with Supply Chain and Financial Sustainability (R-value 0.796), in which 63.4% of each sub-dimension's total variability is explained by the other.

Table 6. Analysis of associations, correlation coefficients and significance levels of sub-dimensions.

	Sub-Dimensions	1	2	3	4	5	6	7	8	9	10	11	12	13
	Sustainability in Society	1 75												
Market	Environmental Issues	0.566 0.000 75	1 75											
	Sustainability in PR Marketing	0.587 0.000 75	0.362 0.001 75	1 75										
	Function of Sustainability	0.605 0.000 75	0.333 0.004 75	0.698 0.000 75	1 75									
Organization	Sustainability of Senior Management	0.551 0.000 75	0.536 0.000 75	0.667 0.000 75	0.718 0.000 75	1 75								
	Sustainability of HR	0.472 0.000 75	0.429 0.000 75	0.558 0.000 75	0.590 0.000 75	0.642 0.000 75	1 75							
	Management of Sustainability	0.560 0.000 75	0.413 0.000 75	0.611 0.000 75	0.717 0.000 75	0.813 0.000 75	0.619 0.000 75	1 75						
Project	Sustainability in Construction and/or Design	0.516 0.000 75	0.480 0.000 75	0.593 0.000 75	0.537 0.000 75	0.686 0.000 75	0.638 0.000 75	0.773 0.000 75	1 75					
	Sustainability in Facilities	0.451 0.000 75	0.435 0.000 75	0.516 0.000 75	0.432 0.000 75	0.554 0.000 75	0.624 0.000 75	0.655 0.000 75	0.816 0.000 75	1 75				
	Financial Sustainability	0.595 0.000 75	0.405 0.000 75	0.552 0.000 75	0.600 0.000 75	0.551 0.000 75	0.456 0.000 75	0.520 0.000 75	0.485 0.000 75	0.394 0.000 75	1 75			
Processes	Sustainability of Acquisition and Supply Chain	0.527 0.000 75	0.376 0.001 75	0.515 0.000 75	0.533 0.000 75	0.534 0.000 75	0.464 0.000 75	0.488 0.000 75	0.506 0.000 75	0.423 0.000 75	0.796 0.000 75	1 75		
	Sustainability in Services	0.549 0.000 75	0.436 0.000 75	0.559 0.000 75	0.527 0.000 75	0.634 0.000 75	0.572 0.000 75	0.642 0.000 75	0.522 0.000 75	0.453 0.000 75	0.561 0.000 75	0.671 0.000 75	1 75	
Opinion	Assessment of Sustainability	0.413 0.000 75	0.341 0.003 75	0.563 0.000 75	0.527 0.000 75	0.497 0.000 75	0.535 0.000 75	0.517 0.000 75	0.476 0.000 75	0.412 0.000 75	0.476 0.000 75	0.379 0.001 75	0.500 0.000 75	1 75

3.6. Overall Opinion About Company Sustainability

When respondents were asked about "an overall assessment of the degree of implementation of sustainable practices in their company", only 32% answered, "almost always" and "always". When asked, "Do you think the implementation of sustainability strategies could be applicable in your company?" 70.6% responded, "almost always" and "always". However, there is unanimity among the 88% of respondents who answered, "almost always" and "always" regarding the question of whether "they think it is important to implement sustainability strategies in construction". Surprisingly, no one answered "never" and "hardly ever" to this question. Moreover, 36% of the companies stated that they favor technological solutions for energy efficiency, and 32% of the respondents indicated that they favor the reduction of energy consumption as solutions.

4. Discussion

The research question was analyzed and addressed from the perspective of the following dimensions: (1) Market Dimension: Companies that address their strategies mostly around marketing (market) have a lower development of endogenous processes. This shows that there is less progression in obtaining a long-term competitive advantage, but rather in achieving an immediate market position; (2) Organization Dimension: This should be the most relevant dimension because it is in the organization where business strategy is defined. However, this is not evidenced in the case of the Chilean construction industry; (3) Project Dimension: The management of sustainable processes allows evaluating and improving performance or environmental risks of future projects, although in practice this is not addressed; (4) Processes Dimension: An additional precedent is market pressure, which forces a public relations' reaction from the company, with the subsequent disregard of internal processes, especially present in the residential construction industry.

Although mining construction companies are aware of potential environmental damage, they tend to only respond in compliance with either world-class standards or industry sustainability requirements. All business strategies are directly assessed by society and can generate positive or negative effects if decisions are not properly sized; the cost of inaction could destroy the company's social position with significant consequences for the business.

The importance and understanding of the Marketing sub-dimension is observed in all dimensions of the companies surveyed. However, this importance declines in the following order: organization, project and process, which is consistent with the importance given to the market, along with less intra-organizational promotion. Although a great deal of effort is focused on the Market, there is a gap in (a) understanding the social responsibility-related needs and demands raised by the market; and (b) incorporating bottom-up knowledge from workers who already have ideas about the transformation of sustainable processes. To address these issues in practice, it is important to have innovative employees with the ability and vision to redesign products, processes and business models. Above all, these employees must be able to coherently justify why this journey is a meaningful one [74].

Figure 4 shows a gradual change within the organization, starting with the strategic decision and projecting to the process dimension to conceive the market dimension. This ideal situation takes the shape of a snail; this is compared with the survey's results, privileging the market's public relations response, which takes a more diffuse and disorganized shape (stain) in the direction opposite from the snail. This represents a rapid adaptation of the processes needed to meet the project requirements through design, construction and facilities, albeit in the absence of the traceability of the company's strategic decisions (senior management), which seems diminished in relation to these sub-dimensions. It is interesting to pay attention to the points projecting outside the snail, which both indicate the current situation and show Chilean construction firms' lack of a sustainable strategic plan. Decisions that contribute to defining sustainable construction strategies must be conceived as part of a system involving the whole company. One common mistake is to address decisions separately. The policy should be transversal, primarily adopted in the internal process and then projected to the market. Good decisions should not be conceived separately because this makes them more diffuse and increases the risk of inaccurate results; neither should independent decisions be fused together. Synergies should be found so that the result is successful.

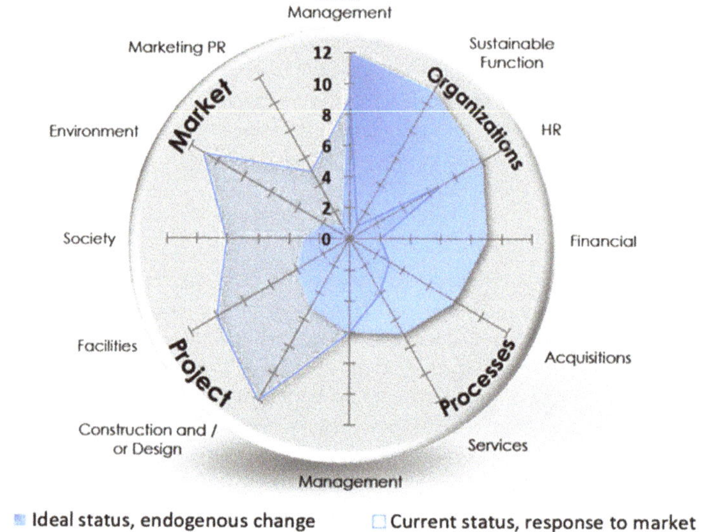

Figure 4. Comparison of endogenous change (ideal, snail-shaped) versus approach based on market response (current situation, irregularly shaped).

Despite the undisputable importance of waste management and recycling, the construction industry takes time to incorporate efficient internal processes, thus placing endogenous strategies at a disadvantage. This is seen in the fact that Construction is the most highly evaluated dimension (average score of 13.25, see Rankings in Table 5), and other dimensions have lower scores, such as Service (9.25), Acquisition and Supply Chain (8.83) and Finance (8.41). Thus, there is greater concern for the delivery of energy-efficient installations to the market. In contrast, water and material recycling from companies' own construction processes are not favored with the same intensity, which reveals the companies' lack of interest in endogenous change.

One of the advantages of endogenous change is the transmission of a company's own knowledge and culture toward the organization and the market, along with the implementation of strategies focused on the project that incorporates sustainable processes. As seen in this study's results, influence is exerted from the company's core. However, it should be exerted from the senior management level in response to the strategic definition.

Respondents not only adhere to the immediate need to rethink strategies but also visualize fertile ground for implementing sustainable strategies; this is indicated by the 79% who responded that the new sustainable business model is "Applicable in the company". Nevertheless, respondents also recognize the implied positive impact on society and the economic benefit for the company, evidenced by the 88% support for the "Importance of Sustainability".

Thirty-six percent of respondents indicated that they favor energy-efficient solutions because of new building requirements or new market demands. The residential construction sector is gradually realizing that has to move towards sustainability because (a) construction operations interfere with society and affect the community's quality of life; (b) certain processes are more beneficial for workers; (c) certain practices are economically viable; and (d) they can achieve greater profitability and social acceptance in the process.

The structured function of sustainability, defined as a permanent role, has a significant impact on both the supply chain and the sustainability of services. The reason is that the Sustainability Function is responsible for disseminating strategy from the senior management level and coordinating all of the company's processes with an economic, social and environmental approach, boosting actions in the primary processes in which practices should be internalized. Notwithstanding, the Sustainability Function is the least valued sub-dimension (7.85, see Table 5). The sustainability role is important in (a) sustaining a plan of activities aimed at involving workers based on different approaches; (b) adequately communicating senior management's vision; (c) transferring intra-organizational knowledge; (d) encouraging practices through incentives; (e) developing policies; (f) managing technical counterparts; (g) consistently articulating processes; and (h) guiding efforts and collecting and transmitting new social, economic and environmental requirements.

Although adaptation to this context seems irreversible, companies can dangerously delay this change until they are not experiencing regulatory pressure. This traditional approach could put the company at a disadvantage compared with competing companies that have accepted the impending change and have decided both to act and to be consistent with the sustainable model. Therefore, conceiving a sustainable vision and acting accordingly will confer a competitive advantage, provided the company reacts in time.

It should be noted that the process of generating strategies is a learning and dynamic process; therefore, it involves a permanent adjustment in which decisions can also be reversed for consistency with the results of other decisions. Practices that are applied as part of a sustainable process should be supported by major decisions involving other overall aspects of the process, probably driven by the market but strengthened by internal processes.

To identify how sustainability has been incorporated into companies operating in Chile, Ernst and Young surveyed 407 business executives from different sectors (e.g., retail, mining, energy, manufacturing, and telecommunications). When asked whether their companies have implemented a sustainability strategy, 51% answered "yes". This is a surprising answer because most of the surveyed companies do not have a sustainability strategy that includes this area in their core business, which should be aligned with their strategy [75]. When comparing this survey with the construction industry survey presented in this paper, it is evident that the construction industry has delayed longer in implementing sustainable strategies. Only 32% of respondents had an overall assessment of their companies' sustainability strategies, thereby demonstrating that the construction industry takes more time than other industries to incorporate sustainable strategies. This is consistent with the degree of importance attributed to sustainability. However, companies do recognize the importance and applicability of providing a high degree of value to sustainability in the company.

Similar results are found worldwide. International studies show agreement about CEOs of different productive sectors about the relevance of corporate sustainability for the business model, while their attitude towards sustainability has increased significantly. For example, the report "A New Era of Sustainability" [76] shows the results of a survey conducted to 766 CEOs of 13 productive sectors in 100 countries. This report reveals that 93% of 766 CEOs believe that taking care of sustainability issues in advance will be critical to their businesses' future success, whereas 96% believe that sustainability issues should be fully integrated into their companies' strategy and operations [76]. This report also shows a significant increase of the CEOs attitude toward sustainability which increases from 72% in 2007 to 96% in 2010. Other survey based-study carried out by the Massachusetts Institute of Technology in 2012 involved 2874 managers and executives from 113 countries [77]. The study shows that 70% of firms have incorporated sustainability on their management agendas, while 66% of respondents indicates that sustainability was crucial to be competitive under today's market and social demands. Moreover, this study found 31% of companies are currently profiting from sustainable business practices. The results of these international and national studies as well as the results shown in this paper agree that companies' commitment toward sustainability is increasing significantly because this is key for competitive advantages and company's long-term permanence.

5. Conclusions

Overall, construction companies implement sustainable construction strategies focused on introducing their construction projects to the market. Despite companies provide timely responses to market pressure, this approach is usually incompatible with long-term organizational changes. Although companies adapt their processes to meet customers' requirements, they lack a transversal strategy involving all of the processes within the organization. This gap is evidenced in management processes that prioritize customer-oriented strategies such as energy-efficient facilities and architectural design strategies instead of other management processes that address waste and materials recycling, acquisitions and supply chains. Although these strategies are visible to the market, they only approach the finished product, not a deep organizational change.

Nevertheless, this study showed that although construction firms need to respond to the sustainable demands of society, they implement more and deeper sustainable construction strategies when regulations are stricter, such as the case for construction companies working in the mining industry. There is also evidence that company size directly affects sustainability. Large companies, as indicated by the number of employees, are more likely to develop and implement strategies to address new social, environmental and economic requirements. Likewise, the greatest effect on sustainability is observed in companies with a higher turnover.

This research showed that the participating companies did not pursue strategies that encourage profound organizational change toward corporate sustainability. The function of sustainability management is not a permanent role in the organizational structure, as evidenced by the lack of a permanent sustainability role such as a Director of Sustainability. The lack of a role and organizational structure related to sustainability is a significant barrier for establishing a corporate sustainable strategy and the dissemination, follow-up, and results of sustainable strategies. This strategy should be conceived from the senior management and extended downwards to the entire organization. Thus, the strategy's fragmentation indicates a gap in the dissemination of the policy and a defective communication of the long-term sustainable vision.

As evidenced by this study, only 32% of construction companies declare that they implement sustainable strategies. However, 71% of the respondents believe these strategies could be applied and 88% indicate that these strategies are important for business sustainability. From this perspective, companies recognize sustainable strategies' positive impact on society and on the environment, along with their inherent economic benefits for the firm. Nevertheless, their efforts are focused on the market's demands. The lack of a long-term vision may affect the transformation of sustainable strategy into a competitive advantage. Therefore, if companies do not proceed opportunely, the delay might represent a disadvantage when competitors have experienced transformation based on a sustainable business model.

Acknowledgments: The authors thank the Chilean Construction Chamber for providing the database of the construction companies registered in their records, and the Vice Chancellor's research (VRI) scholarship of the Pontificia Universidad Católica de Chile. The authors also express their gratitude for the support of the Center for Sustainable Urban Development (CEDEUS), research grant CONICYT/FONDAP 15110020.

Author Contributions: Carlos Giannoni, Luis Fernando Alarcón and Sergio Vera, conceived, designed, and wrote together this paper.

Conflicts of Interest: The authors declare no conflict of interest.

References

1. Matsushita, K. *A Green New Deal as an Integration of Policies towards Sustainable Society, Achieving Global Sustainability: Policy Recommendations*; United Nations: New York, NY, USA, 2011.
2. Wade, J.; Cloutier, R.; Barboza, C. 2015 conference on systems engineering research towards a renewable energy decision making model. *Procedia Comput. Sci.* **2015**, *44*, 568–577.
3. Vyas, S.; Ahmed, S.; Parashar, A. Bee (bureau of energy efficiency) and green buildings. *Int. J. Res.* **2014**, *1*, 23–32.

4. Şener, S.; Sarıdoğan, E.; Staub, S.; Yılmaz, M.; Bakış, A. World conference on technology, innovation and entrepreneurship sustainability in construction sector. *Procedia Soc. Behav. Sci.* **2015**, *195*, 2253–2262.
5. Thore, S.; Tarverdyan, R. The sustainable competitiveness of nations. *Technol. Forecast. Soc. Chang.* **2016**, *106*, 108–114. [CrossRef]
6. Robichaud, L.B.; Anantatmula, V.S. Greening project management practices for sustainable construction. *J. Manag. Eng.* **2011**, *27*, 48–57. [CrossRef]
7. Wu, P.; Low, S.P. Project management and green buildings: Lessons from the rating systems. *J. Prof. Issues Eng. Educ. Pract.* **2010**, *136*, 64–70. [CrossRef]
8. Kibert, C.J. *Sustainable Construction: Green Building Design and Delivery*, 2nd ed.; John Wiley & Sons: Hoboken, NJ, USA, 2008.
9. Wagner, M.; Schaltegger, S.; Wehrmeyer, W. The relationship between the environmental and economic performance of firms. *Greener Manag. Int.* **2001**, *34*, 95–108. [CrossRef]
10. Bolis, I.; Morioka, S.N.; Sznelwar, L.I. When sustainable development risks losing its meaning. Delimiting the concept with a comprehensive literature review and a conceptual model. *J. Clean. Prod.* **2014**, *83*, 7–20.
11. Sfakianaki, E. Resource-efficient construction: Rethinking construction towards sustainability. *World J. Sci. Technol. Sustain. Dev.* **2015**, *12*, 233–242. [CrossRef]
12. Lu, Y.; Zhang, X. Corporate sustainability for architecture engineering and construction (AEC) organizations: Framework, transition and implication strategies. *Ecol. Indic.* **2016**, *61 Pt 2*, 911–922. [CrossRef]
13. Epstein, M.J.; Roy, M.-J. Sustainability in action: Identifying and measuring the key performance drivers. *Long Range Plan.* **2001**, *34*, 585–604. [CrossRef]
14. Baumgartner, R.J. Managing corporate sustainability and CSR: A conceptual framework combining values, strategies and instruments contributing to sustainable development. *Corp. Soc. Responsib. Environ. Manag.* **2014**, *21*, 258–271. [CrossRef]
15. Engert, S.; Rauter, R.; Baumgartner, R.J. Exploring the integration of corporate sustainability into strategic management: A literature review. *J. Clean. Prod.* **2016**, *112 Pt 4*, 2833–2850. [CrossRef]
16. Van Marrewijk, M.; Werre, M. Multiple levels of corporate sustainability. *J. Bus. Ethics* **2003**, *44*, 107–119. [CrossRef]
17. Baumgartner, R.J.; Ebner, D. Corporate sustainability strategies: Sustainability profiles and maturity levels. *Sustain. Dev.* **2010**, *18*, 76–89. [CrossRef]
18. Lee, M.-D.P. Configuration of external influences: The combined effects of institutions and stakeholders on corporate social responsibility strategies. *J. Bus. Ethics* **2011**, *102*, 281–298. [CrossRef]
19. Del Sol, P. Ganar sin Competir (*legítimamente*). Ediciones El Mercurio: 2016. Available online: https://www.casadellibro.com/ebook-ganar-sin-competir-legitimamente-ebook/9789567402502/3008609 (accessed on 28 December 2017).
20. Lloret, A. Modeling corporate sustainability strategy. *J. Bus. Res.* **2016**, *69*, 418–425. [CrossRef]
21. Bansal, P. Envolving sustainably: A longitudinal study of corporate sustainable development. *Strateg. Manag. J.* **2005**, *26*, 197–218. [CrossRef]
22. Commission of the European Communities (CEC). *Promoting a European Framework for Corporate Social Responsibility: Green Paper*; Office for Official Publications of the European Communities: Brussels, Belgium, 2001.
23. Langston, C.; Ding, G. *Sustainable Practices in the Built Environment*; Routledge: Abingdon, UK, 2008.
24. Shen, L.-Y.; Tam, V.W.Y.; Tam, L.; Ji, Y.-B. Project feasibility study: The key to successful implementation of sustainable and socially responsible construction management practice. *J. Clean. Prod.* **2010**, *18*, 254–259. [CrossRef]
25. Stead, J.G.; Stead, E. Eco-enterprise strategy: Standing for sustainability. *J. Bus. Ethics* **2000**, *24*, 313–329. [CrossRef]
26. Jin, Z.; Bai, Y. Sustainable development and long-term strategic management: Embedding a long-term strategic management system into medium and long-term planning. *World Future Rev.* **2011**, *3*, 49–69. [CrossRef]
27. Reefke, H.; Sundaram, D. Key themes and research opportunities in sustainable supply chain management—Identification and evaluation. *Omega* **2017**, *66*, 195–211. [CrossRef]
28. Sarkis, J.; Zhu, Q.; Lai, K.-H. An organizational theoretic review of green supply chain management literature. *Int. J. Prod. Econ.* **2011**, *130*, 1–15. [CrossRef]

29. Seuring, S.; Müller, M. From a literature review to a conceptual framework for sustainable supply chain management. *J. Clean. Prod.* **2008**, *16*, 1699–1710. [CrossRef]
30. Walton, S.V.; Handfield, R.B.; Melnyk, S.A. The green supply chain: Integrating suppliers into environmental management processes. *Int. J. Purch. Mater. Manag.* **1998**, *34*, 2–11. [CrossRef]
31. Al-Saleh, Y.M.; Taleb, H.M. The integration of sustainability within value management practices: A study of experienced value managers in the GCC countries. *Proj. Manag. J.* **2010**, *41*, 50–59. [CrossRef]
32. Richardson, J. The business model: An integrative framework for strategy execution. *Strateg. Chang.* **2008**, *17*, 133–144. [CrossRef]
33. Govindan, K.; Madan Shankar, K.; Kannan, D. Sustainable material selection for construction industry—A hybrid multi criteria decision making approach. *Renew. Sustain. Energy Rev.* **2016**, *55*, 1274–1288. [CrossRef]
34. González, M.J.; García Navarro, J. Assessment of the decrease of CO_2 emissions in the construction field through the selection of materials: Practical case study of three houses of low environmental impact. *Build. Environ.* **2006**, *41*, 902–909. [CrossRef]
35. Lockrey, S. A review of life cycle based ecological marketing strategy for new product development in the organizational environment. *J. Clean. Prod.* **2015**, *95*, 1–15. [CrossRef]
36. Cronin, J.J.; Smith, J.S.; Gleim, M.R.; Ramirez, E.; Martinez, J.D. Green marketing strategies: An examination of stakeholders and the opportunities they present. *J. Acad. Mark. Sci.* **2011**, *39*, 158–174. [CrossRef]
37. Josephine, P.B.; Ritsuko, O. Pro-environmental products: Marketing influence on consumer purchase decision. *J. Consum. Mark.* **2008**, *25*, 281–293.
38. Ken, P.; Andrew, C. Green marketing: Legend, myth, farce or prophesy? *Qual. Mark. Res. Int. J.* **2005**, *8*, 357–370.
39. Van Dam, Y.K.; Apeldoorn, P.A.C. Sustainable marketing. *J. Macromark.* **1996**, *16*, 45–56. [CrossRef]
40. Ibrahim, M.I.M. Estimating the sustainability returns of recycling construction waste from building projects. *Sustain. Cities Soc.* **2016**, *23*, 78–93. [CrossRef]
41. Jacobsen, N.B. Industrial symbiosis in kalundborg, denmark: A quantitative assessment of economic and environmental aspects. *J. Ind. Ecol.* **2006**, *10*, 239–255. [CrossRef]
42. Dolan, P.J.; Lampo, R.G.; Dearborn, J.C. *Concepts for Reuse and Recycling of Construction and Demolition Waste*; USACERL Technical Report 97/58; Construction Engineering Research Laboratories, US Army Corps of Engineers: Champaign, IL, USA, 1999; Available online: http://acwc.sdp.sirsi.net/client/en_US/search/asset/1002266;jsessionid=FAEE3D91AB91CE5DB23877FE64090F19.enterprise-15000, (accessed on 28 December 2017).
43. Serpell, A.; Kort, J.; Vera, S. Awareness, actions, drivers and barriers of sustainable construction in Chile. *Technol. Econ. Dev. Econ.* **2013**, *19*, 272–288. [CrossRef]
44. Sfakianaki, E.; Moutsatsou, K. A decision support tool for the adaptive reuse or demolition and reconstruction of existing buildings. *Int. J. Environ. Sustain. Dev.* **2015**, *14*, 1–19. [CrossRef]
45. Tabassi, A.A.; Roufechaei, K.M.; Ramli, M.; Bakar, A.H.A.; Ismail, R.; Pakir, A.H.K. Leadership competences of sustainable construction project managers. *J. Clean. Prod.* **2016**, *124*, 339–349. [CrossRef]
46. Shriberg, M. Toward sustainable management: The university of Michigan housing division's approach. *J. Clean. Prod.* **2002**, *10*, 41–45. [CrossRef]
47. Pemsel, S.; Müller, R.; Söderlund, J. Knowledge governance strategies in project-based organizations. *Long Range Plan.* **2016**, *49*, 648–660. [CrossRef]
48. Lampel, J.; Scarbrough, H.; Macmillan, S. Managing through projects in knowledge-based environments: Special issue introduction by the guest editors. *Long Range Plan.* **2008**, *41*, 7–16. [CrossRef]
49. Cole, R.J. Building environmental assessment methods: Redefining intentions and roles. *Build. Res. Inf.* **2005**, *33*, 455–467. [CrossRef]
50. Guggemos, A.A.; Horvath, A. Comparison of environmental effects of steel- and concrete-framed buildings. *J. Infrastruct. Syst.* **2005**, *11*, 93–101. [CrossRef]
51. Sydow, J.; Lindkvist, L.; DeFillippi, R. Project-based organizations, embeddedness and repositories of knowledge: Editorial. *Organ. Stud.* **2004**, *25*, 1475–1489. [CrossRef]
52. Hobday, M. The project-based organisation: An ideal form for managing complex products and systems? *Res. Policy* **2000**, *29*, 871–893. [CrossRef]

53. Qi, G.Y.; Shen, L.Y.; Zeng, S.X.; Jorge, O.J. The drivers for contractors' green innovation: An industry perspective. *J. Clean. Prod.* **2010**, *18*, 1358–1365. [CrossRef]
54. Abidin, N.Z. Investigating the awareness and application of sustainable construction concept by malaysian developers. *Habitat Int.* **2010**, *34*, 421–426. [CrossRef]
55. Michael, P.; Matthew, T.; Mike, R.; Jennifer, L. Towards sustainable construction: Promotion and best practices. *Constr. Innov.* **2009**, *9*, 201–224.
56. Porter, M.; Kramer, M. Creating shared value. How to reinvest capitalism—And unleash a wave of innovation and growth. *Harv. Bus. Rev.* **2011**. Available online: https://sharedvalue.org/sites/default/files/community-posts/CREATING%20SHARED%20VALUE%20-%20FOR%20INDIVIDUALS.pdf (accessed on 28 December 2017).
57. Konrad, A.; Steurer, R.; Langer, M.; Martinuzzi, A. Empirical findings on business-society relations in europe. *J. Bus. Ethics* **2006**, *63*, 89–105. [CrossRef]
58. Steurer, R.; Langer, M.; Konrad, A.; Martinuzzi, A. Corporations, stakeholders and sustainable development I: A theoretical exploration of business-society relations. *J. Bus. Ethics* **2005**, *61*, 263–281. [CrossRef]
59. Moran, J.; Brightman, B. Leading organizational change. *J. Workplace Learn.* **2000**, *12*, 66–74. [CrossRef]
60. Armenakis, A.A.; Harris, S.G.; Feild, H.S. Making change permanent a model for institutionalizing change interventions. In *Research in Organizational Change and Development*; Emerald Group Publishing Limited: Bingley, UK, 2000; pp. 97–128.
61. Schneider, B.; Brief, A.P.; Guzzo, R.A. Creating a climate and culture for sustainable organizational change. *Organ. Dyn.* **1996**, *24*, 7–19. [CrossRef]
62. Fergusson, H.; Langford, D. Strategies for managing environmental issues in construction organizations. *Eng. Constr. Architect. Manag.* **2006**, *13*, 171–185. [CrossRef]
63. Thompson, B. *Exploratory and Confirmatory Factor Analysis: Understanding Concepts and Applications*; American Psychological Association: Washington, DC, USA, 2004.
64. Cronbach, L.J. Coefficient alpha and the internal structure of tests. *Psychometrika* **1951**, *16*, 297–334. [CrossRef]
65. Tabachnick, B.G.; Fidell, L.S.; Osterlind, S.J. *Using Multivariate Statistics*, 6th ed.; Pearson Education Limited: Essex, UK, 2014.
66. Kaiser, H.F. The varimax criterion for analytic rotation in factor analysis. *Psychometrika* **1958**, *23*, 187–200. [CrossRef]
67. Henson, R.K.; Roberts, J.K. Use of exploratory factor analysis in published research: Common errors and some comment on improved practice. *Educ. Psychol. Meas.* **2006**, *66*, 393–416. [CrossRef]
68. Kaiser, H.F. A second generation little jiffy. *Psychometrika* **1970**, *35*, 401–415. [CrossRef]
69. Bartlett, M.S. Tests of significance in factor analysis. *Br. J. Math. Stat. Psychol.* **1950**, *3*, 77–85. [CrossRef]
70. Tukey, J.W. Comparing individual means in the analysis of variance. *Biometrics* **1949**, *5*, 99–114. [CrossRef] [PubMed]
71. Pearson, K. Notes on the history of correlation. *Biometrika* **1920**, *13*, 25–45. [CrossRef]
72. Pearson, K.X. On the criterion that a given system of deviations from the probable in the case of a correlated system of variables is such that it can be reasonably supposed to have arisen from random sampling. *Lond. Edinb. Dublin Philos. Mag. J. Sci.* **1900**, *50*, 157–175. [CrossRef]
73. Student. The probable error of a mean. *Biometrika* **1908**, *6*, 1–25.
74. Senge, P.M. La cadena de suministro sustentable. *Harv. Bus. Rev.* **2011**, *89*, 48–50.
75. Serrano, B. ¿qué tanto nos preocupamos por ser sostenibles? In *Poder & Negocios*; Editorial Tiempo Presente: Santiago, Chile, 2011; pp. 16–24.
76. Lacy, P.; Cooper, T.; Hayward, R.; Neuberger, L. A New Era of Sustainability. 2010. Available online: https://www.unglobalcompact.org/docs/news_events/8.1/UNGC_Accenture_CEO_Study_2010.pdf (accessed on 28 December 2017).
77. Kiron, D.; Kruschwitz, N.; Haanaes, K.; Von Streng Velken, I. Sustainability nears a tipping point. *MIT Sloan Manag. Rev.* **2012**, *53*, 69–74. [CrossRef]

 *sustainability*

Article

Performance Evaluation and Life Cycle Cost Analysis Model of a Gondola-Type Exterior Wall Painting Robot

Dong-Jun Yeom [1], Eun-Ji Na [1], Mi-Young Lee [1], Yoo-Jun Kim [1], Young Suk Kim [1,*] and Chung-Suk Cho [2]

[1] Department of Architectural Engineering, Inha University, Incheon 22212, Korea; dj09051@inha.edu (D.-J.Y.); 22131041@inha.edu (E.-J.N.); 22141007@inha.edu (M.-Y.L.); 22162010@inha.edu (Y.-J.K.)
[2] Department of Civil, Infrastructure and Environmental Engineering, Khalifa University, P.O. Box 127788, Abu Dhabi L2017E, UAE; chung.cho@kustar.ac.ae
* Correspondence: youngsuk@inha.ac.kr; Tel.: +82-32-860-7593

Received: 12 September 2017; Accepted: 6 October 2017; Published: 8 October 2017

Abstract: The amount and market size of apartment complex exterior wall painting work continues to increase each year in South Korea. Nevertheless, there are difficulties with the supply and demand of human resources due to the high risks associated with conventional painting work. To resolve these issues, research and development has recently been conducted on a Gondola-type Exterior Wall Painting robot (GEWPro). The aims of this study were to develop a performance evaluation and life cycle cost (LCC) analysis model for a GEWPro and deduce its performance and economic efficiency through a case study. According to the results, the performance of the automated method was 16.8% higher than that of the conventional method, and the economic efficiency was also superior (benefit/cost ratio 6.39). These results show that the proposed performance evaluation and LCC analysis model can predict the productivity and economic efficiency of automated methods.

Keywords: exterior wall painting; automation robot; life cycle cost; performance evaluation

1. Introduction

The number of newly constructed apartment complexes in South Korea continues to increase each year. In particular, apartment complexes containing more than 304,432 dwellings were constructed in 2016 [1]. Because the exterior finishes of these apartment complexes normally require painting, the increase in the supply of newly built apartment complexes will result in an increase in the demand for exterior wall painting. Moreover, in the case of existing apartment complexes to which exterior wall painting has been applied, full-scale repainting on the entire outer wall of an apartment should be performed every five years using a fresh coat of paint according to the long-term repair plan of the Korea Housing Act [2] to enhance the long-term durability and sustainability of the buildings. An analysis of the repainting frequency for 50 apartment complexes in metropolitan areas of South Korea showed that exterior wall repainting was performed every 5.34 years on average. Therefore, it is expected that the market for exterior wall painting of apartment complexes will grow even more in the future considering the enormous number of dwellings in apartment complexes in South Korea.

Exterior wall painting for apartment complexes is performed by laborers working from a suspended scaffold. Therefore, there is a high risk of accidents, such as falls or collisions with the outer wall. Moreover, there is a high risk of safety accidents, such as falling objects hitting pedestrians passing by below. In particular, the intensity of an accident would be very high when it does occur because exterior wall painting work is performed after the safety structures (safety net, flared shield in, etc.) have been removed from the exterior walls. Thus, the inherent problems, such as various risks and the number of insufficient skilled laborers involved in the exterior wall painting work of apartment complexes in South Korea, are expected to worsen in the future.

Recently, a unique Gondola-type Exterior Wall Painting robot (GEWPro) was developed to resolve the above problems related to the conventional exterior wall painting work of apartment complexes in South Korea [3]. To determine if this construction automation technology is attractive and acceptable in the field, it is important to verify quantitatively the benefits gained using the automated method by a comparison with the conventional method. Over the last three decades, a number of automated or semi-automated construction machines or systems have been developed and demonstrated in the construction industry [4–8]. On the other hand, few studies have verified the superiority of such automated methods based on a comparison and analysis of their overall performance in terms of productivity, cost, safety, and quality [9]. As a result, their practical use in the field has been underestimated.

In this study, the conventional process of the exterior wall painting of apartment complexes in South Korea and its problems were identified, and the automated method using the developed GEWPro was assessed. A model for performance evaluation and Life Cycle Cost (LCC) analysis [10–12] of GEWPro was developed in an effort to enhance the sustainability for the practical use of robots on construction sites.

Based on the model and data collected from several field trials, the overall performance of GEWPro was analyzed and its LCC was then compared with that of the conventional method to verify the benefits obtained from automation. Multiple laboratory and field tests were conducted to acquire essential data for the performance evaluation and LCC analysis. Conclusions and recommendations are made regarding the value of implementing and commercializing GEWPro. Finally, the performance evaluation and LCC analysis methodology and the model presented in this study can be adapted to analyze the performance of any other construction automation machine or robot.

2. Background Studies

2.1. The Conventional Exterior Wall Painting Work Process and the Problems

According to an analysis based on a field examination regarding exterior wall painting work currently performed in South Korea, the process of the exterior wall painting work can be categorized into the following: (1) preparation work; (2) installation work; (3) painting work; (4) disassembly work; (5) horizontal moving work; and (6) demobilization work. Figure 1 illustrates the detailed work process.

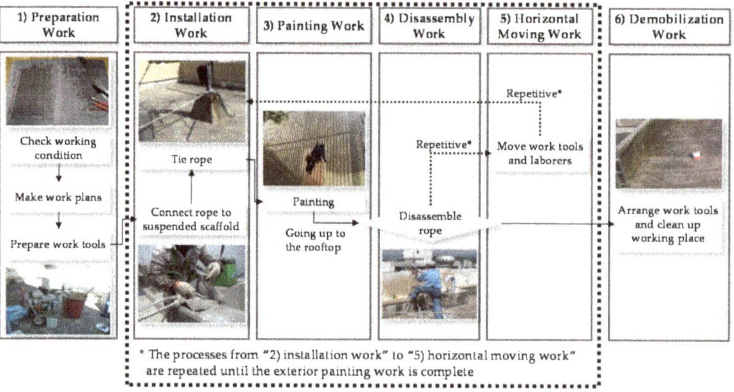

Figure 1. Conventional exterior wall painting work process.

To understand the exterior wall painting work currently performed in South Korea and investigate the problems of the current method, a questionnaire including questions related to the following three major aspects (Sections 2.1.1–2.1.3) was distributed to 69 on-site field managers who work for seven construction companies (including both general and specialty contractors) in South Korea.

2.1.1. Aspect of the Supply and Demand of Human Resources

Conventional exterior wall painting work is strongly dependent on laborers, as shown by the ratio of the laborer cost to the total construction cost of 80% [13]. Therefore, the supply and demand of exterior wall painting laborers with good quality and construction speeds are very important factors for completing exterior wall painting work successfully and economically. On the other hand, there has always been a difficulty in the supply and demand of skilled laborers, because of the inherent high risks involved in the exterior wall painting work of high-rise apartment complexes. In the survey, all respondents also indicated that such laborer issues would continue to become much more serious in the near future considering that it generally takes approximately eight to nine years for an entry-level laborer to become a skilled laborer, and there is an undersupply.

2.1.2. Aspect of Productivity and Quality

According to the survey result regarding the productivity and quality aspects of the conventional exterior wall painting work, 57.8% of respondents perceived the following to be the most urgent problems to be solved in the aspects of productivity and quality: (1) a decrease in productivity due to the work done by insufficiently skilled laborers and unskilled laborers; and (2) inconsistent thickness of the painted surface due to manual work and deviations in the workmanship of the labor force, respectively. Most of the survey respondents also answered that such problems are expected to worsen in the future due to aging of the labor manpower and the phenomena of avoiding construction fields among the younger work force of South Korea.

2.1.3. Aspect of Safety

Conventional exterior wall painting is a high-risk type of work, in which the work is performed by a laborer while suspended in a hanging scaffold. According to the survey regarding the risk and safety aspects of the conventional exterior wall painting work, 66% of respondents answered that the risk of exterior wall painting work is still very high, and alternatives, such as automation (applying robots), need to be considered in an effort to prevent or minimize deadly accidents that can occur in the exterior wall painting work conducted by labor work force.

2.2. Development Status of Gondola-Type Exterior Wall Painting Robot

To resolve aforementioned problems, a GEWPro has recently been developed in South Korea for the automation of the exterior wall painting work in high-rise buildings [3]. In this study, the characteristics of the developed GEWPro based on its composition and specifications were analyzed to develop a performance evaluation and LCC analysis model [10–12]. The GEWPro is equipped with the capability for exterior wall painting of apartment complexes as its major function and is supplemented with a function for cleaning the outer walls as well. To perform painting and cleaning work, the GEWPro is composed of four parts—"main body of the robot", "hanger", "paint and cleaning water supply device", and a "controller". Figure 2 shows the composition of the main body of the developed GEWPro, and lists its detailed specifications.

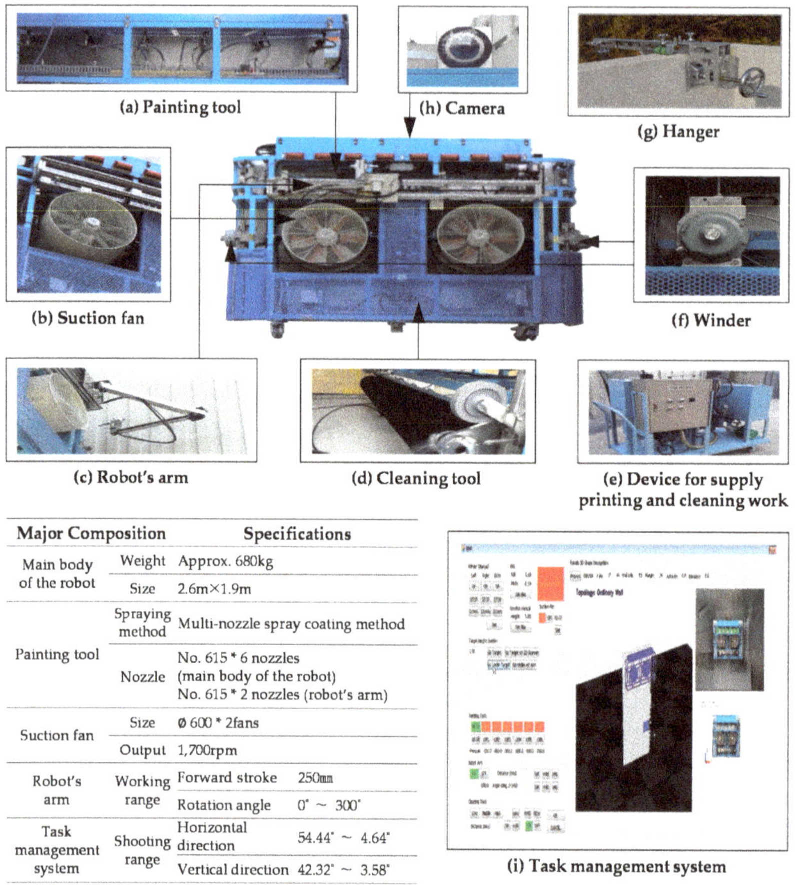

Major Composition		Specifications	
Main body of the robot	Weight	Approx. 680kg	
	Size	2.6m×1.9m	
Painting tool	Spraying method	Multi-nozzle spray coating method	
	Nozzle	No. 615 * 6 nozzles (main body of the robot) No. 615 * 2 nozzles (robot's arm)	
Suction fan	Size	Ø 600 * 2fans	
	Output	1,700rpm	
Robot's arm	Working range	Forward stroke	250㎜
		Rotation angle	0° ~ 300°
Task management system	Shooting range	Horizontal direction	54.44° ~ 4.64°
		Vertical direction	42.32° ~ 3.58°

Figure 2. Major composition of GEWPro.

2.2.1. Main Body of the GEWPro

The main body of GEWPro consists of a camera, painting tool, robot's arms, winder, suction pan, and a cleaning tool. The painting tool is the main device for performing exterior wall painting work and can carry out the exterior wall painting work through six nozzles that are equipped with an anti-scattering cover. The robot arms of GEWPro were installed to enable the simultaneous painting of the protruding areas and blind spots of the external walls without additional work and could perform the painting work through two nozzles. The suction fan was designed to improve wind resistance performance of GEWPro and maintain a consistent distance between the exterior wall and main body of GEWPro during the exterior wall painting. Using the two suction fans installed in the main body of GEWPro, it is possible to perform stable painting work and overcome protruding obstacles while moving up or down. Winder and endless wires are designed to minimize the risk of falling accidents of GEWPro. In addition, a camera attached to the main body of GEWPro allows real-time monitoring of the condition and painting quality of the exterior wall painting work. A cleaning tool has been designed to clean the outer walls, if necessary, by brushing and washing with water.

2.2.2. Hanger

The hanger has been developed to enable the lifting of GEWPro from a rooftop in the form shown in Figure 2g, and it can be easily installed, disassembled, and moved horizontally in the rooftop (parapet) without additional equipment. The system was designed to be small and light to increase the ease of work, and it can be applied to apartment complex rooftops (parapet) of various thicknesses with easy installation and disassembling.

2.2.3. Paint and Water Supply Device

The paint and water supply device was designed to be located on the ground to reduce the weight of GEWPro and the convenience of the work, and it was developed as an integrated device for the airless pump and solution tank (Figure 2e).

2.2.4. Task Management System

The GEWPro was designed to be able to scan the outer wall in three dimensions and automatically detect obstacles and openings on the exterior walls while moving up from the ground. Based on this scanning information, the system performs painting work by automatically making judgments about the areas to be painted while moving down. As shown in Figure 2i, a task management system (controller) was developed to allow an operator to monitor and control the current status of GEWPro and the overall condition of the painting work being conducted in real time using a camera attached to the upper part of the robot (Figure 2h).

2.2.5. Work Process of the GEWPro

According to an analysis based on a field examination regarding the exterior wall painting work that is being performed by GEWPro, the process of the exterior wall painting work by GEWPro can be categorized into the following: (1) preparation work; (2) installation work; (3) painting work; (4) disassembly work; (5) horizontal moving work; and (6) demobilization work. Figure 3 presents the detailed work process.

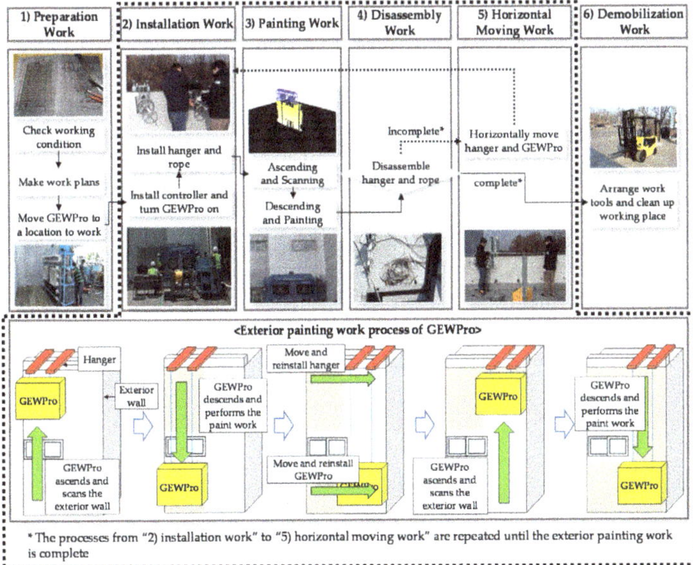

Figure 3. Major composition of GEWPro.

3. Development of Performance Evaluation and LCC Analysis Model of GEWPro

If the economic validity of automated construction robots were not secured, it would be difficult to introduce them to an actual construction field even if their safety and productivity are far superior to the conventional method. Therefore, proving the advantage of an automated method over the conventional method through economic validity analysis is crucial to the research, development, and commercialization of automated construction robots [14]. Thus, this study aims to develop a performance evaluation and LCC analysis model (Figure 4) that reflects the characteristics of building exterior painting work as well as the specifications and performance of GEWPro and verifies the validity of the proposed model and commercialization possibility of GEWPro through a case study.

3.1. Performance Evaluation Model of GEWPro

The performance evaluation model (Figure 4-Step 2) developed in this study appraised the performance of GEWPro by analyzing the resources required for exterior wall painting work per day, work time, and work productivity. As shown in Figures 1 and 3, work processes starting from "preparation work" through "horizontal moving work" are basically identical in the conventional method and automated method using GEWPro. The difference in productivity between the two methods stems from the difference in allocated resource amount (equipment, labor, etc.) and the required hours of work. In this study, the performance evaluation of GEWPro was carried out by comparing the productivity between conventional and automated methods resulting from work processes starting from "preparation work" through to "horizontal moving work" excluding the "demobilization work", which typically occurs after the daily working hours (8 hours/day).

3.2. LCC Analysis Model of GEWPro

The LCC analysis of GEWPro can be carried out by analyzing and calculating the additional cost incurred in exterior wall painting work using GEWPro (compared to that of the conventional method) and the benefits derived from the introduction of this automated method [15]. In this study, reasonable assumptions were made to conduct an economic analysis of GEWPro and the LCC analysis model that reflects the initial cost, maintenance cost, interest rate, number of workable days per year, possible total exterior painting area per year, and net profit per year resulting from the automated method was suggested. In addition, the economic feasibility of GEWPro was verified through a case study. For the means of economic analysis of GEWPro, (1) net present worth; (2) benefit/cost ratio; and (3) break-even point analyses were used; and (4) cost saving effect analysis of the automated method being introduced was conducted.

3.3. Sensitivity Analysis Model

In this study, the major factors that influence the performance evaluation and LCC analysis of GEWPro were identified and sensitivity analysis was performed on those factors to improve the reliability of the performance evaluation and LCC analysis result. Figure 4-Step 4 presents a model for sensitivity analysis.

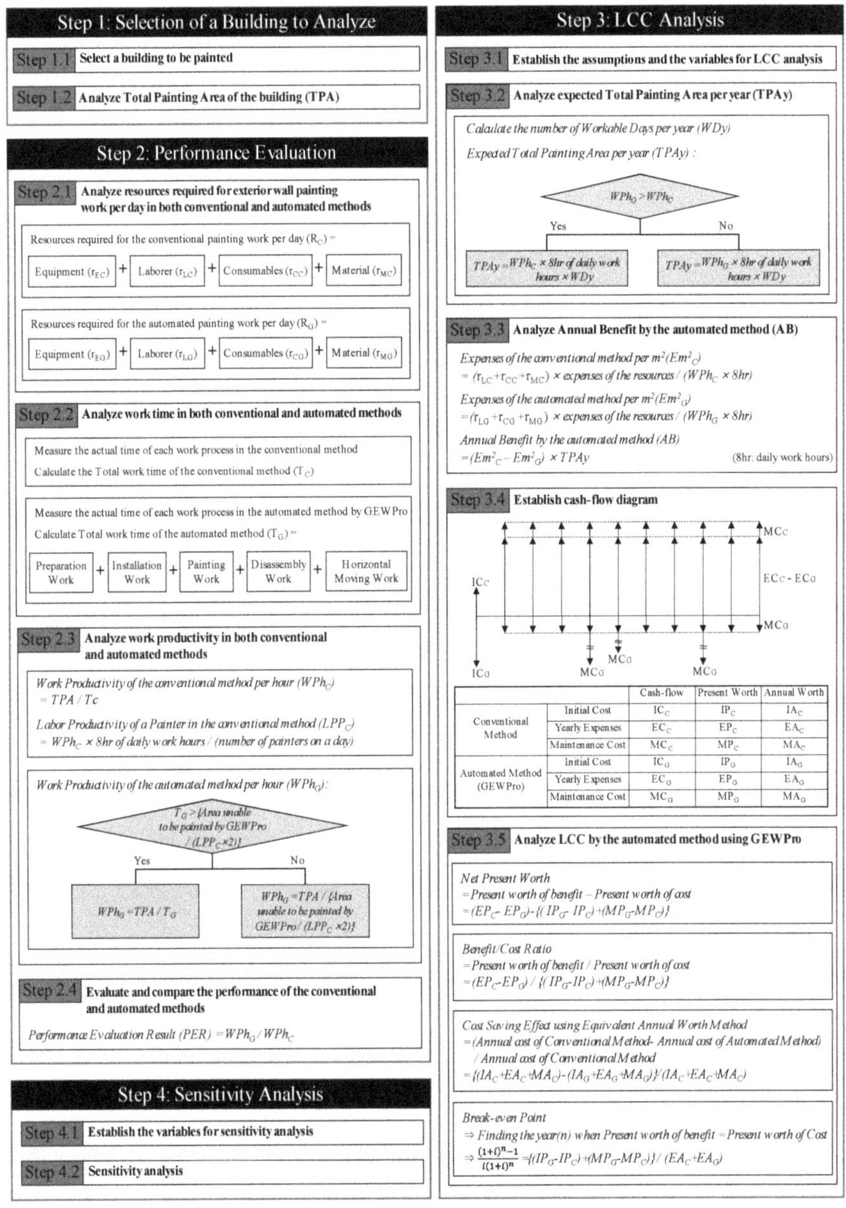

		Cash-flow	Present Worth	Annual Worth
Conventional Method	Initial Cost	IC_C	IP_C	IA_C
	Yearly Expenses	EC_C	EP_C	EA_C
	Maintenance Cost	MC_C	MP_C	MA_C
Automated Method (GEWPro)	Initial Cost	IC_G	IP_G	IA_G
	Yearly Expenses	EC_G	EP_G	EA_G
	Maintenance Cost	MC_G	MP_G	MA_G

Figure 4. Performance evaluation and LCC analysis model of GEWPro.

4. Performance Evaluation and LCC Analysis of GEWPro

4.1. Buildings for the Case Study

For a performance evaluation and LCC analysis of GEWPro, apartment complexes comprised of residences, 99.0~115.5 m² in size, which form the highest proportion (24%) among all apartments being built in South Korea [1], were considered for the case study. Two apartment complex construction sites,

Chungbuk Jincheon (Site A) and Gangwon Wonju (Site B), where exterior wall painting work was in progress at the time of this study, were selected (Table 1). From the floor plans for both sites, the total painting area of Sites A and B was 8275.0 m^2 and 8013.3 m^2, respectively, while the total painting area excluded from the study due to the difficulty of GEWPro access (bold lines of Table 1) was 2446.8 m^2 and 1841.0 m^2 for Sites A and B, respectively (Table 1).

Table 1. Floor plan and outline of the buildings for the case study.

Site	Detailed Summary		Floor Plan
Chungbuk Jincheon Site (Site A)	Floor area	108.9 m^2	
	No. of floors	20 floors above ground	
	Building height	57.8 m	
	Building perimeter	143.2 m	
	Total painting area	8275.0 m^2	
	Inaccessible area to GEWPro *	2446.8 m^2	
Gangwon Wonju Site (Site B)	Floor area	105.6 m^2	
	No. of floors	20 floors above ground	
	Building height	58.8 m	
	Building perimeter	136.3 m	
	Total painting area	8013.3 m^2	
	Inaccessible area to GEWPro *	1841.0 m^2	

* Inaccessible area to GEWPro can be defined as follows: Places where hangers cannot be installed on the parapet of a rooftop because the parapet is not continuously connected or where the width of the area to be painted is more narrow than 2.6 m, the width of GEWPro.

4.2. Performance Evaluation of GEWPro

4.2.1. Daily Resource Analysis of Conventional and Automated Methods Using GEWPro

As summarized in Table 2, daily resources (R_C, R_G) consisting of the equipment, laborer, consumables, material, etc. required for conventional exterior painting work and automated method using GEWPro were identified based on an interview conducted with exterior painting work experts as well as field test results using GEWPro.

Table 2. Daily resource analysis of the conventional and automated methods using GEWPro.

Item		Conventional Method		GEWPro
		Site A	Site B	
Equipment (r_E)	GEWPro	-	-	1 ea.
	Airless pump	4 ea.	3 ea.	2 ea.
Laborer (r_L)	Foreman	1	-	1
	Painter (Spray coating)	4	3	2
	Painter (Roller, Brush)	-	2	1
	Laborer (Helper, Safety)	1	1	1
	Operator	-	-	1
	Total	6	6	6
Consumables (r_C)	Spray gun	4	3	2
	Hose	4	3	2
	Nozzle tip	4	3	2
Material (r_M)	Paint	0.24 ℓ/m^2	0.24 ℓ/m^2	0.24 ℓ/m^2

4.2.2. Total Work Time Analysis of Conventional and Automated Methods Using GEWPro

The total work time required for conventional exterior painting work and the automated method using GEWPro was analyzed using the work sampling technique according to Step 2.2 of Figure 4. The results show that using the conventional method, the total time required for exterior wall painting work (T_C) for a single apartment building on both sites (Site A and B) was 3 days (24 h), whereas the automated method using GEWPro (T_G) took 20.9 h for Site A and 20.2 h for Site B for a single apartment building (Table 3). The total work time using GEWPro excludes painting work for areas not accessible by GEWPro (bold lines of Table 1).

Table 3. Analysis of the total work time required using GEWPro on a single apartment building.

Work Process	Required Work Time		Comment
	Site A	Site B	
Preparation Work	30.0 min.		Performed once initially
Installation Work	42.8 min.		Performed once initially
Painting Work	820.4 min.	796.6 min.	Ascend/descend speed 6.2 m/min.
Disassembly Work and Horizontal Moving Work	361.2 min.	344.4 min.	Dismantling and horizontal movement time 8.4 min/event
Total	1254.4 min (20.9 h)	1213.8 min (20.2 h)	

4.2.3. Work Productivity Analysis of Conventional and Automated Methods Using GEWPro

As summarized in Table 4, the hourly work productivity of the conventional method and automated method using GEWPro (WPh_C, WPh_G, respectively) and single painter's hourly productivity (LPP_C) were analyzed according to Step 2.3 of Figure 4. Regarding the exterior painting area, where GEWPro was not accessible (bold lines of Table 1), two additional painters were deployed to carry out the work. Because the conventional exterior painting work and automated method using GEWPro can be done concurrently without any mutual interference, the total required work time used to calculate GEWPro work productivity was determined by taking the larger value between T_G and $LPP_C \times 2$, which was the time required for two additional painters deployed to work on areas not accessible by GEWPro. Because the time taken for two additional painters to complete the painting work for areas not accessible by GEWPro was less than T_G on both case study buildings, the total work time of GEWPro (T_G) was used in GEWPro work productivity analysis (Figure 4-Step 2.3).

Table 4. Work productivity analysis result.

Each Building	Conventional Method		GEWPro	
	Site A	Site B	Site A	Site B
Painting area	8275.3 m^2	8013.3 m^2	8275.0 m^2	8013.3 m^2
Total time required	24 h	24 h	20.9 h	20.2 h
Hourly work productivity	344.8 m^2/h	333.9 m^2/h	396.2 m^2/h	396.5 m^2/h
Average hourly work productivity for both sites	339.4 m^2/h		396.4 m^2/h	
Work productivity of single painter	86.2 m^2/M·h	66.8 m^2/M·h	86.2 m^2/M·h	66.8 m^2/M·h
Work time required for areas not accessible by GEWPro	-	-	14.2 h	13.8 h

4.2.4. Performance Evaluation of GEWPro

According to Step 2.4 in Figure 4, a performance evaluation of GEWPro was conducted using Equation (1). In calculating and analyzing the difference in productivity between the conventional and automated methods, the average hourly work productivities (WPh_C, WPh_G) obtained from Sites A and B for each method was used for a performance evaluation. Performance evaluation analysis showed that the productivity of an automated method using GEWPro (WPh$_G$) was 16.8% higher than that of the conventional method (WPh$_G$).

$$Performance\ Evaluation\ Result\ (PER)$$
$$= (Work\ Productivity\ of\ the\ automated\ method\ per\ hour\ (WPh_C))$$
$$/\ Work\ Productivity\ of\ the\ conventional\ method\ per\ hour\ (WPh_G)$$
$$= (396.4\ \mathrm{m}^2/\mathrm{h})/(339.4\ \mathrm{m}^2/\mathrm{h}) = 1.168 \tag{1}$$

4.3. LCC Analysis of GEWPro

4.3.1. Establishment of the Assumptions and the Variables for LCC Analysis

Prior to performing LCC analysis, assumptions and variables necessary for LCC analysis were established, as shown in Table 5 below.

Table 5. Assumptions and variables for LCC analysis.

Assumptions and Variables		Value	Value Setting Basis
Interest Rate		2.90%	Average inflation rate of South Korea in recent 10 years
Conventional Method	Initial Cost	$14,494	Airless pump cost
	Maintenance Cost	$544/year	Airless pump maintenance cost
	Expected Service Life	10 year	Airless pump specification
GEWPro	Initial Cost	$60,568	Production cost of GEWPro
	Maintenance Cost	Annually ($716) Every 4 years ($16,099) Every 5 years ($6577)	Replacement costs of detailed parts per varying durability
	Expected Service Life	10 year	Life expectancy of GEWPro

4.3.2. Analysis of Expected Total Painting Area per Year

Set as a standard workload in performing the LCC analysis, the expected total painting area per year (TPAy) was calculated by multiplying the number of workable days per year (WDy: day/year) by the daily exterior wall painting work productivity (WPh$_C$ × 8 h: m^2/day). By establishing the unworkable conditions (i.e., average daily temperature of $\leq$5 °C; highest daily temperature of $\geq$35 °C; daily precipitation of $\geq$10 mm; daily maximum wind speed of $\geq$10 m/s; and national holidays) derived based on the Suspended Scaffold Safety Regulations of Korea Occupational Safety and Health Agency (KOSHA) [16] and the Painting Specifications of Korea Land and Housing Corporation [17] and by analyzing the recent ten years of weather data [18], the number of yearly workable days (WDy) was calculated to be 162 days. As shown in Step 3.2 of Figure 4, this yearly workable days (WDy) is multiplied by the smaller of the WPh$_C$ and WPh$_G$ to obtain the possible total painting area per year (TPAy). In this study, the TPAy required for LCC analysis was calculated to be 439,862.4 m^2/year according to Equation (2).

$$Expected\ Total\ Painting\ Area\ per\ year\ (TPAy)$$
$$= Work\ Productivity\ of\ the\ automated\ per\ hour\ (WPh_C) \times 8\ \mathrm{h/day}$$
$$\times\ the\ number\ of\ Workable\ Days\ per\ year\ (WDy)$$
$$= (339.4\ \mathrm{m}^2/\mathrm{h}) \times 8\ \mathrm{h/day} \times (162\ \mathrm{day/year}) = 439{,}862.4\ \mathrm{m}^2/\mathrm{year} \tag{2}$$

4.3.3. Annual Benefit Analysis of the Automated Method Using GEWPro

According to Step 3.3 in Figure 4, the annual net profit of introducing the automated method using GEWPro was calculated by multiplying the expected total painting area per year (TPAy) by the difference between the expenses of the conventional method per square meter (Em^2_C) and the automated method per square meter (Em^2_G). The expenses per unit area (m^2) for conventional and automated methods were taken from the sum of all the necessary expenses of applied resources (laborer, consumables, material) excluding the equipment expense, which is applied directly to the cash-flow diagram. In measuring the expenses for resources, the unit prices actually utilized in the field were applied. As shown in Table 6, which summarizes the resources and expenses for conventional and automated methods, the expenses of the conventional method per unit area (Em^2_C) were $0.9351/m^2$, whereas the expenses for automated method per unit area (Em^2_G) were calculated to be $0.7998/m^2$. Therefore, the annual present worth benefit (AB) of applying GEWPro was calculated to be \$59,520.1/year according to Equation (3).

$$
\begin{aligned}
\text{Annual Benefit by the automated method (AB)} \\
= \left(\text{Expenses of the conventional method per } m^2 \ \left(Em^2_C\right)\right. \\
- \left.\text{Expenses of the automated method per } m^2 \ \left(Em^2_G\right)\right) \\
\times \text{Expected Total Painting Area per year (TPAy)} \\
= \left(\$0.9351/m^2 - \$0.7998/m^2\right) \times 439{,}862.4 \ m^2/\text{year} = \$59{,}520.1/\text{year}
\end{aligned}
\tag{3}
$$

Table 6. Resources and expenses for the conventional and automated methods.

Category			Conventional Method			GEWPro	
Resources		Unit Price	Resources Applied		Daily Work Expenses ($): Average of Two Sites	Resources Applied	Daily Work Expenses ($)
			Site A	Site B			
Consumables	Spray gun	$13.9/unit	4 items	3 items	$48.5	2 items	$27.7
	Hose						
	Tip						
Laborer	Foreman	$93.4 per person	1	-		1	
	Painter (Spray coating)	$239.9 per person	4	3		2	
	Painter (Roller, Brush)	$219.1 per person	-	2	$1209.8	1	$1009.3
	Laborer	$115.5 per person	1	1		1	
	Driver	$101.5 per person	-	-		1	
Material	Paint	$2.6/ℓ	481.3 ℓ	508 ℓ	$1280.6	578 ℓ	$1496.6
Daily work expenses			$2538.9/day			$2533.6/day	
Daily work productivity			2714.7 m²/day			3167.6 m²/day	
Expenses per unit area (m²)			$0.9351/m²			$0.7998/m²	

4.3.4. Cash-Flow Diagram for LCC Analysis of GEWPro

Using the set assumptions and variables established previously and reflecting the expenses calculated for conventional and automated methods, Figure 5 presents the cash-flow diagram of GEWPro for the durability life of 10 years. Table 7 also lists the net present worth (NPW) and equivalent annual worth (EAW) derived from the annual benefit by the automated method (AB), which was reflected in the cash-flow diagram and necessary expenses (i.e., initial cost, maintenance cost) of equipment resources (r_{EC}, r_{EG}) for conventional and automated methods according to Step 3.4 of Figure 4.

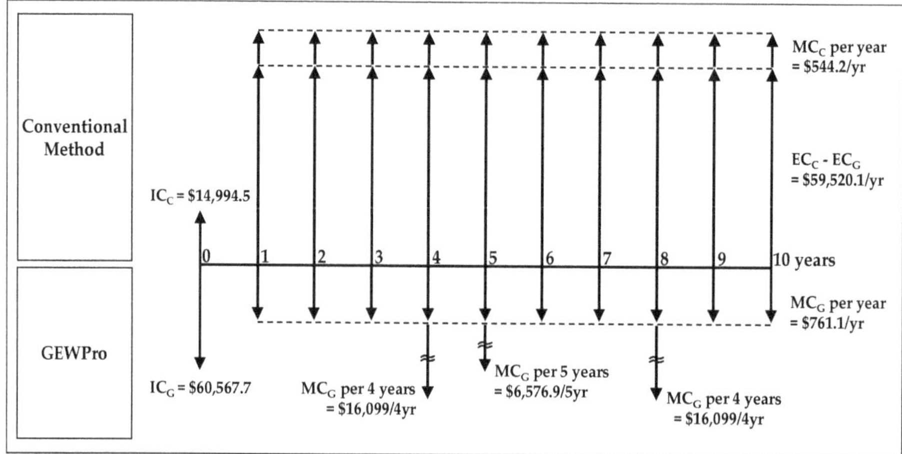

Figure 5. Cash-flow diagram for GEWPro's LCC analysis.

Table 7. Cash-flow analysis (present worth and annual worth).

			Cash-Flow Diagram	Present Worth	Annual Worth
Conventional method	Initial cost		$14,994.5	$14,994.5	$1748.9/year
	Yearly Expenses		$411,340.1/year	$3,526,789.7	$411,340.1/year
	Maintenance cost	Annually	$544.2/year	$4665.5	$544.2/year
GEWPro	Initial cost		$60,567.7	$60,567.7	$7064.0/year
	Yearly Expenses		$351,820.0/year	$3,016,470.2	$351,820.0/year
	Maintenance cost	Annually	$716.1/year	$6139.5	$716.1/year
		Every 4 years (at 4th year)	$16,099.0/4 year	$14,359.4	$1674.8/year
		Every 4 years (at 8th year)		$12,807.8	$1493.8/year
		Every 5 years	$6576.9/5 year	$5701.0	$664.9/year

4.3.5. LCC Analysis of GEWPro

- Net Present Worth; NPW

Net present worth analysis is a method for comparing the cash flow of a subject under economic analysis based on its present value. In this study, the present worth was calculated based on the cost difference incurred over ten years of operation (expected service life of GEWPro) using the conventional method versus automated method using GEWPro. According to the economic analysis using net present worth, the present worth of net profit was calculated to be $430,404, as shown in Equation (4).

$$
\begin{aligned}
&\text{Net Present Worth (NPW)}\\
=~&\text{Present worth of benefit} - \text{Present worth of cos}\,t\\
=~&(\text{EP}_C - \text{EP}_G) - \{(\text{IP}_G - \text{IP}_C) + (\text{MP}_G - \text{MP}_C)\}\\
=~&(\$3{,}526{,}789.7 - \$3{,}016{,}470.2) - \{(\$60{,}567.7 - \$14{,}994.5) +\\
&(\$6139.5 + \$14{,}359.4 + \$12{,}807.8 + \$5701) - \$4665.5\}\\
=~&\$430{,}404
\end{aligned}
\tag{4}
$$

- Benefit/Cost ratio

The B/C ratio is used in economic analysis where the economic payability is determined by calculating the ratio of the present worth of the incurred benefit to the present worth of the incurred cost. The system is economically viable if the B/C ratio is greater than one. In this study, as per Equation (5), the B/C ratio was calculated to be 6.39 based on the ratio of the incurred benefit and incurred cost of using conventional and automated methods over ten years (expected service life of GEWPro).

$$
\begin{aligned}
&\text{Benefit/Cost ratio (B/C)}\\
&= \text{Present worth of benefit / Present worth of } \cos t\\
&= (EP_C - EP_G)/\{(IP_G - IP_C) + (MP_G - MP_C)\}\\
&= (\$3{,}526{,}789.7 - \$3{,}016{,}470.2)/\{(\$60{,}567.7 - \$14{,}994.5)\\
&\quad + (\$6139.5 + \$14{,}359.4 + \$12{,}807.8 + \$5701) - \$4665.5\}\\
&= 6.39
\end{aligned}
\tag{5}
$$

- Break-even point

The break-even point is a point at which the present values of the incurred benefit and cost are the same [19]. The actual gain from using the GEWPro begins to occur after the break-even point. In this study, the break-even point was found to be 1.36 years (16.3 months) according to Equation (6).

$$
\begin{aligned}
&\text{Break} - \text{even point}\\
&\rightarrow \text{Finding the year}(n)\text{when present worth of benefit} = \text{present worth of } \cos t\\
&\rightarrow \frac{(1+i)^n-1}{i(1+i)^n} = \{(IP_G - IP_C) + (MP_G - MP_C)\}/(EA_C - EA_G)\\
&\rightarrow \text{when the interest rate }(i)\text{ is 2.90\%,}\\
&\frac{(1+0.029)^n-1}{0.029(1+0.029)^n}\\
&\qquad = \{(\$60{,}567.7 - \$14{,}994.5)\\
&\qquad + (\$6139.5 + \$14{,}359.4 + \$12{,}807.8 + \$5701) - \$4665.5\}\\
&\qquad /((\$411{,}340.1/\text{year} - \$351{,}820.0/\text{year}))\\
&\qquad = 1.34\\
&\therefore n = 1.36 \text{ year}
\end{aligned}
\tag{6}
$$

- Cost saving effect using equivalent annual worth method

The analysis of the construction cost saving effect utilizes the equivalent annual worth method in obtaining the ratio of the present worth of cost saving using automated method over the conventional method. Applying Equation (7), the cost saving effect of using the GEWPro was 12.2% for this study.

$$
\begin{aligned}
&\text{Cost saving effect using equivalent annual worth method}\\
&= (\text{annual } \cos t \text{ of conventional method} - \text{annual } \cos t \text{ of automated method})\\
&/(\text{annual } \cos t \text{ of conventional method})\\
&= \{(IA_C + EA_C + MA_C) - (IA_G + EA_G + MA_G)\}/(IA_C + EA_C + MA_C)\\
&= \{(\$1748.9/\text{year} + \$411{,}340.1/\text{year} + \$544.2/\text{year})\\
&- (\$7064.2/\text{year} + \$351{,}820.0/\text{year} + \$716.1/\text{year} + \$1674.8/\text{year}\\
&+ \$1493.8/\text{year} + \$664.9/\text{year})\}/(\$1748.9/\text{year} + \$411{,}340.1/\text{year} + \$544.2/\text{year})\\
&= 0.122
\end{aligned}
\tag{7}
$$

4.4. Sensitivity Analysis

4.4.1. Setting Major Variables for Sensitivity Analysis

Among the assumptions and variables previously established, sensitivity analysis was performed on the variables expected to have a significant influence or variables that must be considered for the future commercialization of GEWPro. The three significant variables finally selected for performing sensitivity analysis were (1) work time of each work process in the automated method by GEWPro (preparation and installation work, painting work, disassembly and horizontal moving work); (2) projected expenses of GEWPro equipment (initial cost of GEWPro, maintenance cost of GEWPro); and (3) interest rate. The sensitivity ranges for each variable in performing sensitivity analysis were established, as shown in Table 8.

Table 8. Setting ranges of the significant variables for sensitivity analysis.

(a) GEWPro Work Time of Each Work Process	Preparation and installation work	10~140 [min]
	Painting work	5~12 [m/min]
	Disassembly work and horizontal movement	1~13 [min/each]
(b) GEWPro Equipment Projected Expenses	GEWPro initial cost	−50~+50% [$]
	GEWPro maintenance cost	
	(c) Interest Rate	0.5~6 [%]

4.4.2. Sensitivity Analysis

- GEWPro work time of each work process

Sensitivity analysis regarding the work time of each work process of GEWPro was performed because it has a direct impact on work productivity, which in turn has a decisive effect on the result of LCC analysis. According to sensitivity analysis of the GEWPro work time of each work process, even if the time required for preparation and installation work process varies within the sensitivity range set in Table 8 (a), the use of GEWPro was deemed to be economical (Figure 6a). On the other hand, it was observed to be uneconomical if the speed of painting work drops below 5 m/min (Figure 6b) or if the time for each disassembly and horizontal moving work process exceeds 13 min (Figure 6c). Therefore, to secure the economic viability of GEWPro, it is recommended that the painting work process be maintained at least 5 m/min (currently, 6.2 m/min) while keeping the disassembly and horizontal moving work to be done within 13 min/process (currently, 8.4 min/process).

- GEWPro equipment projected expensess

Sensitivity analysis of the GEWPro equipment cost is essential to enhance the reliability of economic analysis because the projected expenses of GEWPro (i.e., initial cost, and maintenance cost) can vary according to the manufacturing process and the supply/demand of GEWPro (particularly the maintenance cost, because it can be generated randomly during the expected service life of GEWPro). The results of sensitivity analysis on the projected expenses of GEWPro equipment (Figure 6d,e) suggests that even if the GEWPro equipment projected expenses vary within the sensitivity range set in Table 8 (b), the use of GEWPro is economical.

- Interest rate to reflect economic condition

In this study, LCC analysis was conducted using the interest rate derived from the average inflation rate over the last 10 years in South Korea. Because interest rates can fluctuate depending on the economic conditions at home and abroad, it is imperative that sensitivity analysis is performed to enhance the analysis reliability. As shown in Figure 6f, sensitivity analysis showed that even if the interest rate varies within the sensitivity range set in Table 8 (c), the use of GEWPro is economical.

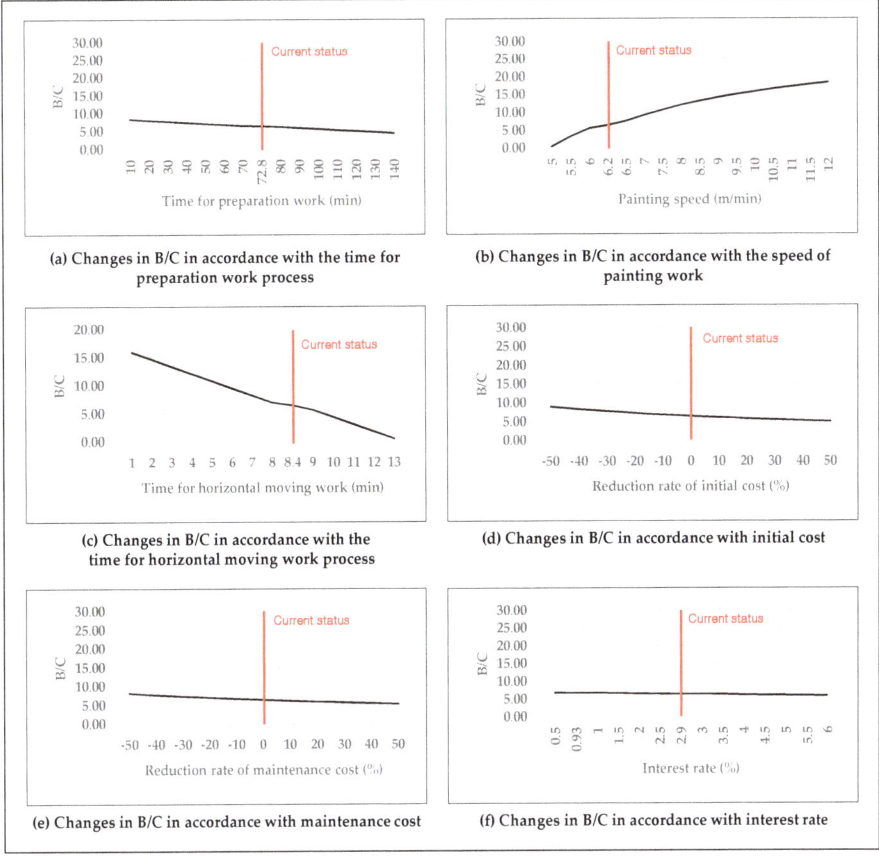

(a) Changes in B/C in accordance with the time for preparation work process

(b) Changes in B/C in accordance with the speed of painting work

(c) Changes in B/C in accordance with the time for horizontal moving work process

(d) Changes in B/C in accordance with initial cost

(e) Changes in B/C in accordance with maintenance cost

(f) Changes in B/C in accordance with interest rate

Figure 6. Results of sensitivity analysis.

5. Conclusions

This study developed a performance evaluation and LCC analysis model and used the model to evaluate the performance and analyze the LCC of a GEWPro, which was developed to improve the problems related to conventional exterior wall painting work. The results of the study are as follows:

(1) According to the survey conducted to analyze the characteristics and problems related to exterior wall painting work, issues with the work productivity and product quality arose due to the aging of skilled exterior wall painters and the difficulty in the supply of a new workforce. In particular, 66% of the survey respondents perceived the risk of exterior painting work to be high and recognized that preparing alternatives, such as automation (utilizing robots, etc.) of exterior painting work was urgent.

(2) To meet these needs and automate the exterior painting work, a Gondola-type Exterior Wall Painting robot (GEWPro) was recently developed in South Korea. This study developed a performance evaluation and LCC analysis model for GEWPro based on an understanding of the conventional exterior work characteristics and an analysis of the specifications and performance of GEWPro.

(3) According to the performance evaluation of GEWPro, productivity of the automated method using GEWPro was 16.8% superior to the conventional method. Moreover, LCC analysis of

GEWPro showed that the present net profit, B/C ratio, break-even point, and annual construction cost saving rate was $430,404, 6.39, 1.36 years, and 12.2%, which showed that the use of GEWPro was economically more efficient and viable compared to the conventional method.

(4) According to sensitivity analysis, which was performed to improve the reliability of a performance evaluation and LCC analysis of GEWPro, among the six major variables identified for sensitivity analysis, four variables (preparation and installation work process, GEWPro initial cost, maintenance cost, and interest rate) had a very low impact on the economic efficiency of GEWPro within the set sensitivity range. To secure the economic efficiency of using GEWPro over the conventional method, however, the speed of the painting work process should be greater than 5 m/min (currently, 6.2 m/min), whereas the disassembly and horizontal moving work should be done within 13 min/process (currently, 8.4 min/process).

Regarding economic analysis, although this study only considered the quantitative resources and expenses, such as equipment and labor cost, it is expected that the economic reliability and validity of the automated method would be furthered if additional consideration is given to qualitative factors, such as safety enhancement and uniformity in the quality of work brought by the automated method. Finally, authors need to consistently experiment and verify the sustainable applicability and operational stability of GEWPro on construction sites. For this, a number of field trials and case studies should be carried out in the near future.

Acknowledgments: This work was supported by INHA UNIVERSITY Research Grant.

Author Contributions: Dong-Jun Yeom conceived the idea for this study and wrote the manuscript. Eun-Ji Na and Mi-Young Lee designed the early methodology. Yoo-Jun Kim contributed to some of the LCC analysis and editing the paper. Young-Suk Kim and Chung-Suk Cho conceived the idea, supervised the research, and revised the manuscript. All authors read and approved the final manuscript.

Conflicts of Interest: The authors declare no conflict of interest.

Nomenclature

AB	Annual Benefit by the automated method ($/year)
B/C	Benefit/Cost
EA_C	Yearly Expenses of the conventional method at Annual worth ($/year)
EA_G	Yearly Expenses of the automated method at Annual worth ($/year)
EC_C	Yearly Expenses of the conventional method at Cash-flow ($)
EC_G	Yearly Expenses of the automated method at Cash-flow ($)
Em^2_C	Expenses of the conventional method per m^2 ($/$m^2$)
Em^2_G	Expenses of the automated method per m^2 ($/$m^2$)
EP_C	Yearly Expenses of the conventional method at Present worth ($)
EP_G	Yearly Expenses of the automated method at Present worth ($)
GEWPro	Gondola-type Exterior Wall Painting robot
IA_C	Initial cost of the conventional method at Annual worth ($/year)
IA_G	Initial cost of the automated method at Annual worth ($/year)
IC_C	Initial cost of the conventional method at Cash-flow ($)
IC_G	Initial cost of the automated method at Cash-flow ($)
IP_C	Initial cost of the conventional method at Present worth ($)
IP_G	Initial cost of the automated method at Present worth ($)
LCC	Life Cycle Cost ($)
LPP_C	Labor Productivity of a Painter in the conventional method (m^2/man)
MA_C	Maintenance cost of the conventional method at Annual worth ($/year)
MA_G	Maintenance cost of the automated method at Annual worth ($/year)
MC_C	Maintenance cost of the conventional method at Cash-flow ($)
MC_G	Maintenance cost of the automated method at Cash-flow ($)
MP_C	Maintenance cost of the conventional method at Present worth ($)
MP_G	Maintenance cost of the automated method at Present worth ($)

PER	Performance Evaluation Result
R_C	Resources required for the conventional painting work per day
r_{CC}	Consumable resources required for the conventional painting work per day
r_{CG}	Consumable resources required for the automated painting work per day
r_{EC}	Equipment resources required for the conventional painting work per day
r_{EG}	Equipment resources required for the automated painting work per day
R_G	Resources required for the automated work per day
r_{LC}	Laborer resources required for the conventional painting work per day
r_{LG}	Laborer resources required for the automated painting work per day
r_{MC}	Material resources required for the conventional painting work per day
r_{MG}	Material resources required for the automated painting work per day
T_C	Total work time of the conventional method (h)
T_G	Total work time of the automated method (h)
TPA	Total Painting Area of the Building (m^2)
TPAy	Total Painting Area per year ($/year)
WDy	The number of Workable Days per year (day/year)
WPh_C	Work Productivity of the conventional method per hour (m^2/h)
WPh_G	Work Productivity of the automated method per hour (m^2/h)

References

1. Statistics Korea. Annual Construction Report of Apartment Complex. Available online: http://www.kostat.go.kr (accessed on 15 May 2017).
2. Ministry of Land. Infrastructure and Transport in Korea, Enforcement Regulations of Multi-family Housing Management. Available online: http://www.law.go.kr (accessed on 15 May 2017).
3. Korea Agency for Infrastructure Technology Advancement. Final Report of Maintenance Robot. Available online: https://www.kaia.re.kr/portal/landmark/readTskFinalView.do?tskId=55292&yearCnt=5&menuNo=200100 (accessed on 15 October 2015).
4. Slocum, A.; Schena, B. Blockbot: A Robot to Automate Construction of Cement Block Walls. *Robot. Autonom. Syst.* **1988**, *4*, 111–129. [CrossRef]
5. Pritschow, G.; Dalacker, M.; Kurz, J.; Gaenssle, M. Technological Aspects in the Development of a Mobile Bricklaying Robot. *Autom. Constr.* **1996**, *5*, 3–13. [CrossRef]
6. Lichtenberg, J. The Development of a Robot for Paving Floors with Ceramic Tiles. In Proceedings of the 20th International Association for Automation and Robotics in Construction, Eindhoven, Holland, 2003; pp. 85–88. Available online: http://www.irbnet.de/daten/iconda/CIB13500.pdf (accessed date 11 October 2017).
7. Lee, J.; Yoo, H.; Kim, Y.; Lee, J.; Cho, M. The Development of a Machine Vision-Assisted Tele-operated Pavement Crack Sealer. *Autom. Constr.* **2006**, *15*, 616–626. [CrossRef]
8. Kim, Y.; Lee, J.; Kim, S.; Lee, J. Development of an Automated Machine for PHC Pile Head Grinding and Crushing Work. *Autom. Constr.* **2009**, *18*, 737–750. [CrossRef]
9. Kim, Y.; Lee, J.; Yoo, H.; Lee, J.; Jung, U. A Performance Evaluation of a Stewart Platform based Hume Concrete Pipe Manipulator. *Autom. Constr.* **2009**, *18*, 665–676.
10. Sherif, Y.; Kolarik, W. Life Cycle Costing: Concept and Practice. *Omega* **1981**, *9*, 287–296. [CrossRef]
11. Schau, E.M.; Traverso, M.; Lehmann, A.; Finkbeiner, M. Life Cycle Costing in Sustainability Assessment—A Case Study of Remanufactured Alternators. *Sustainability* **2011**, *3*, 2268–2288. [CrossRef]
12. Özkan, A.; Günkaya, Z.; Tok, G.; Karacasulu, L.; Metesoy, M.; Banar, M.; Kara, A. Life Cycle Assessment and Life Cycle Cost Analysis of Magnesia Spinel Brick Production. *Sustainability* **2016**, *8*, 662. [CrossRef]
13. Cho, J.; Yoo, H.; Choi, S.; Kim, Y. Wind Resistance Performance Analysis of Automated Exterior Wall Painting Robot for Apartment Buildings. *KSCE J. Civ. Eng.* **2015**, *19*, 510–519. [CrossRef]
14. Kim, Y. *Development of Remote Control Device for Automation of Road Surface Maintenance*; Research Report; Ministry of Land, Infrastructure and Transport (MOLIT): Sejong, Korea, 2004.
15. Kim, Y.; Lee, J. A Conceptual Model and Technical—Economical Feasibility Analysis of Apartment Exterior Wall Painting Robot. *KICEM Korea Inst. Constr. Eng. Manag.* **2006**, *22*, 139–150.
16. Korea Occupational Safety & Health Agency. Suspended Scaffold Safety Guidelines. Available online: http://www.kosha.or.kr (accessed on 7 June 2017).

17. Korea Land & Housing (LH) Corporation. Painting Specifications. Available online: http://www.lh.or.kr (accessed on 10 June 2017).
18. Korea Meteorological Administration. Available online: http://www.kma.go.kr (accessed on 11 June 2017).
19. Remer, D.; Tu, J.; Carson, D.; Ganiy, S. The State of the Art of Present Worth Analyses of Cash Flow Distributions. *Eng. Costs Prod. Econ.* **1984**, *7*, 257–278. [CrossRef]

Article

Behaviour of a Sustainable Concrete in Acidic Environment

Salim Barbhuiya * and Davin Kumala

Department of Civil Engineering, Curtin University, Perth 6845, Australia; davin.kumala@student.curtin.edu.au
* Correspondence: salim.barbhuiya@curtin.edu.au; Tel.: +61-892-662-392

Received: 29 July 2017; Accepted: 30 August 2017; Published: 1 September 2017

Abstract: Sustainability has become one of the most important considerations in building design and construction in recent years. Concrete is susceptible to acid attack because of its alkaline nature. The socioeconomic losses associated with infrastructure deterioration due to acid attack exceed billions of dollars all around the world. An experimental investigation was carried out to study the behaviour of sustainable concrete in 3% sulphuric acid and 1.5% nitric acid environment in which cement was replaced by a combination of fly ash and ultra fine fly ash. It was found that the compressive strength loss of concrete in these acid environments was the minimum in which cement was replaced by 30% fly ash and 10% ultra fine fly ash. This mix also showed the lowest mass loss when exposed to these acids.

Keywords: sustainable concrete; fly ash; ultra fine fly ash; sulphuric acid; nitric acid

---

1. Introduction

The impact of concrete, being one of the most commonly used construction materials worldwide, on sustainability can be significant. Concrete, in general, has a relatively low embodied energy compared to other construction materials. Fly ash, a by-product from thermal power stations, has been proven to have a lower embodied energy compared to ordinary Portland cement (OPC) [1]. The use of fly ash as a supplementary cementitious material (SCM) in concrete is well recognised for its economic and performance advantages such as improved workability and durability [2–5]. In fact, fly ash is specified in various Standards for use as a SCM [6] and in General Purpose and Blended Cements [7]. Studies have shown that by using high volumes of fly ash (>50%) it is possible to achieve the desired properties of concrete with a minimized cost [6,7].

The pozzolanic reaction of fly ash is a slow process. Therefore, the early strength of fly ash concrete is much lower than the concrete which does not contain any fly ash [8]. Different approaches have been used to accelerate the pozzolanic reaction of fly ash in concrete [9–12]. One of the approaches studied is the incorporation of very small size pozzolanic materials. In particular, microsilica has been used to improve the early age strength properties of concrete containing fly ash [13–15]. Ultra fine fly ash (UFFA) is a recently developed material. It is produced by a proprietary separation system with a mean particle diameter of 1–5 microns and contains 20% more amorphous silica than typical class F fly ash (particle diameter of 1–300 microns) [16]. Therefore, not only have the benefits of using UFFA in concrete been studied [17–19], but also the effectiveness of UFFA in improving the strength of fly ash concrete at early age has been evaluated [20]. The use of UFFA in concrete also contributes to the sustainability. This is because, compared to cement production, the UFFA production does not require any high energy-intensive process.

It has been recognised that, in general, ordinary Portland cement (OPC) concrete has minimal (almost no) resistance to acid attacks. While some weaker acids can be tolerated if exposed occasionally, OPC is known to be unable to hold up against any solution with a pH of 3 or lower [21]. Sulphuric acid

(H_2SO_4) is one of the most deleterious acids to act on concrete due to the combination of acid and sulphate attack. The deterioration of concrete sewer pipes due to sulphuric acid attack is a global problem all around the world. Moreover, industrial waste often contains a large amount of sulphuric acid. Therefore, concrete structures in industrial areas are exposed to of sulphuric acid attack. Sulphuric acid reacts with calcium hydroxide (CH), hydration product of cement in concrete and produce gypsum. The creation of gypsum in concrete causes volume increase. The gypsum also reacts with calcium aluminate hydrate (C_3A) to produce ettringite. The volume of ettringite is almost seven times more than the initial compounds [22]. Ettringite causes inner pressure in concrete leading to the formation of cracks [23]. Ultimately, the corroded concrete loses its mechanical strength that contributes to more cracking, spalling and finally leads to completely destruction [24].

Nitric acid (HNO_3) is another powerful corrosive acid that is immensely aggressive in nature. Nitric acid occurs in chemical plants producing explosives, artificial manure and similar products. Although nitric acid is not as strong as sulphuric acid, its effect on concrete at brief exposure is more destructive. The nitric acid reacts with CH of concrete and produces a highly soluble calcium nitrate salt. This salt weakens the cement paste structure and reduces the strength of concrete.

Different strategies have been used to enhance the resistance of concrete in acidic environment. One of the strategies, found to be very effective, is the use of various supplementary cementitous materials such as fly ash, slag, microsilica and calcite laterites [25–27]. Although extensive research has been carried out on the use of UFFA in concrete either individually or in combination with fly ash, very few studies evaluated its effectiveness on the durability properties of fly ash concrete. This paper reports the results of an investigation on the behaviour of a concrete in sulphuric acid and nitric acid environment where cement was replaced with fly ash and UFFA.

2. Experimental Programme

2.1. Materials

The cement used in this study was a General Purpose Grey Portland cement (PC) supplied by Cockburn Cement of Western Australia. The commercially available class F fly ash (FA) and ultra fine fly ash (UFFA) were used as partial replacement of cement. The UFFA had 18% more amorphous content compared to FA. The chemical composition and physical properties of all materials used in this study are summarised in Table 1. The aggregates used consisted of coarse aggregates with sizes of 20 mm and 10 mm, while the fine aggregate was natural sand.

Table 1. Chemical composition and physical properties of materials.

Chemical Composition			
Oxides	Cement (%)	Fly Ash (%)	Ultra Fine Fly Ash (%)
SiO_2	21.1	51.8	73.4
Al_2O_3	4.7	26.4	17.7
Fe_2O_3	2.8	13.2	4.4
CaO	63.8	1.61	0.9
MgO	2.0	1.1.7	0.6
MnO	-	0.10	0.1
K_2O	-	0.68	1.03
Na_2O	-	0.31	0.11
P_2O_5	-	1.39	0.20
TiO_2	-	1.44	0.70
SO_3	2.4	0.21	0.20
Physical Properties			
Particle Size	25–40% $\leq 7\ \mu m$	40% of 10 μm	Mean Size 3.4 μm
Specific gravity	2.7–3.2	2.6	2.0–2.55
Surface area (m^2/kg)	352	340	2510
Loss of Ignition (%)	2.4	0.50	0.60

2.2. Mix Proportions

In total, five mixes of concrete were cast. The first mix was a control mix with 100% OPC. The remaining mixes contained OPC with at varying percentages of FA and UFFA. The amount of UFFA was kept constant at 10% based on previous studies [17–19]. Details of mix proportions are shown in Table 2. The water–binder ratio was kept constant at 0.35 for all the mixes. A polycarboxylate-based superplasticiser was used to maintain a constant workability (slump = 100 ($\pm$5) mm). Due to high fineness of UFFA, the water demand in the mixes increased with the increase in the quantity of UFFA. Therefore, to balance the water demand, it was needed to use higher quantity of superplasticizer as the UFFA content in the mixes increased. The target strength of control mix was 35 MPa at 28 days.

Table 2. Mix proportions.

| Mix No. | Mix ID | kg/m^3 | | | | | | SP * (%) |
		OPC	FA	UFFA	Water	Fine Aggregate	Coarse Aggregate	
Mix 1	OPC (control)	400	0	-	140	600	1250	0
Mix 2	20% FA + 10% UFFA	280	80	40	140	600	1250	0.5
Mix 3	30% FA + 10% UFFA	240	120	40	140	600	1250	1.0
Mix 4	40% FA + 10% UFFA	200	160	40	140	600	1250	1.2
Mix 5	50% FA + 10% UFFA	160	200	40	140	600	1250	1.5

* SP: % by mass of total binder.

2.3. Specimen Preparation

A 160 kg capacity pan mixer was used to prepare the concrete. The speed of the mixer was 26 rotations per minute. Cube specimens (100 mm $\times$ 100 mm $\times$ 100mm) were cast in two layers. After casting each layer, the specimens were compacted using a vibrating table. The vibration was carried out until air bubbles stopped appearing on the surface. The frequency of the vibration was 60 Hz. A plastic sheet was used to cover the specimens in moulds, and these were then kept in casting room for 24 h. The temperature of the room was maintained at 20 ($\pm$1) °C. After 24 h the specimens were demoulded and placed in water bath for 3 days. The temperature of water bath was maintained at 20 ($\pm$1) °C. After this, the specimens were sealed in polythene sheets and kept in a storage laboratory until the day of testing. The temperature and relative humidity of the storage laboratory was maintained at 20 ($\pm$1) °C and 65% $\pm$ 1%, respectively.

2.4. Test Methods

2.4.1. Compressive Strength

The compressive strength testing was performed in accordance with the guidelines given in AS (1012.9-2014 [28]) at the age of 7, 14, 28, 56 and 90 days. At each age, three specimens were tested and the mean value of these measurements is reported.

2.4.2. Strength and Mass Loss

After 28 days of curing, the 100 mm cube samples were immersed in sulphuric acid of 3% concentration (H_2SO_4, pH $\approx$ 3) and nitric acid of 1.5% concentration (HNO_3, pH $\approx$ 3) for a period up to 90 days. These concentrations have been taken from existing literatures [29,30]. The solutions of acids were prepared by mixing concentrated acids with a predetermined amount of tap water. The pH level of acid solutions was monitored regularly using a portable digital pH meter (standard error: $\pm$0.05). To maintain the desired pH levels, the concentrated acid was added either weekly or when the pH level went up. It has to be mentioned that the pH value depends on the degree of dissociation of radicals, and it may not be a true indicator of the concentration of acid in the solution [31]. Therefore, in the present study, the concentration was used directly as an indicator of the aggressiveness of the exposure environment.

The samples were removed from the acid solution after the exposure period and brushed carefully to remove the loose particles from the surface. They were then left for drying under room temperature for 1 h before determining the loss in compressive strength and the mass changes. The loss in compressive strength was calculated by determining the strengths at 7, 14, 28, 56 and 90 days. The mass loss was determined at 3, 7, 14, 28, 56 and 90 days.

2.4.3. Scanning Electron Microscope (SEM)

The microstructure was studied using scanning electron microscope, Zeiss EVO-40 (Carl-Zeiss, Germany). The small cut samples were polished using silicon carbide paper and coated with platinum before imaging.

3. Results and Discussion

3.1. Compressive Strength Development

The compressive strength development of concrete containing different amounts of fly ash and ultra fine fly ash is shown in Figure 1. The compressive strength values are summarised in Table 3 along with standard deviation. In Figure 1, it can be seen that the strength development for Mix 1 (OPC) was much faster than the other mixes containing FA and UFFA. The compressive strength of Mix 1 (OPC) at seven days was 31.8 MPa, while this was below 30 MPa for other mixes. This trend also continues at the 14 day of curing. Therefore, it can be said that the strength gain of concrete mixes containing high volumes of FA is much slower than the concrete mixes without FA. This is due to the slower pozzolanic reactions of FA, in which the reaction between FA and water creates a slower hydration rate compared to the reaction between cement and water. However, at later ages (28 days or after), it can be seen that the strength for all the FA concrete mixes begins to develop at an accelerating rate, most notably for Mix 2 (20% FA + 10% UFFA). At 90 days, it can be seen that the compressive strength of mix Mix 2 (20% FA + 10% UFFA) far exceeds that of Mix 1 (OPC) and the rest, with Mix 3 (30% FA + 10% UFFA) coming in at the second. This also conforms to the findings of existing literature [32] that FA concrete has a slower strength gain at early age, but the strength exceeds the OPC concrete without any FA at 90 days.

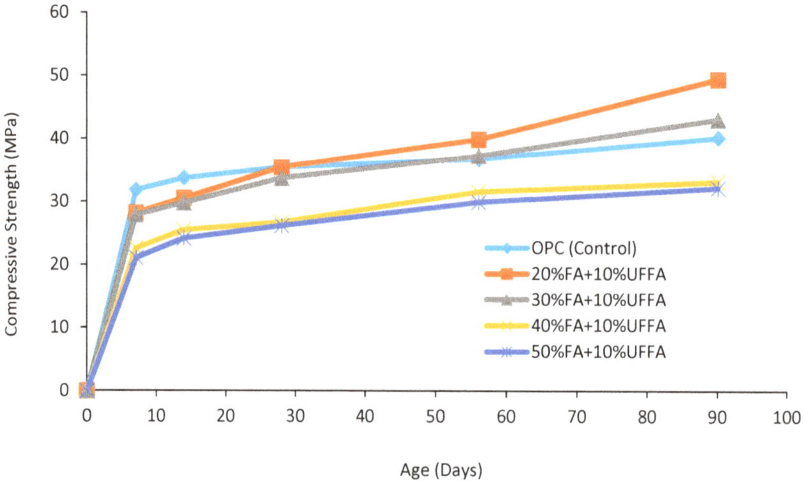

Figure 1. Compressive strength development of concrete.

Table 3. Compressive strength results.

Mix No.	Mix ID	7 Days (MPa) (Avg ± SD)	14 Days (MPa) (Avg ± SD)	28 Days (MPa) (Avg ± SD)	56 Days (MPa) (Avg ± SD)	90 Days (MPa) (Avg ± SD)
Mix 1	OPC (control)	31.67 ± 0.25	33.53 ± 0.59	35.37 ± 1.16	36.23 ± 2.59	40.13 ± 0.15
Mix 2	20% FA + 10% UFFA	28.37 ± 0.32	29.60 ± 1.73	35.3 ± 2.31	40.17 ± 2.48	49.00 ± 1.23
Mix 3	30% FA + 10% UFFA	27.60 ± 0.30	29.53 ± 1.76	33.50 ± 0.53	37.30 ± 1.45	42.73 ± 0.60
Mix 4	40% FA + 10% UFFA	22.43 ± 0.85	25.60 ± 0.35	26.53 ± 0.40	31.47 ± 0.74	33.27 ± 0.91
Mix 5	50% FA + 10% UFFA	20.97 ± 1.72	24.03 ± 0.68	26.07 ± 1.42	29.33 ± 1.64	32.50 ± 1.31

3.2. Behaviour in Sulphuric Acid Environment

Figure 2 shows the compressive strength loss for the five mixes when they are immersed in 3% sulphuric acid for a period of up to 90 days. The losses of compressive strength are summarised in Table 4 along with standard deviation. It can be observed from Figure 2 that Mix 1 (OPC) had the highest loss in the compressive strength at 90 days. Although Mix 1 (OPC) possessed the highest compressive strength initially, it was only able to retain 37.1% of its seven-day compressive strength after 90 days. This indicates that Mix 1 (OPC) was affected the most in 3% sulphuric acidic environment. Mix 2 (20% FA + 10% UFFA) showed the second largest variance, with compressive strength of 18.6 MPa compared to 26.0 MPa at seven days. Mix 3 (30% FA + 10% UFFA) also showed a declining compressive strength trend. However, this was not as severe as Mix 1 (OPC) and Mix 2 (20% FA + 10% UFFA). Mix 4 (40% FA + 10% UFFA) and Mix 5 (50% FA + 10% UFFA) showed minimal changes in strength loss, with less than 15% strength loss.

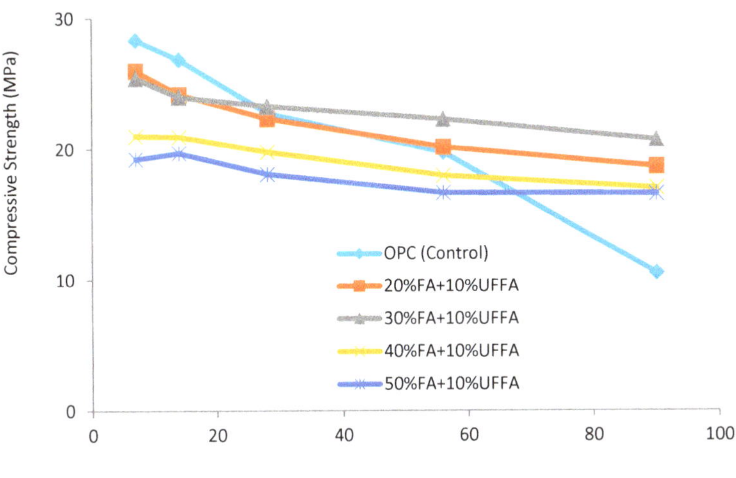

Figure 2. Compressive strength loss of concrete in sulphuric acid (3%).

Table 4. Compressive strength loss of concrete in sulphuric acid (3%).

Mix No.	Mix ID	7 Days (MPa) (Avg ± SD)	14 Days (MPa) (Avg ± SD)	28 Days (MPa) (Avg ± SD)	56 Days (MPa) (Avg ± SD)	90 Days (MPa) (Avg ± SD)
Mix 1	OPC (control)	27.93 ± 2.84	26.70 ± 1.08	22.57 ± 0.42	19.53 ± 1.50	10.73 ± 3.17
Mix 2	20% FA + 10% UFFA	25.80 ± 1.25	24.17 ± 1.08	22.40 ± 0.20	20.17 ± 1.25	18.37 ± 2.37
Mix 3	30% FA + 10% UFFA	25.07 ± 1.61	23.97 ± 0.40	23.23 ± 0.32	22.20 ± 0.10	20.77 ± 0.35
Mix 4	40% FA + 10% UFFA	21.37 ± 2.97	20.70 ± 2.29	19.20 ± 2.08	17.67 ± 0.51	16.70 ± 0.66
Mix 5	50% FA + 10% UFFA	19.20 ± 0.26	19.77 ± 1.01	18.07 ± 1.85	16.73 ± 1.56	16.37 ± 0.32

Figure 3 shows the percentage mass loss of concrete cubes immersed in 3% sulphuric acid for a period of up to 90 days. The mass losses are summarised in Table 5 along with standard deviation. It can be observed that the maximum mass loss occurred in Mix 1 (OPC). The per cent mass reduction increases as the exposure period prolongs, showing an almost linear rate of mass loss. Mix 1 (OPC) showed a mass loss of 1.2% at three days, increasing to 10.3% at 28 days up to 22.7% at 90 days. While not as significant as Mix 1 (OPC), Mix 2 (20% FA + 10% UFFA) presented a mass loss of 2.7% at 28 days up to 8.9% at 90 days. Other mixes showed minimal mass loss, with less than 1% change at the end of 90 days. It can also be seen that the percentage of mass loss decreases as the volume of FA increases in each mix. The minimal mass loss per cent change in Mix 3 (30% FA + 10% UFFA), Mix 4 (40% FA + 10% UFFA) and Mix 5 (50% FA + 10% UFFA) is associated with the greater volume of FA to cement replacement, which provides a higher resistance to sulphuric acid attack. The could also be due to the accumulation of gypsum at the surface, effectively blocking or reducing further reactions from occurring, whilst already possessing a denser matrix.

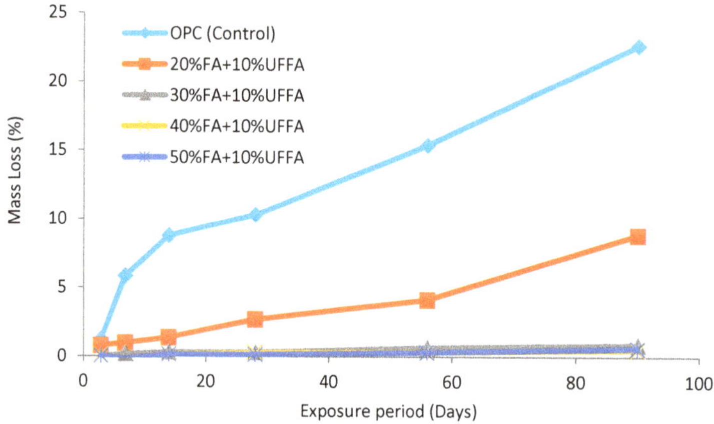

Figure 3. Mass loss of concrete in sulphuric acid (3%).

Table 5. Mass loss of concrete in sulphuric acid (3%).

Mix No.	Mix ID	3 Days (MPa) (Avg ± SD)	7 Days (MPa) (Avg ± SD)	14 Days (MPa) (Avg ± SD)	28 Days (MPa) (Avg ± SD)	56 Days (MPa) (Avg ± SD)	90 Days (MPa) (Avg ± SD)
Mix 1	OPC (control)	1.22 ± 0.08	5.82 ± 0.24	8.78 ± 0.12	10.35 ± 0.49	15.40 ± 0.45	22.64 ± 0.35
Mix 2	20% FA + 10% UFFA	0.77 ± 0.05	0.96 ± 0.12	1.35 ± 0.08	2.68 ± 0.36	4.15 ± 0.17	8.89 ± 1.04
Mix 3	30% FA + 10% UFFA	0.13 ± 0.14	0.17 ± 0.06	0.36 ± 0.15	0.30 ± 0.07	0.65 ± 0.04	0.88 ± 0.08
Mix 4	40% FA + 10% UFFA	0.09 ± 0.08	0.28 ± 0.05	0.17 ± 0.08	0.28 ± 0.06	0.34 ± 0.03	0.52 ± 0.07
Mix 5	50% FA + 10% UFFA	0.04 ± 0.03	0.27 ± 0.09	0.15 ± 0.08	0.14 ± 0.03	0.34 ± 0.05	0.61 ± 0.08

3.3. Behaviour in Nitric Acid Environment

Figure 4 shows the compressive strength loss of concrete immersed in 1.5% nitric acid for a period of up to 90 days. The compressive strength losses are summarised in Table 6 along with standard deviation. In Figure 4, it can be seen that Mix 1 (OPC) had the greatest decline in compressive strength at 90 days, with a compressive strength of 21.3 MPa. Mix 1 (OPC) was only able to retain 72% of its seven-day compressive strength after 90 days. Mix 2 (20% FA + 10% UFFA) showed the second largest variance, with compressive strength of 22.5 MPa compared to 27.4 MPa at seven days. The other mixes showed minimal changes in strength loss, with less than 10% strength loss. Comparing Figures 2 and 4, it can also be observed that the strength loss of concrete in 3% sulphuric acid is much greater than in 1.5% nitric acid.

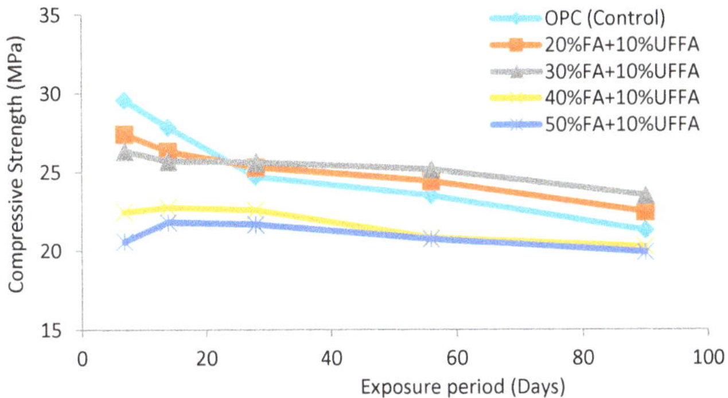

Figure 4. Compressive strength loss of concrete in nitric acid (1.5%).

Table 6. Compressive strength loss of concrete in nitric acid (1.5%).

Mix No.	Mix ID	7 Days (MPa) (Avg ± SD)	14 Days (MPa) (Avg ± SD)	28 Days (MPa) (Avg ± SD)	56 Days (MPa) (Avg ± SD)	90 Days (MPa) (Avg ± SD)
Mix 1	OPC (control)	29.40 ± 0.26	27.60 ± 2.19	24.03 ± 1.40	23.47 ± 0.40	21.23 ± 0.49
Mix 2	20% FA + 10% UFFA	27.30 ± 0.44	26.27 ± 0.21	24.60 ± 1.50	24.53 ± 0.86	22.50 ± 0.10
Mix 3	30% FA + 10% UFFA	26.23 ± 0.25	25.73 ± 0.38	25.53 ± 0.61	25.13 ± 0.15	23.40 ± 2.01
Mix 4	40% FA + 10% UFFA	22.27 ± 1.78	22.70 ± 0.56	22.40 ± 0.46	20.83 ± 0.25	20.13 ± 0.49
Mix 5	50% FA + 10% UFFA	20.60 ± 0.50	21.67 ± 1.42	21.73 ± 0.31	20.73 ± 0.65	20.10 ± 0.46

Figure 5 shows the percentage mass loss of concrete cubes immersed in 1.5% nitric acid for a period of up to 90 days. The mass losses are summarised in Table 7 along with standard deviation. It can be observed in Figure 5 that Mix 1 (OPC) had the most significant loss in mass, from 1.5% at three days, 4% at 28 days and 5% at 90 days. Rest of the mixes showed a much lower rate of loss with less than 0.6% at three days and less than 1% at 28 days. At 90 days, Mix 2 (20% FA + 10% UFFA) reached a mass loss of 2.7% while Mix 3 (30% FA + 10% UFFA) reached 2.2%. Both Mix 4 (40% FA + 10% UFFA) and Mix 5 (50% FA + 10% UFFA) showed a mass loss of 2% at 90 days, indicating the highest resistance. All mixes showed a consistent trend with the mass loss per cent increasing as the exposure period increased. This indicates that the resistance improves as the FA replacement level increases. The reduction of mass loss in mixes containing FA and UFFA can be attributed to the lower traces of CH due to pozzolanic reactions, minimising further reactions from the nitric acid.

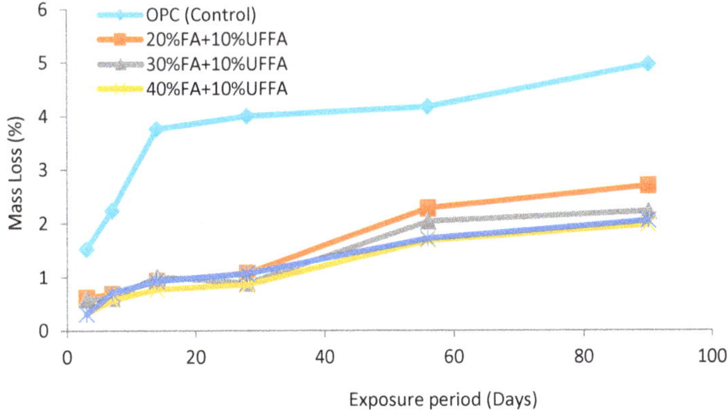

Figure 5. Mass loss of concrete in nitric acid (1.5%).

Table 7. Mass loss of concrete in nitric acid (1.5%).

Mix No.	Mix ID	3 Days (MPa) (Avg ± SD)	7 Days (MPa) (Avg ± SD)	14 Days (MPa) (Avg ± SD)	28 Days (MPa) (Avg ± SD)	56 Days (MPa) (Avg ± SD)	90 Days (MPa) (Avg ± SD)
Mix 1	OPC (control)	1.52 ± 0.08	2.22 ± 0.19	3.64 ± 0.57	4.09 ± 0.23	4.17 ± 0.45	4.79 ± 0.89
Mix 2	20% FA + 10% UFFA	0.62 ± 0.05	0.68 ± 0.08	0.94 ± 0.05	1.08 ± 0.11	2.28 ± 0.43	2.65 ± 0.53
Mix 3	30% FA + 10% UFFA	0.58 ± 0.03	0.62 ± 0.07	1.08 ± 0.24	0.89 ± 0.09	2.09 ± 0.59	2.23 ± 0.29
Mix 4	40% FA + 10% UFFA	0.30 ± 0.020	0.57 ± 0.08	0.92 ± 0.04	0.85 ± 0.06	1.67 ± 0.22	1.96 ± 0.67
Mix 5	50% FA + 10% UFFA	0.31 ± 0.04	0.69 ± 0.08	1.09 ± 0.23	1.08 ± 0.17	1.74 ± 0.24	2.07 ± 0.43

3.4. Visual Inspection

Figure 6 shows the various stages of concrete deterioration in 3% sulphuric acid environment. It can be seen that Mix 1 (OPC) suffered the greatest signs of deterioration at the end of 90 days in compared to the other mixes. Mix 1 (OPC) also showed the signs of peeling and full exposure of the aggregate surface at 28 days. At 90 days, the initial layer was found to be completely disintegrated with some of the initial surface aggregates already falling off. This would also link to the reduction of mass and compressive strength for this mix. Mix 2 (20% FA + 10% UFFA) showed the signs of deterioration, with the formation of gypsum at the surface at 28 days and becoming more porous. It was also observed that the initial layer of the surface started spalling off and exposed aggregates at 90 days. Mix 3 (30% FA + 10% UFFA) and Mix 4 (40% FA + 10% UFFA) showed similar behaviour. However, the deterioration signs were less as the volumes of FA increased. The deterioration was much slower, with the aggregates being slightly exposed at 90 days. The volumes of these two mixes also appeared to have increased at 28 days, which could be as a result of the formation of gypsum. Mix 5 (50% FA + 10% UFFA) appeared to be the most aesthetically resistant, with no major structural changes at the end of 90 days.

Figure 6. Concrete deterioration in 3% sulphuric acid solution.

Figure 7 shows the various stages of concrete deterioration in 1.5% nitric acid environment. Similar to 3% sulphuric acid environment, Mix 1 (OPC) showed the most serious damage in 1.5% nitric acid, with spalling of the surface beginning already at 28 days. At 90 days, larger surfaces of the aggregates can be observed with more severe spalling of the surface. Mix 2 (20% FA + 10% UFFA) showed the higher resistance compared to Mix 1 (OPC). However, Mix 2 (20% FA + 10% UFFA) showed severe spalling and exposed aggregates at 90 days. Other mixes behaved in a similar fashion with structural changes not as severe as Mix 1 (OPC) or Mix 2 (20% FA + 10% UFFA). All mixes showed a browning of colour at 28 days, turning lighter again at 90 days after disintegration and spalling of the initial layer.

Figure 7. Concrete deterioration in 1.5% nitric acid solution.

3.5. Microstructural Observation from SEM

The SEM image of a sample from Mix 1 (OPC) at 28 days is shown in Figure 8. The hexagonal plate-shaped crystals of CH and C-S-H gels are clearly visible in the image. The presence of excess hydrous calcium-sulpho-aluminate hydrate (also known as ettringite) characterised by needle-like structures is also evident. Large number of pores and voids can also be seen in the image. The SEM image of a sample from Mix 3 (30% FA + 10% UFFA) at 28 days is shown in Figure 9. The SEM image shows a denser matrix with much lower trace of the CH crystals. It is considered that the majority of CH content might have reacted with the amorphous silica of FA and UFFA to produce secondary C-S-H gel by the pozzolanic reactions. The denser microstructure is likely to be associated with micro-filing effects of UFFA. The UFFA might have filled the pores and voids between the unreacted particles in the hydrated matrix, effectively densifying the pore structure.

Figure 10 shows the SEM image of a sample from Mix 3 (30% FA + 10% UFFA) exposed to 3% sulphuric acid for 28 days. The surface appears to be highly porous in the image. A large scale of possible micro-cracks and voids can also be observed. A noticeable amount of C-S-H gel appears to have been decomposed into finer particles. Remains of calcium hydroxide crystals and unreacted FA and UFFA also appear to be present. Furthermore, the signs of gypsum can be seen to cover the surface area including particles of FA. The extensive formation of gypsum in the surface regions may have caused the disintegration resulting the spalling of the surface. Figure 11 shows the SEM image of sample from Mix 3 (30% FA + 10% UFFA) immersed in 1.5% nitric acid for a period of 28 days. The surface also appears to be very porous, with the salt by-products on the surface caused by the reaction of the acid with the CH. Small round particles appear are the unreacted FA and UFFA. The broken surface pieces are likely to be the traces of calcium nitrate salt and calcium nitro-aluminate hydrate. It also appears that the ions from the nitric acid have completely disintegrated the C-S-H gel on the outer surface of the sample leading to dissolution and deterioration of the surface layer.

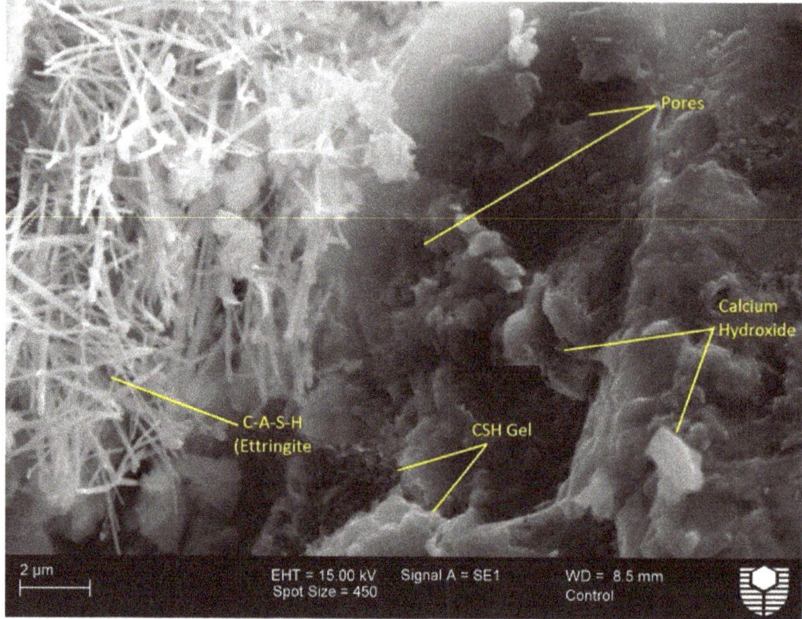

Figure 8. SEM image of Mix 1 (OPC) (28 days of water curing).

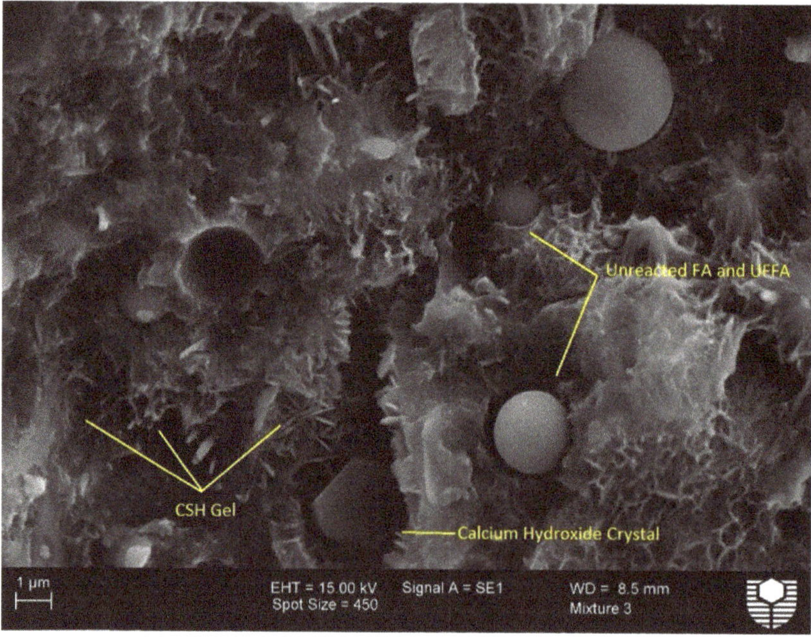

Figure 9. SEM image of Mix 3 (30% FA + 10% UFFA) (28 days of water curing).

Figure 10. SEM image of Mix 3 (30% FA + 10% UFFA) in 3% sulphuric acid (exposure period 28 days).

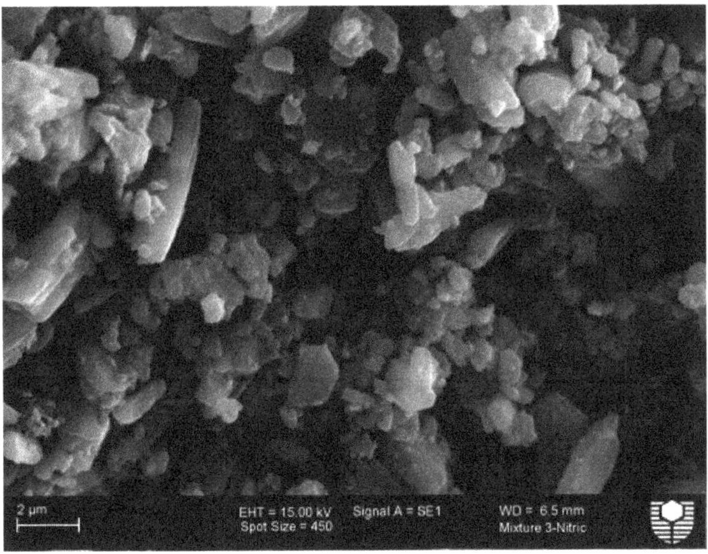

Figure 11. SEM image of Mix 3 (30% FA + 10% UFFA) in 1.5% nitric acid (exposure period 28 days).

Total porosity and the presence of microcracks have significant influence on the permeability of concrete. In general, permeability decreases with an increase in porosity up to a certain level, and then the influence of porosity on permeability is negligible. The presence of microcracks also increases the permeability of concrete, and thus encourages more rapid deterioration.

4. Conclusions

The behaviour of a sustainable concrete containing fly ash and ultra fine fly ash in 3% sulphuric acid and 1.5% nitric acid environment was studied in this research. Based on the results obtained, the following conclusions are made:

- In sulphuric acid environment, the compressive strength loss was minimum for a concrete mix in which cement was replaced with 30% fly ash and 10% ultra fine fly ash. The mass loss was less in this mix compared to the mix without fly ash. However, mass loss was also less in mixes containing higher amounts of fly ash.

- In nitric acid environment, concrete mixes containing 20% fly ash and 10% ultra fine fly ash and 30% fly ash and 10% ultra fine fly ash had the minimum compressive strength loss. However, the mass loss in mix containing 30% fly ash and 10% ultra fine fly ash was less than the mix containing 20% fly ash and 10% ultra fine fly ash.

- The SEM image of concrete mix with 30% fly ash and 10% ultra fine fly ash cured in water for 28 days showed denser microstructure characterised by less amounts of calcium hydride crystals. The SEM image of concrete mix containing 30% fly ash and 10% ultra fine fly ash exposed to sulphuric acid for 28 days showed that the surface is highly porous. A noticeable amount of C-S-H gel appears to have been decomposed into finer particles. When the same mix was exposed to nitric acid for a period of 28 days, the SEM image showed that the surface is very porous, with the salt by-products on the surface caused by the reaction of the acid with the calcium hydroxide.

Author Contributions: Salim Barbhuiya conceived and designed the experiments and wrote the paper. Davin Kumala performed the experiments and analysed the data.

Conflicts of Interest: The authors declare no conflict of interest.

References

1. Flower, D.; Sanjayan, J.; Baweja, D. Environmental impacts of concrete production and construction. In Proceedings of the Concrete Institute of Australia Biennial Conference, Melbourne, Australia, 17–19 October 2005.
2. Han, S.H.; Kim, J.K.; Park, Y.D. Prediction of compressive strength of fly ash concrete. *Cem. Concr. Res.* **2003**, *33*, 965–971. [CrossRef]
3. Malhotra, V.M. Durability of concrete incorporating high-volume of low calcium (ASTM class F) fly ash. *Cem. Concr. Compos.* **1990**, *12*, 487–493. [CrossRef]
4. Bilodeau, A.; Malhotra, V.M. High-volume fly ash system: Concrete solution for sustainable development. *ACI Mater. J.* **2000**, *97*, 41–48.
5. Langey, W.S.; Carette, G.G.; Malhotra, V.M. Strength development and temperature rise in large concrete blocks containing high-volumes of low-calcium (ASTM class F) fly ash. *ACI Mater. J.* **1992**, *89*, 362–368.
6. Standards Australia. *Australian Standard AS3582.1, Supplementary Cementitious Materials for Use with Portland and Blended Cement—Part 1: Fly Ash*; Standards Australia International: Sydney, Australia, 1998.
7. Standards Australia. *Australian Standard AS3972, General Purpose and Blended Cements*; Standards Australia International: Sydney, Australia, 2010.
8. Malhotra, V.M.; Zhang, M.H.; Read, P.H.; Ryell, J. Long-term mechanical properties and durability characteristics of high-strength/high-performance concrete incorporating supplementary cementing materials under outdoor exposure conditions. *ACI Mater. J.* **2000**, *97*, 518–525.
9. Xu, A.; Sarkar, S.L. Microstructural study of gypsum activated fly ash hydration in cement paste. *Cem. Concr. Res.* **1991**, *21*, 1137–1147.
10. Paya, J.; Monzo, J.; Borrachero, M.V. Mechanical treatment of fly ashes: Part III. Studies on strength development of ground fly ashes. *Cem. Concr. Res.* **1997**, *27*, 1365–1377. [CrossRef]
11. Katz, A. Microstructure study of alkali-activated fly ash. *Cem. Concr. Res.* **1998**, *28*, 197–208. [CrossRef]
12. Barbhuiya, S.; Gbagbo, J.; Russell, M.; Basheer, M. Properties of fly ash concrete modified with hydrated lime and silica fume. *Constr. Build. Mater.* **2009**, *23*, 3233–3239. [CrossRef]

13. Nochaiya, T.; Wongkeo, W.; Chaipanich, A. Utilization of fly ash with silica fume and properties of Portland cement fly ash-silica fume concrete. *Fuel* **2010**, *89*, 768–774. [CrossRef]

14. Gesoglu, M.; Gunevisi, E.; Ozbav, E. Properties of self-compacting concretes made with binary, ternary, and quaternary cementitious blends of fly ash, blast furnace slag, and silica fume. *Constr. Build. Mater.* **2009**, *23*, 1847–1854. [CrossRef]

15. Bingol, A.F.; Tohumcu, I. Effects of different curing regimes on the compressive strength properties of self-compacting concrete incorporating fly ash and silica fume. *Mater. Des.* **2013**, *51*, 12–18. [CrossRef]

16. Obla, K.H.; Hill, R.L.; Shashiprakash, S.G.; Perebatova, O. Properties of concrete containing ultra-fine fly ash. *ACI Mater. J.* **2003**, *100*, 426–433.

17. Subramaniam, K.V.; Gromotka, R.; Shah, S.P.; Obla, K.; Hill, R. Influence of ultrafine fly ash on the early age response and the shrinkage cracking potential of concrete. *J. Mater. Civ. Eng.* **2005**, *17*, 45–53. [CrossRef]

18. Chindaprasirt, P.; Jaturapitakkul, C.; Sinsiri, T. Effect of fly ash fineness on compressive strength and pore size of blended cement paste. *Cem. Concr. Compos.* **2005**, *27*, 425–428. [CrossRef]

19. Choi, S.; Lee, S.S.; Monteiro, P.J.M. Effect of Fly ash fineness on temperature rise, setting, and strength development of mortar. *J. Mater. Civ. Eng.* **2012**, *24*, 499–505. [CrossRef]

20. Shaikh, F.U.A.; Supit, S.W.M. Compressive strength and durability properties of high volume fly ash (HVFA) concretes containing ultrafine fly ash (UFFA). *Constr. Build. Mater.* **2015**, *82*, 192–205. [CrossRef]

21. Shi, C.; Stegemann, J.A. Acid corrosion resistance of different cementing materials. *Cem. Concr. Res.* **2000**, *30*, 803–808. [CrossRef]

22. Monteny, J.; Vincke, E.; Beeldeens, A.; De Belie, N.; Taerwe, L.; Gemert, D.; Verstraete, W. Chemical, microbiological, and in situ test methods for biogenic sulfuric acid corrosion of concrete. *Cem. Concr. Res.* **2000**, *30*, 623–634. [CrossRef]

23. Monteny, J.; Belie, N.D.; Vincke, E.; Verstraete, W.; Taerwe, L. Chemical and microbiological tests to simulate sulfuric acid corrosion of polymer-modified concrete. *Cem. Concr. Res.* **2001**, *31*, 1359–1365. [CrossRef]

24. Lee, S.; Hooton, R.; Jung, H.; Park, D.; Choi, C. Effect of limestone filler on the deterioration of mortars and pastes exposed to sulfate solutions at ambient temperature. *Cem. Concr. Res.* **2008**, *38*, 68–76. [CrossRef]

25. Roy, D.M.; Arjunan, P.; Silsbee, M.R. Effect of silica fume, metakaolin and low-calcium fly ash on chemical resistance of concrete. *Cem. Concr. Res.* **2001**, *31*, 1809–1813. [CrossRef]

26. Monteny, J.; Belie, N.D.; Taerwe, L. Resistance of different types of concrete mixtures to sulphuric acid. *Mater. Struct.* **2003**, *36*, 242–249. [CrossRef]

27. Baret, G.; Poppe, A.M.; Belie, N.D. Strength and durability of high-volume fly ash concrete. *Struct. Concr.* **2008**, *9*, 101–108.

28. Standards Australia. *AS 1012.9-2014 Methods of Testing Concrete Compressive Strength Tests—Concrete, Mortar and Grout Specimens*; Standards Australia: Sydney, Australia, 2014.

29. Hewayde, E.; Nehdi, M.; Allouche, E.; Nakhla, G. Using concrete admixtures for sulphuric acid resistance. *Constr. Mater.* **2007**, *160*, 25–35. [CrossRef]

30. Goyal, S.; Kumar, M.; Sidhu, D.S.; Bhattacharjee, B. Resistance of mineral admixture concrete to acid attack. *J. Adv. Concr. Technol.* **2009**, *7*, 273–283. [CrossRef]

31. Zivica, V.; Bajza, A. Acidic attack of cement based materials—A review Part 2. Factors of rate of acidic attack and protective measures. *Constr. Build. Mater.* **2002**, *16*, 215–222. [CrossRef]

32. Mehta, P.K.; Monteiro, P.J.M. *Concrete: Microstructure, Properties, and Materials*, 3rd ed.; McGraw-Hill Comapnes Inc.: New York, NY, USA, 2006; p. 659.

 *sustainability*

Article

Remote Sensing Techniques for Urban Heating Analysis: A Case Study of Sustainable Construction at District Level

Stefania Bonafoni *, Giorgio Baldinelli, Paolo Verducci and Andrea Presciutti

Department of Engineering, University of Perugia, via G. Duranti 93, 06125 Perugia, Italy;
giorgio.baldinelli@unipg.it (G.B.); paolo.verducci@unipg.it (P.V.); andrea.presciutti@unipg.it (A.P.)
* Correspondence: stefania.bonafoni@unipg.it; Tel.: +39-075-585-3663

Received: 20 June 2017; Accepted: 24 July 2017; Published: 26 July 2017

Abstract: In recent years, many new districts in urban centres have been planned and constructed to reshape the structure and functions of specific areas. Urban regeneration strategies, planning and design principles have to take into account both socioeconomic perspectives and environmental sustainability. A district located in the historical city centre of Terni (Italy), Corso del Popolo, was analysed to assess the construction effects in terms of surface urban heat island (SUHI) mitigation. This district is an example of urban texture modification planned in the framework of the regeneration of the ancient part of the town. The changes were realised starting from 2006; the new area was completed on June 2014. The analysis was carried out by processing Landsat 7 ETM+ images before and after the interventions, retrieving land surface temperature (LST) and albedo maps. The map analysis proved the SUHI reduction of the new area after the interventions: as confirmed by the literature, such SUHI mitigation can be ascribed to the presence of green areas, the underground parking, the partial covering of the local roadway and the shadow effect of new multi-storey buildings. Moreover, an analysis of other parameters linked to the impervious surfaces (albedo, heat transfer and air circulation) driving LST variations is provided to better understand SUHI behaviour at the district level. The district regeneration shows that wisely planned and developed projects in the construction sector can improve urban areas not only economically and socially, but can also enhance the environmental impact.

Keywords: urban regeneration; sustainable construction; environmental effects; surface urban heat island; albedo

1. Introduction

The urbanization processes and related environmental effects are increasingly steered to the study of the urban heat island (UHI) phenomenon. The UHI is a typical form of anthropogenic climate modification mainly due to the growing urbanization process and accompanied by an increase of air pollution and anthropogenic heat sources [1]. The city expansion generally reduces the presence of green areas, with building structures and materials trapping solar radiation, determining significant temperature differences between urban and rural areas [1–4]. Urban texture and construction materials can cause an increase of air and surface temperatures of several degrees with respect to the surrounding rural areas. The UHI effect worsen during the hot season: high temperatures may affect human health, the quality of life of urban residents, energy consumption and air pollution. The UHI effect has been a concern for several decades, affecting not only large metropolitan areas, but also smaller cities; therefore, their planning and design represent the main step for the solution to the problem. The UHI typologies may be grouped in three categories [2]: the surface urban heat island (SUHI); the boundary layer heat island, and; the canopy layer heat island. The canopy layer UHI is affected by the air temperature heating in the atmosphere extending up to the top of buildings, while the boundary layer UHI extends up to 1 km [5]. The SUHI concerns the land surface temperature (LST) pattern, having greater variability and higher thermal peaks with respect to the air temperature. In the paradigm of urban sustainable development, the mitigation of the urban heat island is a key point [6]. Different strategies to reduce UHI effects can be adopted, based on properly designing the urban texture in order to obtain energy savings and health benefits [7–9], such as the increase of urban surface reflectivity [10–12] and urban vegetation (green roofs, street trees, and green spaces) [13–15].

Urbanised areas are characterised by an irregular thermal pattern, with the LST variations generally linked to the impervious surfaces reflectivity: albedo describes quite completely the surface behaviour in terms of reflectivity, as it represents its hemispherical reflection in the wavelength range of the solar spectrum. For instance, a simple way to increase the albedo of building roofs, and consequently to mitigate the UHI, is the adoption of cool roof paint.

Earth observation data from space-borne sensors have been widely exploited to examine the SUHI effects. Unlike in situ measurements, providing uneven distributed data, satellite observations have the advantages of covering large areas at the same time, and during different temporal intervals, ensuring a more effective analysis of the intra-urban SUHI spatial variability, closely related to the building distribution, surface materials and vegetation density. Different space-borne platforms, such as AVHRR [16–18], MODIS [19,20], ASTER [21–23], and Landsat [24–26], were used to retrieve the SUHI. Furthermore, satellite sensor measurements of surface reflectivities make it possible to retrieve albedo maps, both at the local and global spatial scale.

In this work, the retrieval from Landsat 7 satellite data of the urban LST and albedo is carried out over the city of Terni (Central Italy), characterized by several urban changes during the last 10–15 years. In particular, a district located in the historical city centre, Corso del Popolo, is analysed, where a clear construction intervention was realized starting from 2006; the intervention ended in June 2014. This district is an example of urban texture modification in the construction sector planned for the regeneration of the ancient part of the town. An analysis of the SUHI maps of this district before and after the interventions using different satellite images is provided. The 60 m pixel size of Landsat 7 ETM+ thermal channel proved to be suitable to monitor SUHI changes at the district level, making it possible to point out if urban construction changes move towards an urban sustainable criteria. Moreover, an analysis of the correspondent albedo maps is proposed, with a study of the parameters affecting the LST variations through an analytical model. The synergy of satellite techniques and analytical studies is aimed at assessing whether the planning and design principles fulfil the sustainability requirements in terms of urban heat island mitigation, and the more evident reasons for the heating variations.

2. Study Area

The study area is the Corso del Popolo district, located in the historical city centre of Terni, Central Italy (Figure 1). Terni (city centre coordinates 42.56° N, 12.64° E) has about 112,000 inhabitants, covers an overall area of 212 km^2, and its average altitude is 130 m above mean sea level. The city is quite far from the sea, and is located in a flat valley surrounded by mountainous areas; it has a temperate climate typical of mid-latitude regions, with hot, humid and poorly ventilated summers. The selected zone of Corso del Popolo (Figure 2) covers a surface of 300 × 300 m: The Universal Transverse Mercator (UTM) WGS84 coordinates (33T zone) are: longitude range (m E): 306,690–306,990 and latitude range (m N): 4,714,300–4,714,600.

Terni is an average Italian city characterized by an industrial texture at the top of the national table in the manufacturing of iron and steel and the production of electricity. From the second half of the 19th century, Terni underwent a transformation from a historic town, still structured along the lines of the urban layout of the ancient Roman town, into a dynamic industrial city.

The analysed district is an example of urban texture change planned in the framework of the regeneration of the ancient part of Terni following urban sustainable criteria (projected by Mario Ridolfi and Wolfgang Frankl architects) [27]. The construction interventions, between the Nera riverside and Corso del Popolo Street, started in 2006 and were completed in June 2014. Before 2006, a paved car park (with light-coloured asphalt) with perimetral trees was present. The new zone consists of a private area (seven buildings for residential, commercial and managerial end use) and a public area (a three-level underground car park with 1036 parking spaces, a building of 7000 m^2 with parking used as municipal offices, subway, green areas, underground roadway and a pedestrian bridge).

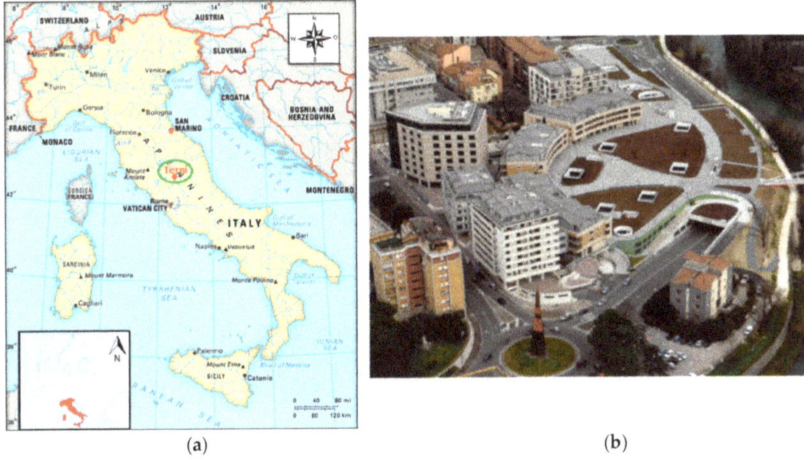

(a)　　　　(b)

Figure 1. (**a**) Terni location (city centre coordinates 42.56° N, 12.64° E); (**b**): the new built-up area of Corso del Popolo, picture from [27].

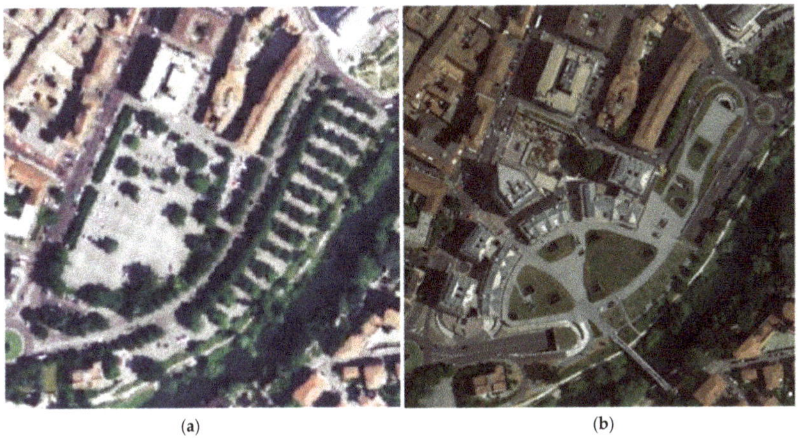

(a)　　　　(b)

Figure 2. (**a**) Regione Umbria (RU) image of Corso del Popolo district on 2005 [28] selected area of 300 × 300 m; (**b**) Google Earth (GE) image on 2015. Universal Transverse Mercator (UTM) WGS84 coordinates (33T zone): longitude range (m E): 306,690–306,990, latitude range (m N): 4,714,300–4,714,600.

The main idea of this operation, embracing an area approximately 4 acres large, is the regeneration of the historical centre of the town integrating it with the neighbour river zone (named "Città Giardino") through both the partial covering of the local roadway and the construction of a pedestrian bridge. The Municipality of Terni chose the project financing system as the most appropriate way to realize the interventions, applied for the first time to an urban area (with both private and public buildings) instead of a single public work. Some environmentally friendly materials were used in the building construction [27]. Figure 3 reports the main external materials employed in the new area.

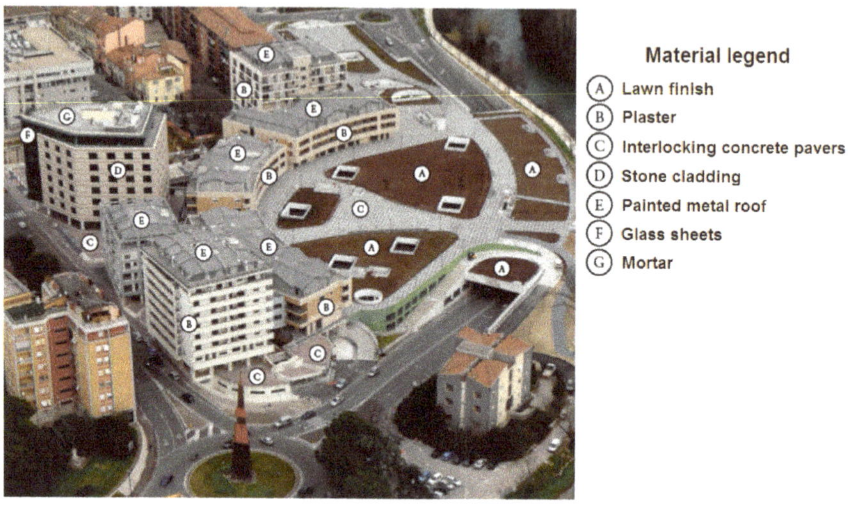

Figure 3. Main external materials used in the new study area.

The brown zones in Figure 3 (lawn finish) correspond to the green spots of Figure 2b.

3. Methods

3.1. Landsat 7 ETM+ Data and Processing

The LST maps, from which the SUHI ones are obtained, and the corresponding albedo maps, were retrieved using observations from the Enhanced Thematic Mapper Plus (ETM+) sensor on board the Landsat 7 satellite. ETM+ has six reflective bands in the visible (VIS), near infrared (NIR) and short-wavelength infrared (SWIR), a band in the thermal infrared region (TIR) and a panchromatic band, as reported in Table 1. The six reflective channels have a spatial resolution of 30 m, the thermal channel of 60 m, and the panchromatic one of 15 m. TIR band data are also delivered at 30 m by the U.S. Geological Survey [29], by means of a cubic convolution resampling method.

Table 1. Landsat 7 ETM+ channels and spectral range. Visible (VIS); near infrared (NIR); short-wavelength infrared (SWIR); thermal infrared region (TIR).

ETM + Channels	Spectral Range (μm)
Ch_1—VIS (blue)	0.45–0.51
Ch_2—VIS (green)	0.52–0.60
Ch_3—VIS (red)	0.63–0.69
Ch_4—NIR	0.77–0.90
Ch_5—SWIR	1.55–1.75
Ch_6—TIR	10.31–12.36
Ch_7—SWIR	2.06–2.35
Ch_8—panchromatic	0.52–0.90

In this work, nine Landsat 7 scenes over Terni before and after the intervention on Corso del Popolo were considered, as reported in Table 2, downloaded from USGS at 30 m resolution. The native

60 m pixel size of ETM + TIR channel, the highest resolution of current satellite-based TIR channels, allows monitoring SUHI changes at the district level with sufficient details. Summer images in clear-sky condition were selected, since it is the period with the more intense SUHI effects, generally with the maximum intensity during the month of July in Italy, as detailed in [20]. The satellite passages are at around 11:50, Central European Summer Time (CEST).

Table 2. Landsat 7 ETM + scenes analysed.

Period	Date
Pre-Intervention	27 June 2005, 24 June 2004, 16 August 2000, 31 July 2000
Post-Intervention	28 August 2016, 27 July 2016, 25 July 2015, 7 June 2015, 6 July 2014

The images were calibrated according to [30], in order to convert digital number values to at-sensor spectral radiances. The surface reflectances of the ETM+ reflective bands were computed through the scheme reported in [31], where only satellite scenes are needed, without the support of ground measurements.

3.2. LST and Albedo Retrieval from Landsat 7 Data

ETM + channel in the TIR band (Ch_6) was employed to retrieve LST by inverting the following radiative transfer equation [32]:

$$L_{sens,\lambda} = \left[\varepsilon_\lambda B_\lambda(T_s) + (1 - \varepsilon_\lambda)L_\lambda^\downarrow\right]\tau_\lambda + L_\lambda^\uparrow \tag{1}$$

where $L_{sens,\lambda}$ is the radiance (W·m^{-2}·sr^{-1}·μm^{-1}) received by the sensor, ε_λ is the emissivity of the surface, $B_\lambda(T_s)$ is the spectral radiance (W·m^{-2}·sr^{-1}·μm^{-1}) of a black body at temperature T_s (K) expressed by the Planck's law, T_s represents the LST, $L_\lambda^\downarrow$ and $L_\lambda^\uparrow$ are, respectively, the downwelling and upwelling atmospheric radiances (W·m^{-2}·sr^{-1}·μm^{-1}), and τ_λ is the transmissivity of the atmosphere. If the surface emissivity is known, LST is retrieved from Equation (1) by the inversion of the Planck's law [33]. The terms τ_λ, $L_\lambda^\downarrow$ and $L_\lambda^\uparrow$ were calculated through the tool available at [34]: it makes use of the atmospheric profiles of the National Centers for Environmental Prediction as input for the code of radiative transfer MODTRAN [35]. The threshold method described in [36,37] and applied to the normalized difference vegetation index (NDVI) was used to estimate the land surface emissivity.

Overall, both thermal and reflective data, selected in cloud-free conditions, were atmospherically corrected.

The total broadband albedo is obtained from the reflective ETM + channels using the following relation [38]:

$$\alpha = 0.356 \times Ch\_1 + 0.130 \times Ch\_3 + 0.373 \times Ch\_4 + 0.085 \times Ch\_5 + 0.072 \times Ch\_7 - 0.0018 \tag{2}$$

Since the albedo is defined as a bi-hemispherical reflectance, whilst the Landsat spectral reflectance is the ratio of the hemispherical incoming radiation to the conical reflected radiation [39], assuming that the surfaces are Lambertian, the broadband albedo can be retrieved from the spectral reflectances of Landsat, regarded as narrowband albedos [38–40].

3.3. SUHI Computation and LST Model

The SUHI intensity is the parameter used to assess the heating effects in urban areas: in this case, it is the difference in physical surface temperature between urban (LST$_{urban}$) and rural (LST$_{rural}$) pixels within a given image:

$$SUHI = LST_{urban} - LST_{rural} \tag{3}$$

It was chosen to compute the LST$_{rural}$ as the mean of LST associated with the rural pixels in the Terni area not affected by land cover changes over the years of the study.

In order to understand deeply the LST variations, an analytical model function of construction material behaviour, air temperature and heat transfer is considered. In fact, the LST variation is not simply driven by the albedo values, but further physical parameters have to be taken into account. Therefore, starting from the energy balance of a built-up surface when exposed to solar radiation, the LST is estimated as a function of different parameters beyond the albedo [10]:

$$(1 - \alpha)I = \sigma\varepsilon\left(T_s^4 - T_{sky}^4\right) + h_c(T_s - T_a) \tag{4}$$

where α is the albedo, I the incident solar radiation on the surface (W/m^2), ε the emissivity of the surface, σ the Stefan–Boltzmann constant (5.67×10^{-8} W/m^2·K^4), T_s is the LST (K), T_{sky} the radiating temperature of the sky (K), h_c the convection (and radiation) coefficient (W/m^2·K), and T_a the air temperature (K).

4. Results and Discussion

The changes of the area under investigation (Figure 2) is monitored in terms of SUHI intensity maps as shown in Figure 4 (before the intervention) and Figure 5 (after the intervention).

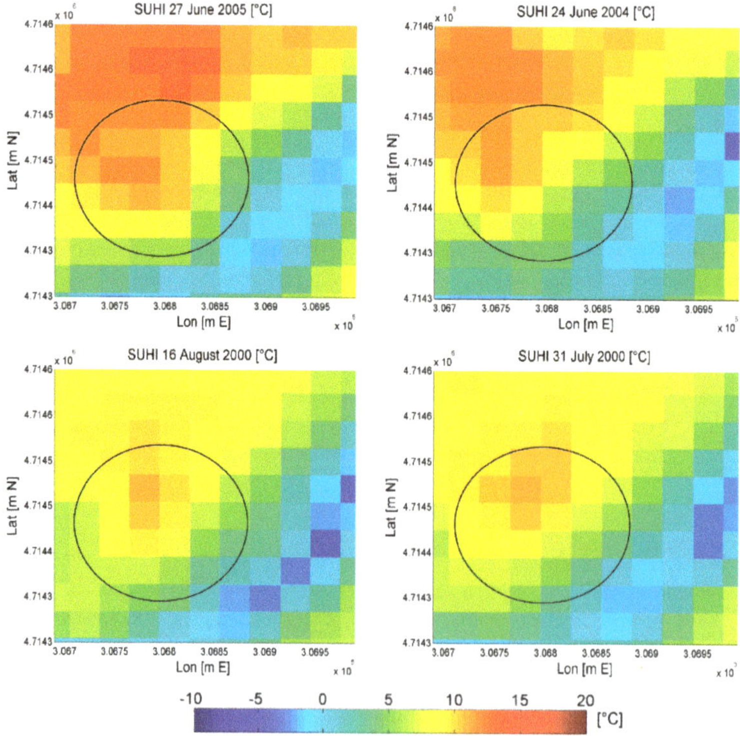

Figure 4. Pre-intervention surface urban heat island (SUHI) maps (°C): 27 June 2005; 24 June 2004; 16 August 2000, and; 31 July 2000. Landsat pixels are 30 × 30 m from the USGS. Lat/lon are in UTM WGS84 coordinates, 33T zone.

The 60 m pixel size of Landsat 7 TIR channel, resampled at 30 m, allows assessing SUHI pattern variations at district level (the selected area is 300 m × 300 m). The level of detail of data derived from satellite is coarser than that of Figure 3, but looking at Figure 3 itself, it emerges that there are areas with the same characteristics (lawn finish, or painted metal roofs) covering a surface comparable or larger than the satellite pixel size. Even though sharp thermal variations are averaged with the surroundings inside the pixel area, the trend of the SUHI pattern can be delineated, revealing its behaviour inside the district.

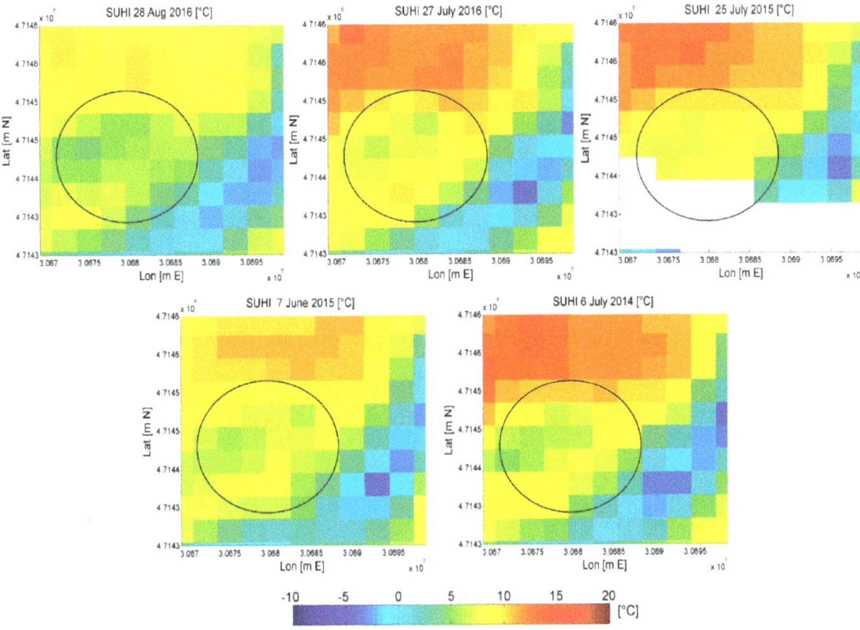

Figure 5. Post-intervention SUHI maps [°C]: 28 August 2016; 27 July 2016; 25 July 2015; 7 June 2015, and; 6 July 2014.

It is clear from Figure 5 how the new construction area (pixels inside the circle) produces a mitigation of the SUHI with respect to the heat island of the built-up area located in the upper part of the reported images (the latter area is present both pre- and post-interventions, as shown in Figure 2). The SUHI reduction is evident also during July, when a strong heating effect is found in the upper built-up area, and a difference of about 10 °C is observed between the two zones. In Figure 4, where the paved car park was not yet replaced by the new area, the heat island is similar or greater than the upper area, and the tree cooling effect is detectable only in the right part of the image, near the Nera River.

The SUHI diminution of the zone under analysis after June 2014 (around 5–6 °C) could be linked to various factors. As widely confirmed from literature [13–15,41], urban greenery mitigates SUHI: in the new district, from summer 2000 to summer 2016, the mean NDVI computed from satellite data increased from 0.19 to 0.23. Furthermore, a three-level underground car park was constructed and a significant part of the roadway transferred underground, thereby avoiding the temperature increase linked to the heat generated by the circulating vehicles on the corresponding area. This benefit is confirmed in a simulated study [42] reporting how the underground space development improves the urban microclimate. Finally, the construction of relatively high buildings with respect to the previous flat car park lot produced a shading effect and a probable improvement of the local air circulation.

The impact of building design on the shadow effect was assessed in [43], showing how shorter urban buildings cause higher surface and near-surface temperatures during the daytime. Hu and Liao [44] suggest how a suitable space distance of buildings contributes to improve ventilation between them, and the new building position in Corso del Popolo seems to match the favourable conditions in [44].

Albedo Maps and Analytical Analysis

The previous SUHI patterns may be also analysed in conjunction with the albedo pattern variations pre- and post-intervention. The albedo retrieved by Landsat data proved to be a useful tool to detect surfaces with different albedo behaviour in a built environment [40,45], without expensive in situ measurements unevenly distributed in space, and therefore unable to provide a global pattern. A high reflective surface (typically light in colour) absorbs less solar radiation than a conventional low reflective surface (dark in colour). Therefore, it is expected that high albedo surfaces lead to lower LST and that low albedo ones lead to higher LST. It is important to underline that vegetation pixels are characterized by low albedo values, but the effect on local overheating is very different with respect to the low albedo of impervious surfaces. In fact, the solar energy stored by vegetation is mainly used for evapotranspiration and life processes, producing a cooling effect of the green areas. Figure 6 shows an example of two albedo maps pre- and post-intervention.

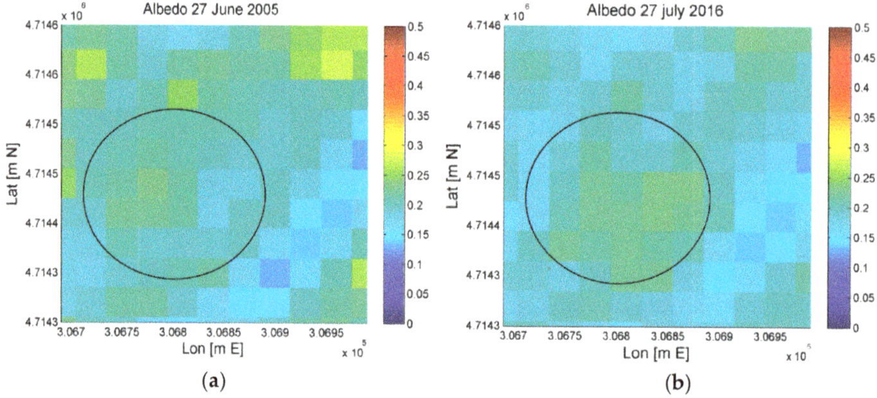

Figure 6. Albedo maps; (a) 27 June 2005; (b) 27 July 2016.

The purpose of this analysis is the evaluation of the albedo pattern in relation with the heating/cooling of the urban area. Figure 5 points out that the albedo variation in the changed area is not remarkable (pixel values are in the range of 0.18–0.24 for both images in the circled zone, with a slight increase post-intervention). The same pattern is roughly found in all the nine images. The satellite remote sensing using Landsat 7 reflective band observations provides a map of albedo values with a 30 m pixel size, averaging intrinsic reflectance heterogeneities at a finer scale. For instance, the presence of vegetation spots (low albedo) inside a built-up texture having high albedo produces a decreased value in the satellite pixel, even though both surfaces have a cooling effect. The shade generated by the new tall buildings may produce a retrieved albedo lower than its correct value, even if the Landsat 7 passages at around 12:00 reduce this effect, since midday sun position causes a reduction of the buildings shadow extension. Overall, taking into account the impact of the area with vegetation on albedo measurements, the new building system and impervious surfaces present an undiminished albedo pattern [46] with respect to the pre-intervention scene.

Before 2006 the paved parking had also perimetral trees that should have been a potential mitigation effect on summer heating of the area. Since the SUHI trend of Figure 3 and Figure 4 shows a clear enhancement post-intervention, the synergy of different urban design solutions, as greenery, parking/roadway covering and building characteristics (dimension and materials) proves to be a more impacting feature. Therefore, to understand deeply the SUHI variations, an overall assessment by means of Equation (4), depending on construction material behaviour, air temperature and heat transfer, is useful.

Equation (4) suggests that the most practical parameter to be changed on a large scale is albedo [10]. At the same time, the emissivity of a surface affects LST. For SUHI mitigation, it is necessary to avoid low-emissivity surfaces since they maintain a higher surface temperature when exposed to the sun than a high-emissivity material. Low-emissivity materials include many unpainted/polished metal surfaces, whereas most not metallic urban surfaces exhibit high emissivity. High albedo roofs and walls can be obtained with white surface coatings, while high albedo pavement includes concrete and conventional asphalt with white aggregate. Considering the thermal storage, the convection coefficient describes the convective heat transfer between the surface and the air flowing over, depending on thermal properties of the medium, characteristics of its flow, surface geometry, thermal boundary conditions and the variability of the surface temperature. Furthermore, Equation (4) is valid in steady-state conditions: since satellite passages are at around midday, the thermal storage phenomenon is nearly terminated, considering that the surface warming by the sun started some hours before in summertime.

An example of the impact of each single parameter variation on the LST is reported in Table 3. The aim of this analysis is the evaluation of the most significant key factors inducing SUHI variation for impervious surfaces. The values for the electromagnetic parameters α and ε are chosen in a typical range for construction materials [47], as well as the h_c range follows the characteristic behaviour in a built environment. The parameters T_a and T_{sky} are chosen in a range following the atmospheric conditions of continental Central Italy summer around midday [48] (T_a in the range 25–35 °C and T_{sky} in the range 278–288 K, the latter computed considering the relative humidity of air ranging from 30% to 70%). The incident solar radiation on the surface I is settled at 900 W/m^2, an average value measured in the Central Italy in July 2016 at around 12:00 CEST.

Table 3. Land surface temperature (LST) variation from Equation (4) (I = 900 W/m^2).

Varying Parameter	Fixed Parameters	LST Variation (°C)	LST Variation (%)
α: 0.13 ÷ 0.40	ε = 0.93 T_{sky} = 283 K T_a = 30°C h_c = 15 W/m^2·K	60.9 ÷ 50.2	−17%
ε: 0.85 ÷ 0.95	α = 0.25 T_{sky} = 283 K T_a = 30°C h_c = 15 W/m^2·K	57.3 ÷ 55.9	−2%
T_{sky}: 278 ÷ 288 K	α = 0.25 ε = 0.93 T_a = 30°C h_c = 15 W/m^2·K	55.2 ÷ 57.3	+3%
T_a: 25 ÷ 35 °C	α = 0.25 ε = 0.93 T_{sky} = 283 K h_c = 15 W/m^2·K	52.8 ÷ 59.5	+13%
h_c: 10 ÷ 20 W/m^2·K	α = 0.25 ε = 0.93 T_{sky} = 283 K T_a = 30 °C	66.3 ÷ 55.1	−17%

It is evident that the LST in correspondence to impervious surfaces turns out to be strongly sensitive not only to the surface albedo, but also to the local convection coefficient, and to the surrounding air temperature. The increase of α shows a clear benefit on urban heating mitigation, as well as the h_c increase and T_a decrease. The emissivity variation impact, in the typical range of built-up materials (significant unpainted or polished metallic surfaces are not present in the study area), is negligible.

Overall, the analysis confirms how the understanding of the SUHI variation requires the evaluation of both city planning factors (green surface, roadway and building design) and physical parameter behaviour (albedo, heat transfer and air circulation) [49].

5. Conclusions

Corso del Popolo district is an example of urban texture modification planned as regeneration of the ancient part of the town of Terni, and the assessment of the sustainable construction in terms of SUHI mitigation is an important practice. Remote sensing data provided by space-borne sensors confirm their usefulness in the monitoring of SUHI spatial pattern and its temporal evolution, making it possible to understand if the urban changes move towards a sustainable development. The advantage of using satellite observation techniques relies, beyond the wide area covered and multiple passages, on the ability to convey information without the need of ground measurements, the latter being unevenly distributed. Even though sharp thermal variations are averaged inside the pixel area, the trend of the SUHI pattern inside the district can be well delineated.

The adopted solutions for mitigating environmental impacts prove their efficiency in terms of SUHI reduction, as verified by the comparative analysis of space-borne LST summer images before and after the interventions. The positive effect on the district microclimate can be ascribed to the not diminished albedo values of the new impervious surfaces and materials, the green areas, the underground parking, the partial covering of the local roadway and the combined effects of shading and air circulation enhancement connected to the new multi-storey buildings.

Acknowledgments: The work has been supported by "Fondo di Ricerca di Base di Ateneo", funded by the Department of Engineering of University of Perugia. Authors wish to thank Eng. Roberta Anniballe for the data processing support.

Author Contributions: Bonafoni, S., Baldinelli, G., Verducci, P. and Presciutti, A. conceived and designed the work and analyzed the data; Bonafoni, S. and Baldinelli, G. wrote the paper.

Conflicts of Interest: The authors declare no conflict of interest.

References

1. Oke, T.R. The energetic basis of the urban heat island. *Q. J. R. Meteorol. Soc.* **1982**, *108*, 1–24. [CrossRef]
2. Voogt, J.A. Urban Heat Island: Hotter Cities. ActionBioscience.org 2004. Available online: http://www.actionbioscience.org/environment/voogt.html (accessed on 5 October 2016).
3. Ogashawara, I.; Bastos, V.S.B. A Quantitative Approach for Analyzing the Relationship between Urban Heat Islands and Land Cover. *Remote Sens.* **2008**, *4*, 3596–3618. [CrossRef]
4. Rizwan, A.M.; Dennis, L.Y.C.; Chunho, L.I.U. A review on the generation, determination and mitigation of Urban Heat Island. *J. Environ. Sci.* **2008**, *20*, 120–128. [CrossRef]
5. Pichierri, M.; Bonafoni, S.; Biondi, R. Satellite air temperature estimation for monitoring the canopy layer heat island of Milan. *Remote Sens. Environ.* **2012**, *127*, 130–138. [CrossRef]
6. Icaza, L.E.; Van Den Dobbelsteen, A.; Van Der Hoeven, F. Integrating urban heat assessment in urban plans. *Sustainability* **2016**, *8*, 320. [CrossRef]
7. Mackey, C.W.; Lee, X.; Smith, R.B. Remotely sensing the cooling effects of city scale efforts to reduce urban heat island. *Build. Environ.* **2012**, *49*, 348–358. [CrossRef]
8. Smith, C.; Levermore, G. Designing urban spaces and buildings to improve sustainability and quality of life in a warmer world. *Energy Policy* **2008**, *36*, 4558–4562. [CrossRef]
9. Shi, Z.; Zhang, X. Analyzing the effect of the longwave emissivity and solar reflectance of building envelopes on energy-saving in various climates. *Sol. Energy* **2011**, *85*, 28–37. [CrossRef]
10. Bretz, S.; Akbari, H.; Rosenfeld, A. Practical issues for using solar-reflective materials to mitigate urban heat islands. *Atmos. Environ.* **1998**, *32*, 95–101. [CrossRef]
11. Taha, H.; Akbari, H.; Rosenfeld, A.; Huang, J. Residential cooling loads and the urban heat island: The effects of albedo. *Build. Environ.* **1988**, *23*, 271–283. [CrossRef]
12. Suehrcke, H.; Peterson, E.L.; Selby, N. Effect of roof solar reflectance on the building heat gain in a hot climate. *Energy Build.* **2008**, *40*, 2224–2235. [CrossRef]
13. Alavipanah, S.; Wegmann, M.; Qureshi, S.; Weng, Q.; Koellner, T. The role of vegetation in mitigating urban land surface temperatures: A case study of Munich, Germany during the warm season. *Sustainability* **2015**, *7*, 4689–4706. [CrossRef]

14. Givoni, B. Impact of planted areas on urban environmental quality: A review. *Atmos. Environ.* **1991**, *25*, 289–299. [CrossRef]

15. Gusso, A.; Cafruni, C.; Bordin, F.; Veronez, M.R.; Lenz, L.; Crija, S. Multi-temporal patterns of urban heat Island as response to economic growth management. *Sustainability* **2015**, *7*, 3129–3145. [CrossRef]

16. Lee, H.Y. An application of NOAA AVHRR thermal data to the study of urban heat islands. *Atmos. Environ.* **1993**, *27*, 1–13. [CrossRef]

17. Gallo, K.P.; McNab, A.L.; Karl, J T.R.; Brown, F.; Hood, J.J.; Tarpley, J.D. The use of NOAA AVHRR data for assessment of the urban heat island effect. *J. Appl. Meteorol.* **1993**, *32*, 899–908. [CrossRef]

18. Streutker, D.R. Satellite-measured growth of the urban heat island of Houston, Texas. *Remote Sens. Environ.* **2003**, *85*, 282–289. [CrossRef]

19. Pu, R.; Gong, P.; Michishita, R.; Sasagawa, T. Assessment of multi-resolution and multi-sensor data for urban surface temperature retrieval. *Remote Sens. Environ.* **2006**, *104*, 211–225. [CrossRef]

20. Anniballe, R.; Bonafoni, S.; Pichierri, M. Spatial and temporal trends of the surface and air heat island over Milan using Modis data. *Remote Sens. Environ.* **2014**, *150*, 163–171. [CrossRef]

21. Lu, D.; Weng, Q. Spectral mixture analysis of ASTER images for examining the relationship between urban thermal features and biophysical descriptors in Indianapolis, Indiana, USA. *Remote Sens. Environ.* **2006**, *104*, 157–167. [CrossRef]

22. Hartz, D.A.; Prashad, L.; Hedquist, B.C.; Golden, J.; Brazel, A.J. Linking satellite images and hand-held infrared thermography to observed neighbourhood climate conditions. *Remote Sens. Environ.* **2006**, *104*, 190–200. [CrossRef]

23. Weng, Q.; Hu, X.; Quattrochi, D.A.; Liu, H. Assessing Intra-Urban Surface Energy Fluxes Using Remotely Sensed ASTER Imagery and Routine Meteorological Data: A Case Study in Indianapolis, U.S.A. *IEEE J. Sel. Top. Appl. Earth Obs. Remote Sens.* **2014**, *7*, 4046–4057. [CrossRef]

24. Xian, G.; Crane, M. An analysis of urban thermal characteristics and associated land cover in Tampa Bay and Las Vegas using Landsat satellite data. *Remote Sens. Environ.* **2006**, *104*, 147–156. [CrossRef]

25. Yuan, F.; Bauer, M.E. Comparison of impervious surface area and normalized difference vegetation index as indicators of surface urban heat island effects in Landsat imagery. *Remote Sens. Environ.* **2007**, *106*, 375–386. [CrossRef]

26. Xu, H.; Ding, F.; Wen, X. Urban Expansion and Heat Island Dynamics in the Quanzhou Region, China. *IEEE J. Sel. Top. Appl. Earth Obs. Remote Sens.* **2009**, *2*, 74–79. [CrossRef]

27. Urbanpromo. Available online: urbanpromo.it/2014-en/progetti/the-project-financing-of-corso-del-popolo-in-terni/ (accessed on 15 November 2016).

28. Regione Umbria—Giunta Regionale—SIAT Sistema Informativo Regionale Ambientale e Territoriale. Available online: siat.regione.umbria.it/paesaggineltempo/ (accessed on 1 December 2016).

29. U.S. Geological Survey, Earth Explorer. Available online: earthexplorer.usgs.gov (accessed on 15 December 2016).

30. Chander, G.; Markham, B.L.; Helder, D.L. Summary of Current Radiometric Calibration Coefficients for Landsat MSS, TM, ETM+, and EO-1 ALI sensors. *Remote Sens. Environ.* **2009**, *113*, 893–903. [CrossRef]

31. Chavez, P.S. Image-based atmospheric corrections—Revisited and improved. *Photogramm. Eng. Remote Sens.* **1996**, *62*, 1025–1036.

32. Jimenez-Munoz, J.C.; Sobrino, J.A. A generalized single-channel method for retrieving land surface temperature from remote sensing data. *J. Geophys. Res.* **2003**, *108*, 1–9. [CrossRef]

33. Bonafoni, S. Downscaling of Landsat and MODIS Land Surface Temperature over the heterogeneous urban area of Milan. *IEEE J. Sel. Top. Appl. Earth Obs. Remote Sens.* **2016**, *9*. [CrossRef]

34. Atmospheric Correction Parameter Calculator. Available online: https://atmcorr.gsfc.nasa.gov/ (accessed on 10 January 2017).

35. Barsi, J.A.; Barker, J.L.; Schott, J.R. An atmospheric correction parameter calculator for a single thermal band earth-sensing instrument. In Proceedings of the IEEE International Geoscience and Remote Sensing Symposium, Toulouse, France, 21–25 July 2003; pp. 3014–3016.

36. Sobrino, J.A.; Jimenez-Munoz, J.C.; Paolini, L. Land surface temperature retrieval from LANDSAT TM 5. *Remote Sens. Environ.* **2004**, *90*, 434–440. [CrossRef]

37. Bonafoni, S. The spectral index utility for summer urban heating analysis. *J. Appl. Remote Sens.* **2015**, *9*. [CrossRef]

38. Liang, S. Narrowband to broadband conversions of land surface albedo—I Algorithms. *Remote Sens. Environ.* **2001**, *76*, 213–238. [CrossRef]
39. Schaepman-Strub, G.; Schaepman, M.E.; Painter, T.H.; Dangel, S.; Martonchik, J.V. Reflectance quantities in optical remote sensing-definitions and case studies. *Remote Sens. Environ.* **2006**, *103*, 27–42. [CrossRef]
40. Baldinelli, G.; Bonafoni, S.; Anniballe, R.; Presciutti, A.; Gioli, B.; Magliulo, V. Spaceborne detection of roof and impervious surface albedo: Potentialities and comparison with airborne thermography measurements. *Sol. Energy* **2015**, *113*, 281–294. [CrossRef]
41. Park, J.-H.; Cho, G.-H. Examining the Association between Physical Characteristics of Green Space and Land Surface Temperature: A Case Study of Ulsan, Korea. *Sustainability* **2016**, *8*, 777. [CrossRef]
42. Yang, X.; Chen, Z.; Cai, H.; Ma, L. A Framework for Assessment of the Influence of China's Urban Underground Space Developments on the Urban Microclimate. *Sustainability* **2014**, *6*, 8536–8566. [CrossRef]
43. Loughner, C.P.; Allen, D.J.; Zhang, D.; Pickering, K.E.; Dickerson, R.R.; Landry, L. Roles of Urban Tree Canopy and Buildings in Urban Heat Island Effects: Parameterization and Preliminary Results. *J. Appl. Meteorol. Climatol.* **2012**, *51*, 1775–1793. [CrossRef]
44. Hu, W.; Liao, X. Practice on improvement of urban thermal environment with evaporation combined CFD simulation. In *Modeling and Computation in Engineering II*; Liquan, X., Ed.; Taylor and Francis Group: London, UK, 2013.
45. Bonafoni, S.; Baldinelli, G.; Verducci, P. Sustainable strategies for smart cities: Analysis of the town development effect on surface urban heat island through remote sensing methodologies. *Sustain. Cities Soc.* **2017**, *29*, 211–218. [CrossRef]
46. Rossi, F.; Castellani, B.; Presciutti, A.; Morini, E.; Anderini, E.; Filipponi, M.; Nicolini, A. Experimental evaluation of urban heat island mitigation potential of retro-reflective pavement in urban canyons. *Energy Build.* **2016**, *126*, 340–352. [CrossRef]
47. Kotthaus, S.; Thomas Smith, E.L.; Wooster, M.J.; Grimmond, C.S.B. Derivation of an urban materials spectral library through emittance and reflectance spectroscopy. *SPRS J. Photogramm. Remote Sens.* **2014**, *94*, 194–212.
48. Weather Underground 2017. Available online: http://www.wunderground.com (accessed on 15 February 2017).
49. Echevarría Icaza, L.; van den Dobbelsteen, A.; van der Hoeven, F. Integrating Urban Heat Assessment in Urban Plans. *Sustainability* **2016**, *8*, 320. [CrossRef]

Article

A Sustainable Graphene Based Cement Composite

Sardar Kashif Ur Rehman [1,*], Zainah Ibrahim [1,*], Shazim Ali Memon [2,*],
Muhammad Faisal Javed [1,3] and Rao Arsalan Khushnood [4]

[1]   Department of Civil Engineering, University of Malaya, 50603 Kuala Lumpur, Malaysia; arbabf1@gmail.com
[2]   Department of Civil Engineering, School of Engineering, Nazarbayev University, 010000 Astana, Kazakhstan
[3]   Department of Civil Engineering, COMSATS Institute of Information Technology,
      22060 Abbottabad, Pakistan
[4]   NUST Institute of Civil Engineering, School of Civil and Environmental Engineering, National University of
      Sciences and Technology, 44000 Islamabad, Pakistan; arsalan.khushnood@nice.nust.edu.pk
*   Correspondence: kashif@engineersdaily.com (S.K.U.R.); zainah@um.edu.my (Z.I.);
     shazim.memon@nu.edu.kz (S.A.M.)

Received: 2 May 2017; Accepted: 4 July 2017; Published: 13 July 2017

Abstract: The rheological properties of fresh cement paste with different content of graphene nanoplatelets (GNPs), different shear rate cycles and resting time was investigated. The rheological data were fitted by the Bingham model, Modified Bingham model, Herschel–Bulkley model and Casson model to estimate the yield stress and plastic viscosity, and to see trend of the flow curves. The effectiveness of these rheological models was expressed by the standard error. Test results showed that yield stress and plastic viscosity increased with the increase in the content of graphene in the cement based composite and resting time while the values of these parameters decreased for higher shear rate cycle. In comparison to control sample, the GNP cement based composite showed 30% increase in load carrying capacity and 73% increase in overall failure strain. Piezo-resistive characteristics of GNP were employed to evaluate the self-sensing composite material. It was found that, at maximum compressive load, the electrical resistivity value reduced by 42% and hence GNP cement based composite can be used to detect the damages in concrete. Finally, the practical application of this composite material was evaluated by testing full length reinforced concrete beam. It was found that graphene–cement composite specimen successfully predicted the response against cracks propagation and hence can be used as self-sensing composite material.

Keywords: rheological properties; self-sensing; piezo-resistivity; graphene nanoplatelets; structural health monitoring

---

1. Introduction

Concrete is the most widely used construction material and usually placed in the plastic form. Fluidity, homogeneity, consistency and workability are the key elements that greatly depend on viscosity of concrete and need to be considered during the mixing and placing of concrete. Any shortcoming in these parameters will lead to bleeding, segregation, laitance and cracking of the concrete [1]. Therefore, the rheological properties of concrete are of utmost significance to achieve homogeneity and good workability and have a strong influence on the overall fresh properties of the concrete [2]. With the advancement in nanotechnology, researchers are emphasizing more on the effect of nanomaterials on cement composite [3]. Numerous researchers studied the effect of various engineered nanomaterials on flow characteristics of the cement paste [4–6] and most of the research was focused on the rolled sheets of graphene and its derivatives, i.e., CNTs (carbon nanotubes) and graphene oxide [6–8]. However, the rheological properties of graphene cement paste remained unexplored and rarely reported. Therefore, an in-depth knowledge of the connection of graphene and rheology of cement paste is required. The flow properties of the cement paste are usually acquired by shear stress and shear rate. Later from the flow curves, viscosity and other flow parameters are calculated by using mathematical models. These rheological models statistically determine the yield stress (shear stress at zero shear rate), plastic viscosity (generalized viscosity for a range of shear rate) and predict the specific trend of the flow. As, these mathematical models possess statistical errors; therefore, one model cannot predict accurately the deformation of cement paste [9]. Therefore, in this research, four rheological models were used to investigate the flow properties of the graphene cement based composite.

Graphene possesses some amazing and extraordinary properties such as huge specific surface area (2630 $m^2 \cdot g^{-1}$), high intrinsic strength (130 GPa), firm Young's module (~1.0 TPa) and high electrical transport properties [10]. Therefore, incorporation of graphene in cement composite will not only alter the rheological characteristics but also affect the electrical properties of the composite. The electrical properties of the GNP–cement composite are important and can be used to monitor the damage in a concrete structure for the purpose of maintaining safe, reliable and sustainable civil infrastructure. It is known that non-destructive test offers skills for speedily and effectively monitoring these structures [11,12]. However, Self-sensing concrete, which can monitor its own strain, is the need of this era. The cement based composite reinforced with conducting fillers can observe its own strain by monitoring the changes in the electrical resistivity values [12]. Self-sensing ability is related to the breaking of conducting fibers when cracks are initiating in the cement based composite consequently enhancing the resistivity of the overall sample. If cracks are opening up due to tensile or fracture loading then resistivity values will be positive while it will be negative when subjected to compressive loading. Newly developed engineering nanocarbons are capable of fabricating new kind of high-performance tailored multifunctional cement-based composite that is capable of self-sensing the real-time damage [13–16]. Carbon nanotubes (CNTs) are composed of sp^2 hybridized carbon atoms, sheets of single layered graphene sheet rolled up in a cylindrical tube [17]. If these rolled sheets of CNTs are opened up in one plane, then it forms the two-dimensional sheet-like structure. These two-dimensional sheets, i.e., graphene nanoplatelets (GNPs) have even greater surface area and aspect ratio. Sixuan [18] investigated the effect of crack depth of the GNP reinforced mortar on the change in the electrical resistance. It was revealed that the specimens (cube and prism shapes) responded to an increase in electrical resistance as the depth of crack became larger. For the same relative crack depth, the change in electrical resistance for the cube was more significant as compared to that of the prism. However, as per authors' knowledge, the response of the GNP–cement composite to various damage levels has not been investigated.

Therefore, in this study, the flow properties of the GNP cement paste were investigated by using Bingham, Modified Bingham, Herschel–Bulkley and Casson models. Variation in flow curves of cement paste with different percentages of graphene nanoplatelets was determined. Rheological properties of graphene cement paste with various resting time (time between sample preparations to casting) and shear rate cycles were also evaluated. Later, the self-sensing properties of the GNP-based cementitious material were determined. The four-probe method was used for electrical resistance measurement purpose. Strain-sensing and fractional change in resistance were observed and used for determining self-sensing characteristics. Finally, application of GNP–cement composite specimen was evaluated on the reinforced beam.

2. Rheological Models

The rheological properties of the cement paste are directly related to the workability, flowability and consistency of the concrete. Cement paste has the complex rheological behavior as it depends on several parameters like water cement ratio, chemical admixtures, shear rate and supplementary cementitious materials [19]. The rheological models consider several factors which have great influence on cement paste rheology, and are necessary to be considered to achieve better and realistic approaches. Due to the influence of several factors in the rheological model for cement paste, the flow behavior cannot be predicted with best fitting curves by using single model [20]. It is also known that the accuracy and efficiency of the mathematical model depend on absolutely fitting the experimental data. Hence, four rheological models have been selected from various standards and researchers [9].

$$\text{Bingham Model} \qquad \tau = \tau_0 + \mu_p \ddot{Y} \qquad (1)$$

$$\text{Modified Bingham Model} \qquad \tau = \tau_0 + \mu_p \ddot{Y} + c\ddot{Y}^2 \qquad (2)$$

$$\text{Herschel-Bulkley Model} \qquad \tau = \tau_0 + K\ddot{Y}^n \qquad (3)$$

$$\text{Casson Model} \qquad \sqrt{\tau} = \sqrt{\tau_0} + \sqrt{\mu_p} \cdot \sqrt{\ddot{Y}} \qquad (4)$$

In these models, τ is considered as shear stress, τ_0 as yield shear stress, μ_p as plastic viscosity, $\ddot{Y}$ as shear rate, K as consistency, n as power rate index and c as regression constant.

Bingham model has been most widely used by researchers to determine the yield stress and plastic viscosity of the cement paste [21]. Bingham mathematical equation is linear (Equation (1)) and comparatively convenient to use for an analytical solution [22]. However, it fails to fit into the nonlinear portion of the flow curve at low shear rate and cannot predict yield shear stress accurately especially for shear thickening behavior [20]. To overcome this deficiency, the Bingham equation was modified and was used to fit the model in pseudo-plastic or shear thickening behavior. The mathematical expression of the modified Bingham model is given by Equation (2) [7]. However, it restricts the response of the cement paste to second order polynomial. Shear thickening behavior of the cement paste was further quantified by using Herschel–Bulkley model. This model has the characteristics of both Bingham and Power models and is given by Equation (3) [22]. This model is based on power rate index (n), which can predict the shear thinning and shear thickening behavior of the cement paste. With the increase in power rate index (n) value more than 1, the shear thickening behavior of the paste will be more prominent. The Casson model has two adjustable parameters, i.e., yield shear stress and plastic viscosity (Equation (4)). It uses the square root of these values, which makes its relation complex and difficult to explain. According to [19], it can predict the viscosity at a very high shear rate (infinite shear rate). However, this equation has a limitation in predicting the flow parameters for very concentrated suspensions [19]. As per [23] observation, Casson equation fits very well to various types of fluid and is more appropriate to use when compared with Herschel–Bulkley equation, yet, it is difficult to explain in most cases.

The ability of any analytical model to accurately match the nonlinear regression at low shear rate will define its accuracy. As this ability varies with each mathematical expression, therefore, the calculated rheological parameters especially yield stress values can offer different values for different models. Standard error for each rheological model has been determined using Equation (5). It depends on the normalized standard deviation. Finally, a comparison has been drawn amongst the calculated values to determine the effective and best-fitted model.

$$Standard\ error = \frac{1000 * \left\{ \frac{\sum (measured\ value - calculated\ value)^2}{(number\ of\ data\ points - 2)} \right\}^{1/2}}{(Maximum\ measured\ value - Minimum\ measured\ value)} \qquad (5)$$

3. Experimental Methodology

3.1. Materials

Ordinary Portland cement (Tasek Corporation Berhad) CEM I 42.5N (Type I) which was based on MS EN 197-1 having specific gravity 3.14 and Blaine surface area 0.351 (m^2/g) was used. The properties of GNP, purchased from Graphene Laboratories, Inc. USA are enlisted in Table 1. Deionized water was used for making cement based composite.

Table 1. The properties of GNPs used for this experiment.

Property	Description/value
Average Flake Thickness (nm)	12 (30–50 Monolayers)
Average particle (lateral) size (nm)	~4500 (1500–10,000 nm)
Specific surface area, SSA (m^2/g)	80
Purity	99.2%
Color	Black

3.2. Preparation of Samples

For specimen's preparation cement, GNP and deionized water were used. The details of mix proportions are given in Table 2. Here, M0 represents the control sample of cement paste while GM3, GM5 and GM10 represent samples containing 0.03%, 0.05% and 0.10% GNP by weight of cement, respectively. The GNP samples (GM3, GM5 and GM10) were used to study the influence of graphene on rheological characteristics of cement paste while for determining the strain-sensing characteristics of graphene–cement based composite, only GM3 having least percentage of GNP was considered. All measurements were repeated with identical samples for three times and averaged. For sample preparation, firstly, GNP (by weight of the cement) was added to the deionized water. To exfoliate nanoparticles, the sample was ultrasonicated (Fisher Scientific™ Model 505 Sonic Dismembrator) for 3 min followed by magnetic stirring for 1 h to ensure uniform dispersion of GNP. Thereafter, cement was added to this aqueous solution of GNP and mixed thoroughly for 5 min in spar mixer (SP-800A). Initially, the speed of the mixer was kept low for 2 min then static for 10 s and finally high for another 2 min. Finally, rheological characteristics of cement paste were measured. The workability of cement paste was determined by ASTM C 1437-15 [24] while for determining electrical properties, prism having dimensions of 160 × 40 × 40 mm was casted. Four-stainless-steel wire meshes (11 × 11 mm cross section) with an average thickness of 1.3 mm and dimensions of 40 × 70 mm was inserted in the samples to measure the electrical properties as shown in Figure 1a. The outer two wire meshes were positioned at 10 mm from the ends of the prism while the inner two wire meshes were positioned at 40 mm from the outer wire mesh. The detailed dimensions are given in Figure 1a. The surface morphology of GNP–cement composite was investigated using Field emission scanning electron microscope (FE-SEM, AJSM-7600F with semi-in-lens) while the elemental composition was acquired using FESEM coupled with an energy dispersive X-ray spectrometer (EDS).

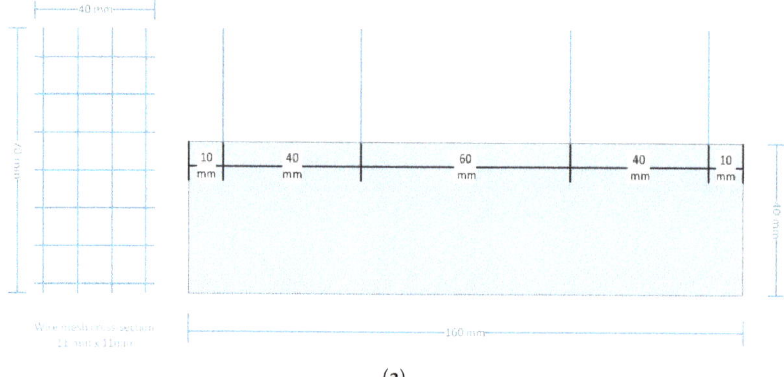

(a)

Figure 1. *Cont.*

Figure 1. Description of specimen and test setup (**a**) Dimensions of graphene–cement composite specimen with wire mesh details; (**b**) placement of specimen on reinforcement; (**c**) casting of reinforced concrete (RC) beam; and (**d**) overall test setup for measuring the electrical properties of graphene–cement composite specimen in the RC beam.

Table 2. Mix Proportions.

Mix Design	W/C Ratio	Cement (g)	Water (mL)	GNP (mg)	GNP/Cement (%)
M0	0.4	100	40	0	0
GM3	0.4	100	40	30	0.03
GM5	0.4	100	40	50	0.05
GM10	0.4	100	40	100	0.10

3.3. Rheological Measurements

The rheological properties were characterized by using Rheometer MCR302 (Anton-Paar) while Rheoplus software was used to monitor and analyze the data points. To determine the rheological characteristics of the GNP–cement paste, smooth parallel plates with a gap thickness of 0.6 mm were used. At first, approximately 15 mL of GNP–cement paste was added on the plate having diameter 25 mm at a fixed temperature of 25 °C. The cement paste was held at rest and after 10 min it was pre-sheared at a shear rate of 100 s^{-1} for 60 s. This was done to re-homogenize the sample, as cement paste has thixotropic character [25]. Then, after 5 min, shear rate was applied from 0.6 s^{-1} to 100 s^{-1} and then back from 100 s^{-1} to 0.06 s^{-1} in 20 decreasing steps. The downslope curve data was used for calculation of rheological parameters. Similarly, GNP–cement paste was also run for two different shear rates cycles, i.e., from 200 s^{-1}–0.6 s^{-1} with 40 steps and 300 s^{-1}–0.6 s^{-1} with 60 steps. The flow properties were also determined at different time interval, i.e., 0 min, 30 min and 60 min. For 30 and 60 min resting time, the samples were manually stirred for 15 s before measuring the rheological data. It is pertinent to mention here that the effect of graphene and its content on the rheology of cement paste was investigated by using mix M0, GM3, GM5 and GM10, while to examine the influence of different shear rate cycle range (100–0.6 s^{-1}, 200–06 s^{-1} and 300–0.6 s^{-1} represented by symbols a, b and c respectively) and different resting times (0, 30 and 60 min time between sample preparation and casting) on rheological properties, only M0 and GM3 mix were used. The detailed description about the shear rate cycle range and resting time of various samples for measuring rheological data is given in Table 3. All measurements were repeated with identical samples for three times and averaged. The apparent viscosity, shear stress and shear rate of GNP–cement paste with different proportions were measured by the rheometer while the yield stress and plastic viscosity of the GNP–cement paste was determined by using the Bingham, Modified Bingham, Herschel–Bulkley and Casson models.

Table 3. Description of rheological samples.

Sample	Sher Rate Cycle Range (s^{-1})	Resting Time (min)
Effect of graphene percentage		
M0 (Control)	200–0.6	0
GM3	200–0.6	0
GM5	200–0.6	0
GM10	200–0.6	0
Effect of shear rate cycle		
M0a	100–0.6	0
M0b	200–0.6	0
M0c	300–0.6	0
GM3a	100–0.6	0
GM3b	200–0.6	0
GM3c	300–0.6	0
Effect of resting time		
M0–0	200–0.6	0
M0–30	200–0.6	30
M0–60	200–0.6	60
GM3–0	200–0.6	0
GM3–30	200–0.6	30
GM3–60	200–0.6	60

3.4. Test Setup for Electrical Properties

The four-probe method was used to determine the piezo-resistive properties of the GNP–cement composite specimens. Four-probe method involves four electrical contact points in which voltage is measured through the inner two electrical contacts while the current is measured using outer two electrical contacts. [26] Four-probe method is better than the two- probe method because measured resistance does not include the contact resistance [27]. According to Han et al. [27], the space between the current pole and voltage pole is very important. However, its influence is marginal if space is more than 7.5 mm. Various researchers used different spacing in their experimental work. Li et al. [28] used the spacing of 10 mm while Le et al. [29] used 40 mm spacing between current and voltage measuring probes. Similarly, the distance between the two voltage measuring probes is not fixed and its value varies. Li et al. [28] used the 40 mm gap between two measuring probes while Le et al. [29] used 80 mm gap. As resistivity values remain unaffected provided that the spacing is above threshold value [27], in this research, we used 40 mm spacing between the current measuring probe and voltage measuring probes and 60 mm between two voltage measuring probes. Pang et al. [30] explored the effect of percentage of graphene nanoplatelets on the strain and damage sensing of the composites. They found that with increase in the percentage of graphene nanoplatelets in cement based composite, the fractional change in resistance values was also increasing. Therefore, based on the study of Pang et al. [30], only M0 and GM3 specimens were considered to explore the electric resistivity of cement based composite materials.

The setup for measuring the electrical resistivity consist of Instron 600 kN machine used for applying a compressive load, TDS-530 data logger to record the voltage measurements, a DC power supply and 10-ohm resistance. Specimens were placed in Instron machine and a constant voltage of 15 V through DC power supply was applied to the samples (Figure 1a). Inner wire meshes were connected to the data logger and measures the voltage drop in V. Meanwhile, one of the outer wire meshes was connected to the negative terminal of the power supply and the remaining outer wire mesh was connected to the resistor followed by the connection to the second channel of the data logger and positive terminal of the power supply forming a series circuit. Finally, the practical application of these GNP–cement composite specimen was investigated by testing the reinforced concrete (RC) beam having cross-sectional dimension 200 × 300 (mm) and span length 3200 mm. For this purpose, GM3 specimen was placed inside the beam in center at the time of beam casting as shown in Figure 1b,c. The beam was tested under flexural loading using Instron 600 kN and response of the GNP–cement composite specimen was recorded using 15 V DC power supply and TDS 530 data logger. The test setup is shown in Figure 2c.

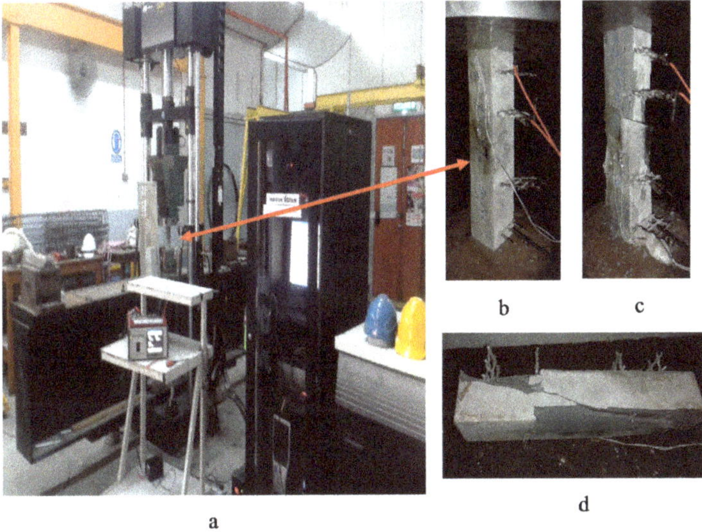

Figure 2. Measurement of electrical properties (**a**) The overall set up for the four-probe method for the electrical properties testing; (**b**) placing of the specimen into the Instron 600 kN machine; (**c**) sample failure under compression; and (**d**) propagation of cracks in failed sample.

4. Results and Discussion

4.1. Rheological Characteristics of GNP–Cement Composite

Variation in rheological characteristics of both cement paste and graphene–cement paste were observed, which originates due to mix composition of the paste, shear rate cycle, resting time and rheological model. As these rheological methods determine the flow property values from statistical trials, several values can be obtained. However, in this research, best curve fitting and minimizing the standard error was considered in determining the rheological flow values. The effect of mix composition mainly includes water to cement ratio, admixture, mixing and testing technique [31]. As these parameters were studied in detail [9,31], all these factors were kept constant and only effect of graphene/cement ratio, shear rate range, and resting time was investigated. Moreover, the parallel plates, whose results are close to reality, were used to determine the rheological characteristics of cement based composites [2]. However, particle sedimentation or "creaming" occurs more in the parallel plate, which may contribute to increase the viscosity [32].

4.1.1. Yield Stress

Yield stress values were determined by Bingham model, Modified Bingham model, Herschel–Bulkley (HB) model and Casson model. The results are presented in Table 4. Due to slippage effect and smooth surface of parallel plates, low yield stress values were found [9]. Generally, the values of yield stress obtained by modified Bingham model were highest followed by Bingham, HB and Casson model. For all rheological models, the values of yield stress increased with the increase in the percentage of graphene in the cement matrix. This increase in yield stress values is in line with the available literature on the derivatives of graphene. The increase in yield stress values for graphene–cement paste might be due to higher surface area of graphene due to which it requires more amount of water for lubrication of graphene [33]. Shang et al. [6] carried out experimental investigation to study the effect of graphene oxide on rheological properties of cement paste and found that in comparison to plain cement, the value of yield stress increased by approximately four times for 0.08% of graphene oxide cement paste. It was noticed that negatively charged graphene nanoplatelets in water interacted with cement particles by electrostatic interaction and formed flocs. Wang et al. [7] also detected the increase in flocculation structures having higher dosage of graphene oxide (GO) in cement paste. According to literature [6], these large agglomerates entrap some water molecules, which reduce the content of free available water. Therefore, it can be concluded that by keeping the water cement (w/c) ratio constant, the values of rheological parameters will increase.

Table 4. Yield stress values determined by various rheological models.

Sample	Bingham (Pa)	Modified BM (Pa)	HB (Pa)	Casson (Pa)
\multicolumn{5}{c}{Effect of graphene percentage}				
M0 (Control)	1.53	1.76	1.22	0.52
GM3	1.67	1.79	1.30	0.62
GM5	1.84	1.89	1.48	0.79
GM10	1.90	1.93	1.53	0.80
\multicolumn{5}{c}{Effect of shear rate cycle}				
M0a	1.60	1.83	1.27	0.55
M0b	1.53	1.76	1.22	0.52
M0c	1.46	1.46	1.17	0.45
GM3a	1.74	1.86	1.34	0.64
GM3b	1.67	1.79	1.30	0.62
GM3c	1.51	1.61	1.21	0.50
\multicolumn{5}{c}{Effect of resting time}				
M0-0	1.53	1.76	1.22	0.52
M0-30	1.65	1.85	1.38	1.75
M0-60	2.54	2.88	3.64	1.93
GM3-0	1.67	1.79	1.30	0.62
GM3-30	1.72	1.92	1.42	1.81
GM3-60	3.78	2.19	1.65	4.91

For the case of shear rate cycle, the values of yield stress decreased when the shear rate cycle increases from 100–0.6 s^{-1} to 300–0.6 s^{-1}. This is possibly due to destruction of potential agglomerated structures in cement paste. For high shear rate cycle, the resistance to flow was reduced and hence shear thinning behavior was dominant. Finally, for both control and GNP–cement paste, the yield stress values increased with the increase in resting time as shown in Table 4. The increase in rheological parameters may be attributed to: (a) hydration reaction of cement paste; and (b) presence of suspended particles (graphene).

4.1.2. Viscosity

Plastic viscosity represents the deformation of cement pastes due to external loading [7]. Plastic viscosities were calculated from the flow curve using the Bingham, Modified Bingham and Casson models. The results are presented in Figure 3. Generally, for control sample (M0), Bingham model showed the highest value of plastic viscosity followed by the Casson and Modified Bingham (Figure 3a). However, for the GM3 sample, Modified Bingham showed the highest value of plastic viscosities followed by the Casson and Bingham model (Figure 3a). Furthermore, direct relationship between graphene and plastic viscosity was observed, as plastic viscosity increased with the increase in the amount of graphene in cement composite. These results are in line with Shang et al. [6], who found that the plastic viscosity values increased by 78% with the addition of 0.04% graphene oxide in plain cement mix. When samples were subjected to high shear rate cycle, the plastic viscosity reduced as shown in Figure 3b. Shang et al. [6] reported that the apparent viscosities are dependent on the shear rate, i.e., at low shear rate, the values of apparent viscosities would be higher while at high shear rate, the values of viscosities would be lower. A possible reason for this may be due to the breaking of the agglomerates of cement paste which in turn resulted in lower apparent viscosity. It was noticed that the values of plastic viscosity increased with the increase in resting time as shown in Figure 3c. It might be related to the hydration of cement particles and the fractional resistance between cement and graphene sheets. The influence of hydration of cement particles and fractional resistance was dominant for the resting time of 60 min, in which plastic viscosity was very high.

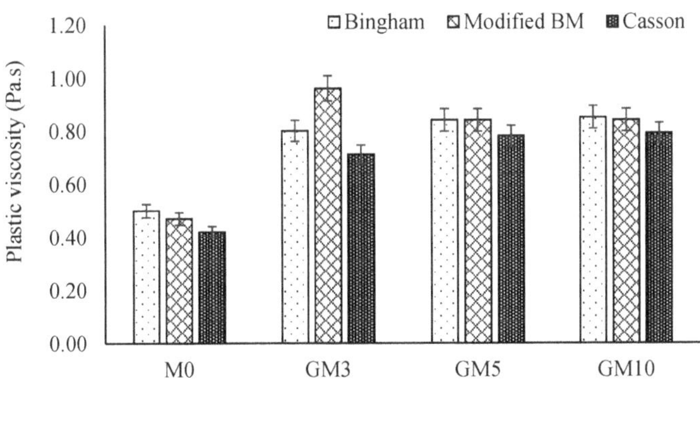

(**a**)

Figure 3. *Cont.*

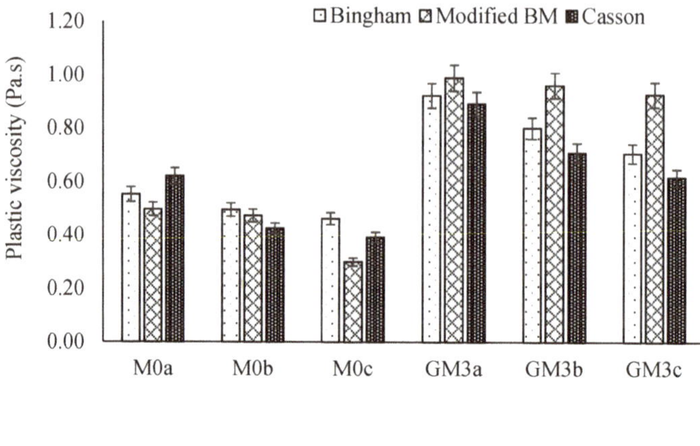

(b)

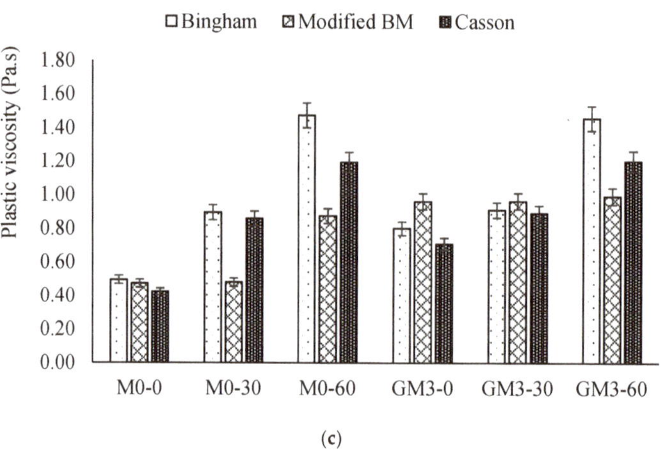

(c)

Figure 3. Plastic viscosity values (Pa.s) for different mixes calculated by Bingham, Modified Bingham and Casson models using smooth parallel plate: (**a**) effect of graphene percentage; (**b**) effect of shear rate cycle range; and (**c**) effect of resting time.

4.1.3. Consistency and Power Rate Index

The trend of the viscosity data was determined using Herschel–Bulkley (HB) model, which considers two factors, i.e., consistency (K) and the power rate index (n). By considering these factors, the relationship between viscosity trend and the shear rate of the flow curve can be determined [9]. Based on power rate index values "n", it also provides the information about the shear deformation, i.e., shear thickening (n > 1) or shear thinning (n < 1) [7] The consistency (K) has no physical meaning and difficult to compare because of its dimension (Pa.s^n) which is dependent on "n" [34]. Wang et al. [7] studied the shear deformation for graphene oxide in cement paste and found that cement paste curve can be divided into shear thinning and shear thickening stage based on the inflection point. They suggested that cement paste with higher graphene oxide content shows the shear thinning effect at high shear rates. The values of the consistency and power rate index for various samples are given in Table 5. It can be seen that the "n" values are less than one and hence cement paste behaves as shear thinning. However, for 60 min resting, the value of "n" for samples M0–60 and GM3-60 exceeded 1, indicating shear thickening behavior. When cement paste comes in contact with water then coagulations and links are formed between two cement particles. With the increase in resting time, these links become strong and provide resistance to flow of cement paste [35]. Therefore, hydration reactions and presence of permanent links between cement particles will result in high resistance to flow.

Table 5. Consistency and rate index values calculated by Herschel–Bulkley model.

Sample	Consistency, K	Rate Index, n
Effect of graphene percentage		
M0 (Control)	0.60	0.96
GM3	0.97	0.97
GM5	0.99	0.98
GM10	0.97	0.98
Effect of shear rate cycle range		
M0a	0.51	0.90
M0b	0.60	0.96
M0c	0.29	0.60
GM3a	0.96	0.99
GM3b	0.97	0.97
GM3c	0.96	0.98
Effect of resting time		
M0–0	0.60	0.96
M0–30	0.96	0.98
M0–60	0.22	1.21
GM3–0	0.97	0.97
GM3–30	0.92	0.99
GM3–60	0.99	1.10

4.1.4. Standard Error

Standard error values for the various rheological models were calculated using Equation (5) and are given in Table 6. A lower value of standard error represents the best-fitting of the mathematical model to the flow curve. It was observed that HB model and Modified Bingham model showed less standard error while Bingham model and Casson model showed higher values of standard error. As HB and Modified Bingham models are nonlinear, they predicted the flow behavior more accurately. Bingham model has a linear mathematical relation and hence it showed higher standard error. Casson model has a limitation in predicting the concentrated suspension [19] and hence it is believed that it showed larger values of standard error. It was noted that, for M0c sample, all of the rheological models showed large values of standard error. It may be due to some calculation and experimental error regarding the flow curve data, as, for the same high shear rate cycle range, GM3c specimen showed comparatively low standard error values. GM10 mix showed the maximum standard error values for all mathematical models. As higher dosage of nanomaterials, i.e., graphene and graphene oxide, results in the formation of flocculation structures in cement paste [7]. Therefore, it is believed that the higher values of standard error may be due to the formation of these flocculated suspensions in the cement paste.

Table 6. Values of standard error for various rheological models.

Sample	Bingham	Modified BM	HB	Casson
Effect of graphene percentage				
M0 (Control)	74.53	74.39	74.39	73.63
GM3	89.01	34.29	70.54	82.93
GM5	30.58	30.67	45.07	57.29
GM10	233.26	223.55	261.57	283.10
Effect of shear rate cycle range				
M0a	163.63	244.69	280.44	140.10
M0b	74.53	74.39	74.39	73.63
M0c	344.36	285.05	319.14	318.37
GM3a	39.39	221.50	28.71	111.43
GM3b	89.01	34.29	70.54	82.93
GM3c	154.96	88.57	94.39	164.24
Effect of resting time				
M0-0	74.53	74.39	74.39	73.63
M0-30	117.72	65.26	123.06	111.40
M0-60	80.53	52.44	51.58	95.21
GM3-0	89.01	34.29	70.54	82.93
GM3-30	91.37	85.60	95.89	97.71
GM3-60	92.42	69.05	73.16	111.50

4.2. Fresh and Hardened Properties of GNP–Cement Composite

The results of workability test are presented in Table 7. The flow diameter of GM3 was found to be 8.5% less when compared to M0. Pan et al. [36] used 0.05% of graphene oxide in cement paste (w/c = 0.5) and observed 41.7% reduction in the slump diameter in comparison to control cement paste. Similarly, for 2% carbon nanotubes in cement paste (w/c = 0.5), 48.9% reduction in slump diameter was found in comparison to control cement paste [37]. Possible reason for reduction in flow diameter could be the large surface area of the graphene sheets, which require more amount of water for lubrication and in turn decrease the free available water. Therefore, the overall workability of the cement paste was reduced by addition of graphene in cement paste.

Table 7. Flow diameter, maximum compressive load and corresponding resistivity values.

Specimen	Flow Diameter (mm)	Maximum Compressive Load (kN)	Four Probe Resistivity at Maximum Compressive Loading (kΩ cm)
M0	175	61	32.93
GM3	160	80	19.12

Stress–strain curves for the M0 and GM3 are given in Figure 4. It can be seen that the addition of GNP greatly enhances the load carrying capacity of the cement paste (Table 7). The enhancement in compressive strength was about 30% as compared to mix M0. It was also observed that GM3 specimen showed more ductile behavior as compared to M0. The percentage strain produced in GM3 increased up to 73%, showing ductile nature of graphene–cement paste. The increase in compressive strength and strain may be attributed to the higher strength of graphene, template effect, and crack bridging by graphene sheets [38]. These results are in agreement with the study of Pan et al. [36] performed on graphene oxide based cement composites. For cement paste containing 0.05% graphene oxide, an improvement of about 22% in compressive strength was observed when compared to control specimen.

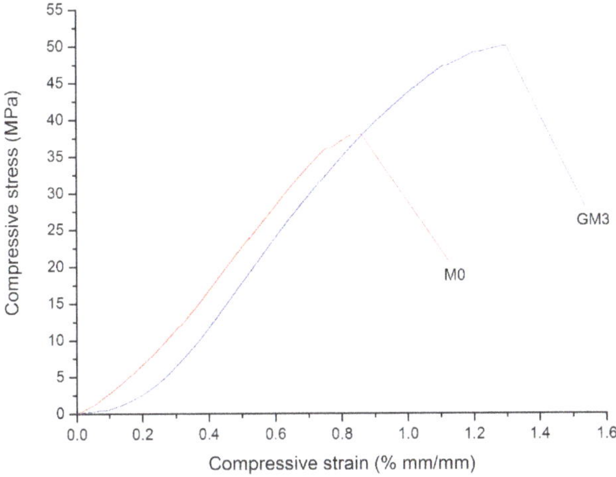

Figure 4. Stress–strain curve for cement paste (M0) and graphene–cement paste (GM3).

To study the possible reasons for the increase in compressive strength and the improvement in ductile behavior, the morphology of cement based composite was investigated. Figure 5 shows the morphology of the GNP–cement composite at 7 days and 28 days. In FESEM images of GM3 captured at seven days (Figure 5a), needle form of calcium silicate hydrate (CSH), hexagonal plates of portlandite, graphene nano particle can be observed. ESB or backscattered image was used to distinguish the carbon materials. It can be seen in Figure 5b that graphene nanoparticle are completely black. EDX image was further employed to confirm the presence of graphene and hydrated cement products. The EDX of graphene sheets (Figure 5c) for the location indicated in Figure 5a shows maximum carbon content, clearly confirming the presence of graphene. The EDX for the needle shape CSH in Figure 5d for the location marked in Figure 5b show maximum content of oxygen followed by the silicon and carbon. At 28 days, the needle form of CSH transformed into honeycomb structure of CSH as shown in Figure 5e. The backscattered image or ESB of Figure 5e shows that the hydrated products grow over graphene (Figure 5f). EDX was also employed in Figure 5g,h to confirm the presence of graphene and honeycombed CSH structure. Hence, based on the above information, it can be deduced that, due to the addition of graphene, hydrated products grow in uniform and ordered way [38], which significantly improved the ductile behavior and compressive strength.

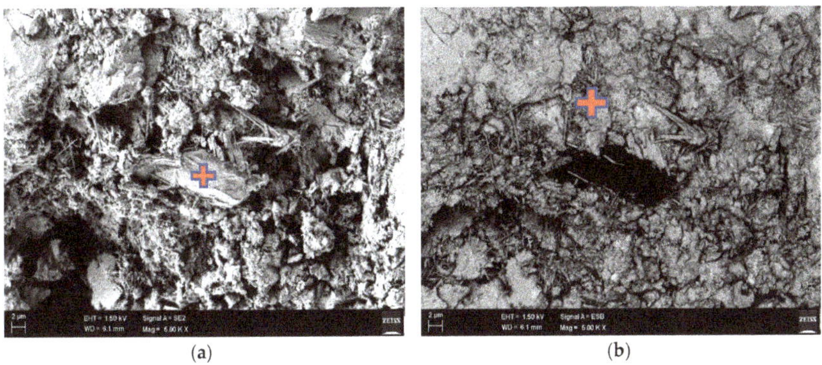

(a) (b)

Figure 5. *Cont.*

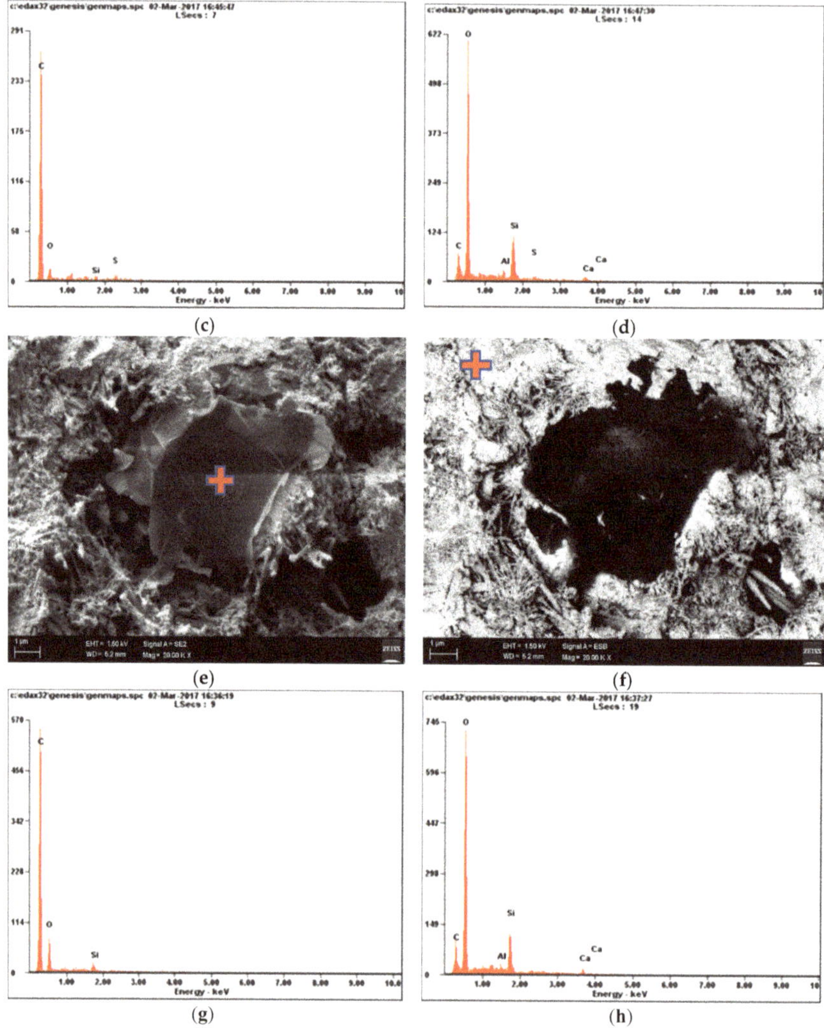

Figure 5. FESEM images of GNP–cement paste at seven days and 28 days: (**a**) FESEM image of GM3 at seven days; (**b**) ESB or backscattered image of (**a**); (**c**) EDX of graphene in GM3 specimen on a point indicated in (**a**); (**d**) EDX for the hydrated cement product at cross hair location in (**b**); (**e**) FESEM image of GM3 at 28 days; (**f**) ESB or backscattered image of (**e**); (**g**) EDX for the marked point in (**e**); and (**h**) EDX at cross-hair location in (**f**).

Generation and growth of cracks for M0 sample were identified using FESEM images, as shown in Figure 6a. These are the nano size cracks; later, due to externally applied forces, they become micro size cracks without any interference and play their role in failure mechanism of material. It can be seen in Figure 6b that graphene successfully interrupted these cracks at nano level and were discontinuous. This is also verified in Figure 6c,d, which shows that graphene platelets are not only holding the micro cracks but are preventing their further growth. Longitudinal growth of cracks is highlighted using green lines while the location of graphene in encircled in red color in Figure 6d. Due to these reasons, the GM3 sample showed more ductile behavior as compared to M0 mix. Pan et al. [36] made the comparison of crack patterns for plain and graphene oxide based cement composite. The authors found that in plain cement matrix cracks passed straight through the dense hydrated product. However, in graphene oxide cement paste, cracks were fine and discontinued. Therefore, it can be deduced that the presence of graphene sheets would make the cracks fine, the crack pattern discontinuous and provide hindrance to their growth, which, in turn, would result in enhancing the ductility and compressive strength of the graphene cement composite.

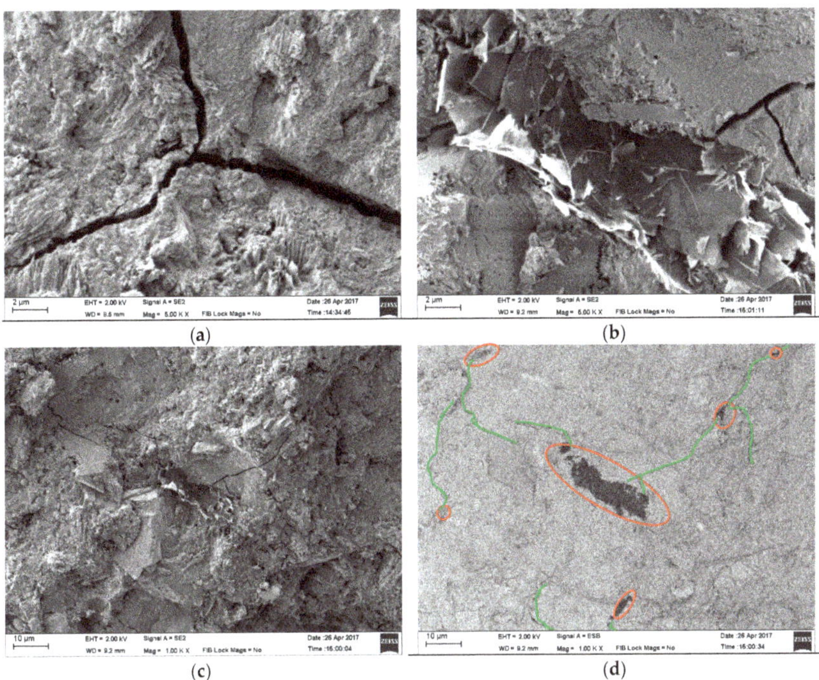

Figure 6. Effect of graphene on propagation of cracks (**a**) Crack propagation in control specimen; (**b**) blockage of the cracks by graphene; (**c**) crack bridging phenomena by graphene; and (**d**) backscattered electron image to identify the graphene in paste.

4.3. Electrical Resistivity Values of the GNP–Cement Composite

Piezo-resistive properties of the M0 and GM3 samples were investigated using the four-probe method. Electrical resistivity for unequal spacing between the probes was calculated using Equation (6).

$$\rho = \frac{V}{I} \times 2\pi \times \frac{1}{\left(\frac{1}{S1} + \frac{1}{S3} - \frac{1}{S1+S2} - \frac{1}{S2+S3} \right)} \tag{6}$$

where V is the floating potential difference between inner two probes, I is current measured by outer two probes, ρ is the resistivity in ohm-cm and S is the spacing in cm and was calculated from current carrying probe to voltage measuring probe. S1 = S3 = 40 cm and S2 = 60 cm.

The results of electrical resistivity are presented in Table 7. The electrical resistivity value at maximum compressive load for GM3 sample was found to be 42% less as compared to M0. To observe the sensing ability of the specimens, normalized compressive loading (NCL) values were calculated. It is the ratio between the applied loads to the maximum compressive load before specimen failure. For electrical resistance values, the fractional change in resistance (FCR) was used. Equations (7) and (8) present the calculation procedure for the NCL and FCR values.

$$\mathrm{NCL} \; = \; \frac{P}{P_{max}} \tag{7}$$

$$\mathrm{FCR} \; = \; \frac{\rho_t - \rho_o}{\rho_o} \times 100\% \tag{8}$$

where ρ_t is the electrical resistivity at the given time during the test; ρ_o is the electrical resistivity at the start of the test; P is the compressive loading at the given time during the test; and P_{max} is the maximum compressive loading for the specimen.

Figure 7 shows the fractional change in resistance (in percentage) for M0 and GM3 against the normalized compressive load. It can be seen that in M0 sample the fractional change in resistance is very less as compared to GM3. The results are in line with Li et al. [39]. As graphene is expensive, therefore, it would be very costly to use GNP–cement composite material in a mega project. However, because of superior self-sensing properties, it can be used as a substitute to do health monitoring of the structure. In order to strengthen this view point and show its practical application, full length RC beam was tested in the laboratory by placing GM3 specimen at the time of casting as shown in Figure 2. This beam was subjected to flexural loading and, due to applied loading, cracks were generated in the region of maximum bending moment. Figure 8 shows the fractional change in resistance of GM3 specimen in RC beam. It can be seen that FCR values were varied with the increase in the applied loading on the beam and a sharp response was noted at the time of beam failure. As the beam was subjected to the flexural loading, resistance values were positive. The in-set figure (enlarged view) shows that with the increase in flexural loading, the fractional change in resistance values are also increasing. It is important to mention here that this response in not linear, however, due to the occurrence of damage and propagation of cracks, the FCR values varied as shown in in-set figure (enlarged view of Figure 8). A sudden drop in electrical resistivity value was observed at 22 kN load. The possible reason may be the occurrence of tensile cracks. Initially, tensile cracks started to occur and electrical resistivity values increased linearly. Thereafter, the steel reinforcement started to carry stresses and the significant variation in resistance was noted in GM3 specimen. This effect was more significant as the specimen (GNP cement based composite) was placed in the tensile region, i.e., just above the reinforcement bars. Finally, at the failure stage abrupt increase in FCR was observed, which shows that widening and irreversible crack opening has occurred inside the beam. Hence, structural member is not capable of carrying additional load.

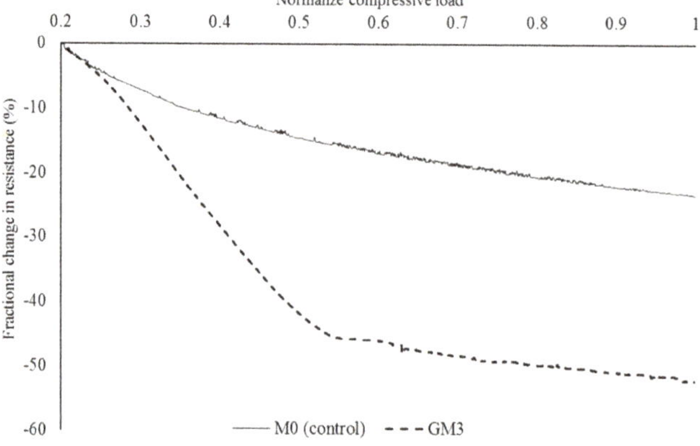

Figure 7. Fractional change in resistance against normalized compression load.

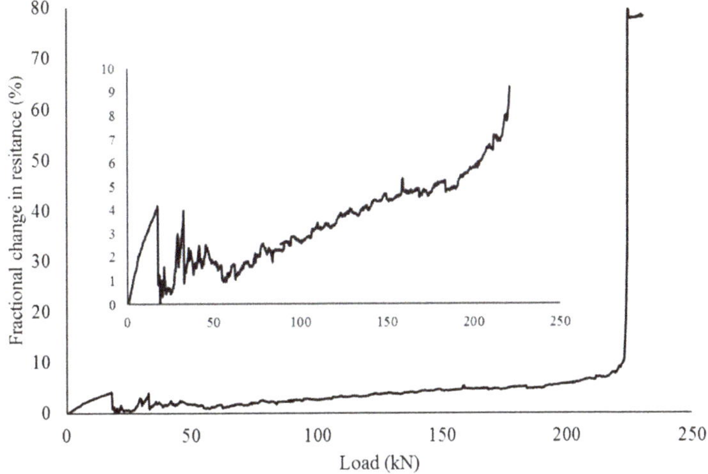

Figure 8. Fractional change in resistance of GM3 specimen against applied compressive loading on the RC beam.

A comparison of strain produced strain against the applied load in reinforcement bars and graphene–cement composite sample is shown in Figure 9. In reinforcement bars, the strain increased linearly and slight variation was observed from 60 kN to 100 kN load as shown in Figure 9a. This variation is very minute and can be neglected for the reinforcement as steel bar was in the linear and elastic region. The GM3 sample also showed variation in strain from 60 kN to 100 kN load as marked in Figure 9a. However, in comparison to reinforcement bar, the variation was large and the strain values for the GM3 sample decreased significantly. This may be related to the redistribution of stresses in the RC concrete beam. Similarly, when the applied loading exceeded 200 kN, the reinforcement bar showed significant variance in strain values. At the same point, an increase in strain values in graphene cement composite sample was noted. It may be related to crushing of concrete in RC concrete beam It is pertinent to mention here that the cost of 5 g of GNPs as per Graphene Laboratories, Inc. USA is 50 USD and 25 cement based composite samples can be casted with GM3 specimen. Hence, GM3 specimen can be used in an efficient and economical way to predict the damages in concrete structures.

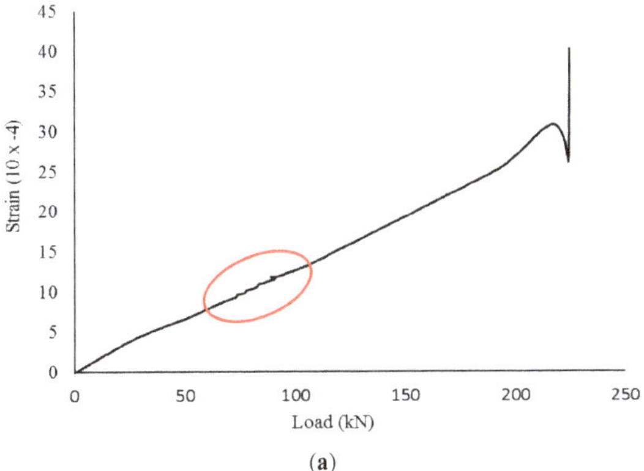

(**a**)

Figure 9. *Cont.*

213

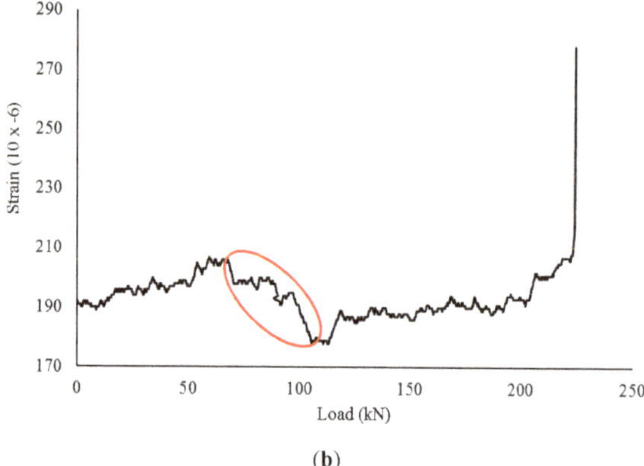

(**b**)

Figure 9. Strain produced by applied compressive load in: (**a**) reinforcement; and (**b**) GNP–cement composite specimen.

5. Conclusions

In this research, cement pastes with different graphene/cement ratio were prepared. Their rheological characteristics of the paste samples were determined from the flow curves of paste using a variety of rheological models, and attained values were analyzed and discussed. After that, hardened properties of the paste were investigated and discussed. Piezo-resistive characteristics of the graphene cement composite were also explored and its practical application was investigated in the laboratory. The following are the conclusions of this research:

1. Generally, for all mixes, the yield stress values estimated by modified Bingham model were the highest while the lowest yield stress values were predicted by Casson model. For all models, the values of yield stress increased with the increase in the content of GNP in cement based composite as well as resting time. For higher shear rate cycle, the yield stress values decreased.

2. For graphene based mixes, Modified Bingham model predicted the higher values of plastic viscosities, while, for control cement paste, Bingham model estimated higher values of plastic viscosities. In general, the plastic viscosities increased with the increase in the content of graphene in the mix and resting time. Reduction in plastic viscosities was observed for higher shear rate cycle.

3. The Standard error resulting from fitting experimental flow curves to various rheological models varied with the content of GNP in cement based composite, shear rate and resting time. Generally, Modified Bingham model best fitted the experimental flow curves while Casson model showed the higher standard error values.

4. Incorporation of GNP showed strong influence on the hardened properties of the cement paste. In comparison to control sample, the GNP cement based composite showed 30% increase in load carrying capacity and 73% increase in overall failure strain.

5. Four-probe method was used to determine the piezo-resistive characteristics of the graphene–cement composite specimen. At maximum compressive load, the electrical resistivity value reduced by 42%. From the results of fractional change in resistance, it can be concluded that GNP cement based composite can be used to detect the damages in concrete.

6. Practical application of graphene–cement composite specimen was evaluated on RC beam. It was found that electrical resistance values varied with the increase in the applied load. Moreover, the graphene–cement composite specimen successfully predicted the response against crack propagation.

Based on above findings, graphene cement composite provides an economical and efficient solution for monitoring the structural health of the concrete members throughout their life span.

Acknowledgments: This research was supported by University Malaya Research Grant (UMRG-Project No. RP004A/13AET), University Malaya Postgraduate Research Fund (PPP-Project No. PG217–2014B) and Fundamental Research Grant Scheme, Ministry of Education, Malaysia (FRGS-Project No. FP004–2014B).

Author Contributions: Sardar Kashif Ur Rehman and Zainah Ibrahim conceived and designed the experiments; Muhammad Faisal Javed and Sardar Kashif Ur Rehman performed the experiments; Sardar Kashif Ur Rehman and Shazim Ali Memon analyzed the data; Zainah Ibrahim contributed reagents/materials/analysis tools; and Sardar Kashif Ur Rehman, Rao Arsalan Khushnood and Shazim Ali Memon wrote the paper.

Conflicts of Interest: The authors declare no conflict of interest.

References

1. Zhang, Y.; Kong, X.; Gao, L.; Lu, Z.; Zhou, S.; Dong, B. In-situ measurement of viscoelastic properties of fresh cement paste by a microrheology analyzer. *Cem. Concr. Res.* **2016**, *79*, 291–300. [CrossRef]
2. Ferraris, C.F. Measurement of the rheological properties of cement paste: A new approach. In *International RILEM Conference on the Role of Admixtures in High Performance Concrete*; National Institute of Standards and Technology: Gaithersburg, MD, USA, 1999; pp. 333–342.
3. Kawashima, S.; Hou, P.; Corr, D.J.; Shah, S.P. Modification of cement-based materials with nanoparticles. *Cem. Concr. Compos.* **2013**, *36*, 8–15. [CrossRef]
4. Ormsby, R.; McNally, T.; Mitchell, C.; Halley, P.; Martin, D.; Nicholson, T. Effect of MWCNT addition on the thermal and rheological properties of polymethyl methacrylate bone cement. *Carbon* **2011**, *49*, 2893–2904. [CrossRef]
5. Konsta-Gdoutos, M.S.; Metaxa, Z.S.; Shah, S.P. Highly dispersed carbon nanotube reinforced cement based materials. *Cem. Concr. Res.* **2010**, *40*, 1052–1059. [CrossRef]
6. Shang, Y.; Zhang, D.; Yang, C.; Liu, Y.; Liu, Y. Effect of graphene oxide on the rheological properties of cement pastes. *Constr. Build. Mater.* **2015**, *96*, 20–28. [CrossRef]
7. Wang, Q.; Wang, J.; Lv, C.-X.; Cui, X.-Y.; Li, S.-Y.; Wang, X. Rheological behavior of fresh cement pastes with a graphene oxide additive. *New Carbon Mater.* **2016**, *31*, 574–584. [CrossRef]
8. Wang, Q.; Cui, X.; Wang, J.; Li, S.; Lv, C.; Dong, Y. Effect of fly ash on rheological properties of graphene oxide cement paste. *Constr. Build. Mater.* **2017**, *138*, 35–44. [CrossRef]
9. Nehdi, M.; Rahman, M.A. Estimating rheological properties of cement pastes using various rheological models for different test geometry, gap and surface friction. *Cem. Concr. Res.* **2004**, *34*, 1993–2007. [CrossRef]
10. Lee, C.; Wei, X.; Kysar, J.W.; Hone, J. Measurement of the elastic properties and intrinsic strength of monolayer graphene. *Science* **2008**, *321*, 385–388. [CrossRef] [PubMed]
11. Rehman, S.K.U.; Ibrahim, Z.; Memon, S.A.; Jameel, M. Nondestructive test methods for concrete bridges: A review. *Constr. Build. Mater.* **2016**, *107*, 58–86. [CrossRef]
12. Chung, D. Piezoresistive cement-based materials for strain sensing. *J. Intell. Mater. Syst. Struct.* **2002**, *13*, 599–609. [CrossRef]
13. Sanchez, F.; Sobolev, K. Nanotechnology in concrete—A review. *Constr. Build. Mater.* **2010**, *24*, 2060–2071. [CrossRef]
14. Raki, L.; Beaudoin, J.; Alizadeh, R.; Makar, J.; Sato, T. Cement and concrete nanoscience and nanotechnology. *Materials* **2010**, *3*, 918–942. [CrossRef]
15. Sobolev, K.; Gutiérrez, M.F. How nanotechnology can change the concrete world. *Am. Ceram. Soci. Bull.* **2005**, *84*, 14.
16. Mukhopadhyay, A.K. Next-generation nano-based concrete construction products: A review. In *Nanotechnology in Civil Infrastructure*; Springer: Berlin/Heidelberg, Germany, 2011; pp. 207–223.
17. Vajtai, R. *Springer Handbook of Nanomaterials: Springer Science & Business Media*; Springer: Berlin/Heidelberg, Germany, 2013.
18. Sixuan, H. Multifunctional Graphite Nanoplatelets (GNP) Reinforced Cementitious Composites. Ph.D. Thesis, Tsinghua University, Beijing, China, 2012.
19. Papo, A. Rheological models for cement pastes. *Mater. Struct.* **1988**, *21*, 41–46. [CrossRef]

20. Yahia, A.; Khayat, K.H. Analytical models for estimating yield stress of high-performance pseudoplastic grout. *Cem. Concr. Res.* **2001**, *31*, 731–738. [CrossRef]
21. Rao, M.A. Flow and functional models for rheological properties of fluid foods. In *Rheology of Fluid, Semisolid, and Solid Foods*; Springer: Berlin/Heidelberg, Germany, 2014; pp. 27–61.
22. Yahia, A.; Khayat, K. Applicability of rheological models to high-performance grouts containing supplementary cementitious materials and viscosity enhancing admixture. *Mater. Struct.* **2003**, *36*, 402–412. [CrossRef]
23. Scott Blair, G. The success of Casson's equation. *Rheol. Acta* **1966**, *5*, 184–187. [CrossRef]
24. American Society for Testing and Materials. *Standard Test Method for Flow of Hydraulic Cement Mortar*; ASTM International: West Conshohocken, PA, USA, 2015.
25. Roussel, N.; Ovarlez, G.; Garrault, S.; Brumaud, C. The origins of thixotropy of fresh cement pastes. *Cem. Concr. Res.* **2012**, *42*, 148–157. [CrossRef]
26. Han, B.G.; Han, B.Z.; Ou, J.P. Experimental study on use of nickel powder-filled Portland cement-based composite for fabrication of piezoresistive sensors with high sensitivity. *Sens. Actuators A Phys.* **2009**, *149*, 51–55. [CrossRef]
27. Han, B.; Guan, X.; Ou, J. Electrode design, measuring method and data acquisition system of carbon fiber cement paste piezoresistive sensors. *Sens. Actuators A Phys.* **2007**, *135*, 360–369. [CrossRef]
28. Li, G.Y.; Wang, P.M.; Zhao, X. Pressure-sensitive properties and microstructure of carbon nanotube reinforced cement composites. *Cem. Concr. Compos.* **2007**, *29*, 377–382. [CrossRef]
29. Le, J.-L.; Du, H.; Dai Pang, S. Use of 2D Graphene Nanoplatelets (GNP) in cement composites for structural health evaluation. *Compos. Part B Eng.* **2014**, *67*, 555–563.
30. Dai Pang, S.; Gao, H.J.; Xu, C.; Quek, S.T.; Du, H. Strain and damage self-sensing cement composites with conductive graphene nanoplatelet. In *SPIE Smart Structures and Materials+ Nondestructive Evaluation and Health Monitoring*; International Society for Optics and Photonics; SPIE: San Diego, CA, USA, 2014; p. 906126.
31. Shaughnessy, R.; Clark, P.E. The rheological behavior of fresh cement pastes. *Cem. Concr. Res.* **1988**, *18*, 327–341. [CrossRef]
32. Barnes, H.A. A handbook of elementary rheology. In *Institute of Non-Newtonian Fluid Mechanics*; University of Wales: Wales, UK, First published 2000; ISBN 0-9538032-0-1.
33. Chuah, S.; Pan, Z.; Sanjayan, J.G.; Wang, C.M.; Duan, W.H. Nano reinforced cement and concrete composites and new perspective from graphene oxide. *Constr. Build. Mater.* **2014**, *73*, 113–124. [CrossRef]
34. Vikan, H.; Justnes, H.; Winnefeld, F.; Figi, R. Correlating cement characteristics with rheology of paste. *Cem. Concr. Res.* **2007**, *37*, 1502–1511. [CrossRef]
35. Wallevik, J.E. Rheological properties of cement paste: Thixotropic behavior and structural breakdown. *Cem. Concr. Res.* **2009**, *39*, 14–29. [CrossRef]
36. Pan, Z.; He, L.; Qiu, L.; Korayem, A.H.; Li, G.; Zhu, J.W. Mechanical properties and microstructure of a graphene oxide–cement composite. *Cem. Concr. Compos.* **2015**, *58*, 140–147. [CrossRef]
37. Collins, F.; Lambert, J.; Duan, W.H. The influences of admixtures on the dispersion, workability, and strength of carbon nanotube–OPC paste mixtures. *Cem. Concr. Compos.* **2012**, *34*, 201–207. [CrossRef]
38. Cao, M.-L.; Zhang, H.-X.; Zhang, C. Effect of graphene on mechanical properties of cement mortars. *J. Cent. South Univ.* **2016**, *23*, 919–925. [CrossRef]
39. Li, H.; Xiao, H.-G.; Ou, J.-P. A study on mechanical and pressure-sensitive properties of cement mortar with nanophase materials. *Cem. Concr. Res.* **2004**, *34*, 435–438. [CrossRef]

MDPI

St. Alban-Anlage 66

4052 Basel

Switzerland

Tel. +41 61 683 77 34

Fax +41 61 302 89 18

www.mdpi.com

Sustainability Editorial Office

E-mail: sustainability@mdpi.com

www.mdpi.com/journal/sustainability

9 783036 504827